Geotechnical Modeling and Intelligent Systems

Gao-Feng Zhao
Editor

Geotechnical Modeling and Intelligent Systems

Editor
Gao-Feng Zhao
State Key Laboratory of Hydraulic
Engineering Intelligent
Construction and Operation
School of Civil Engineering
Tianjin University
Tianjin, China

ISBN 978-981-96-6924-0   ISBN 978-981-96-6925-7 (eBook)
https://doi.org/10.1007/978-981-96-6925-7

This work was supported by Guangzhou KEO Info Technology Co., Ltd.

© The Editor(s) (if applicable) and The Author(s) 2026. This book is an open access publication.

Open Access This book is licensed under the terms of the Creative Commons Attribution 4.0 International License (http://creativecommons.org/licenses/by/4.0/), which permits use, sharing, adaptation, distribution and reproduction in any medium or format, as long as you give appropriate credit to the original author(s) and the source, provide a link to the Creative Commons license and indicate if changes were made.
The images or other third party material in this book are included in the book's Creative Commons license, unless indicated otherwise in a credit line to the material. If material is not included in the book's Creative Commons license and your intended use is not permitted by statutory regulation or exceeds the permitted use, you will need to obtain permission directly from the copyright holder.
The use of general descriptive names, registered names, trademarks, service marks, etc. in this publication does not imply, even in the absence of a specific statement, that such names are exempt from the relevant protective laws and regulations and therefore free for general use.
The publisher, the authors and the editors are safe to assume that the advice and information in this book are believed to be true and accurate at the date of publication. Neither the publisher nor the authors or the editors give a warranty, expressed or implied, with respect to the material contained herein or for any errors or omissions that may have been made. The publisher remains neutral with regard to jurisdictional claims in published maps and institutional affiliations.

This Springer imprint is published by the registered company Springer Nature Singapore Pte Ltd.
The registered company address is: 152 Beach Road, #21-01/04 Gateway East, Singapore 189721, Singapore

If disposing of this product, please recycle the paper.

Preface

This book named *Geotechnical Modeling and Intelligent Systems* partly gathers selected papers accepted and presented during the 2024 7th International Symposium on Traffic Transportation and Civil Architecture (ISTTCA 2024), which is aimed at promoting research and developmental activities related to geotechnical engineering and artificial intelligence, and facilitating scientific information exchange among individuals and research groups.

The complexity and variability of geotechnical bodies, especially their mechanical behavior, deformation characteristics, and hydrogeological response under different environmental conditions, have always been a major challenge in engineering design and construction. With the rapid development of information technology and the increasing maturity of intelligent science, the integration of geotechnical modeling and intelligent systems is becoming a new key to crack this challenge.

The birth of the book is based on the profound insight and forward-looking layout of the future development direction of the geotechnical engineering field in the context of this era. This book aims to explore and explain how to integrate advanced intelligent technology into all aspects of geotechnical modeling, from data acquisition, model construction, parameter optimization to predictive analysis, and to enhance the scientific and intelligent level of geotechnical engineering in all aspects.

This content of the book is mainly related to geotechnical simulation and is focused on the modeling and simulation in geotechnical engineering, computational mechanics, and geotechnical simulation algorithms and systems. It provides rigorous discussions, case studies, and recent developments in geotechnical projects, which are expected to provide a reference for solving real-world engineering problems.

In conclusion, this book is an academic work integrating scientific, prospective, and practicality, which is not only an important reference and teaching material for research students, experts and scholars in the field of geotechnical engineering, but also an invaluable resource for the majority of engineers and technicians to enhance their professional ability and broaden their horizons. Let's work hand in hand to create a new chapter in the development of geotechnical engineering intelligence in the vast world of geotechnical modeling and intelligent systems!

Tianjin, China
September 2024

The Editor in Chief
Gao-Feng Zhao

Contents

Study on the Influence of Pile Arrangement of Wet Sinking Loess Foundation on the Settlement of Outlet Tower Foundation 1
Haichen Zhang, Zihao Ren, Jiaming Li, Yahui Zi, and Xu Xu

Research on Intelligent Mining of Super-High Core Wall Dam Complex Soil Stockyard 17
Chao Hu, Jialin Yu, Guike Zhang, and Zhongyao Wang

Three-Dimensional Numerical Simulation Study on the Influence of Reservoir Water Level Variation on the Stability of Multiple Taluses 31
Zelin Zhou, Jie Zhang, Chao Feng, and Maoyi Liu

Compaction Quality Control of Soil-Rock Mixture Subgrade Based on DEM 43
Zhuoling He, Junyun Zhang, Xiaolong Luo, Tao Yang, Xiaofei Wu, and Le Zhang

Analysis of the Influence of Deep Soft Soil Foundation Cofferdam Filling on Lateral Compression of Bridge Pile Foundations 53
Ruiqi Zhang, Aimin Liu, Guoliang Ye, and Xu Liu

Drilling Optimization Using Digital Image Stratigraphic Modeling 67
Zhengqiang Zeng and Yongchang Cai

Research on the Influence of Ultra-Deep Foundation Pit Excavation on Adjacent Existing Tunnels and Deformation Control 81
Youpeng Wen, Mingxing Zhu, Kunpeng Wu, Xiaocong Liang, and Xiaozhou Yan

Analysis of Stability of Surrounding Rock in the Blasting Excavation of Pumped Storage Power Station Powerhouse 97
Xiji Li, Shuangquan Xu, Chuncheng Ma, and Fan Zhang

Discussion of the Collapsibility of Loess after Pre-Immersion Water Treatment 109
Bai He, Bin Zhi, and Tiantian Wei

Application of Building Information Modeling + Algorithm Model in the "Four Preparations" for Seepage Safety of Concrete Face Rockfill Dam 125
Yi Hou, Bin Mei, Jian Zhang, Wenyang Lin, and Hong Yu

Study on the Secondary Lining Force of Soft Rock Tunnel with Large Deformation 139
Guize Liu, Changyi Yu, Shigang Liu, Yonghua Cao, and Binbin Xu

Construction Risk Assessment of Biogas Layer under Shield of Subway Tunnel 147
Dongyin Qi and Fei Yu

Microstructure-Based Dynamic Characterization of Remodelled Loess under Traffic Loading 159
Hongtao Pei, Kebing Wen, Zezhan Shao, Shenqin Sun, Tianlu Xu, and Junmin Pan

Analysis of Settlement Displacement of Tunnels Traversing Soft Soil Strata Containing Hazardous Gases 175
Jie He

Analysis of Deformation Influence of Gravel Shield Tunneling Under Existing Tunnel 183
Hongtao Pei, Rimei Han, Zezhan Shao, Shenqin Sun, Tianlu Xu, and Junmin Pan

Damage Characteristics of Fractured Rock Under Freeze–Thaw Cycle 205
Yuanqiang Lv, Jingang Zhao, and Haibo Jiang

Research on a Machine Learning-Based Subgrade Compaction Degree Prediction Model 229
Feng Li, Jianfei Zhao, Hongzhao Li, Bing Hui, Zhenkun Wang, Wenjun Zhang, and Guangbo Liu

Distribution Characteristics of Ground Stress Field in the Underground Caverns Under Complex Geological Conditions ... 245
Peiyang Yu, Jun He, Yang Qin, and Jianhua He

Research and Application of Advance Bolt Support Mechanism of Highway Tunnel Under Complex Geological Conditions 259
Zhou Qiao

Experimental Study on Deformation Characteristics of Mud–Stone Mixture 269
Jian-bao Fu, Jian Yu, He-wen Liu, and Yu-bin Guo

Statistical Characterization of the Geotechnical Properties of Changshou Rock 279
Jian-gong Chen and Bing Lu

A Development Method of Geotechnical Testing Instrument Based on 3D Printing Technology 293
Hungchou Lin, Mengyue Wang, Xiaodong Zhou, Yanbo Cao, and Jianbing Peng

Centrifuge Model Test on the Seismic Response of the Pile-Supported Wharf Slope 305
Jiarui Zhang, Yu Shao, Long Lü, Xianlin Liu, and Xilin Lü

Study on Mechanical Experimental Characteristics of Loess Under Different Water Content and Confining Pressure 313
Dehuan Sun

Deformation Analysis of Subway Pit Excavation Under Diaphragm Wall and Steel Support 321
Qi Chen, Xuezhu Li, and Lingfeng Wan

Experimental Study on Water Resistance of Microbial-Magnesium Oxide Improved Red Clay 331
Haodong Qin, Shangbin Wu, Yicong Wang, Yueguang Yang, Wenrong Li, Xiaoqing Wang, and Yuqin Liao

Research on Anti-deformation Technology of High-rise Buildings in Coal Mining Subsidence Areas 349
Keyi Guo, Xing Li, and Yue Li

Multi-criteria Decision Analysis of TGS360Pro Advance Geological Forecasting Results Based on Deviation Maximization 369
Mingcai Zhang, Guanghong Ju, Dazhou Zhang, Zonggang Chen, Dong Li, and Song Han

Ultimate Bearing Capacity of Rigid Foundation Near Embankment Slope Subjected to Water Drawdown 383
Qi Wang, Boyang Xia, Gang Zheng, Zheng Wang, Boyao Zhao, and Xin Yin

Rock Mass Deformation Analysis and Local Collapse Treatment of Powerhouse Slope of Wukuo Power Station 393
Hainian Shan, Feng Zhang, Han Zhang, and Jianhua Deng

Research on Deformation Control Technology for Filling and Mining of High and Steep Slopes in Guizhou Mountainous Areas 409
Yu Wu, Xiaohu Zheng, Qing Liu, Yao Zhong, Qianyong Lv, Jie Huang, and Dandan Liu

Study on the Settlement Change Law of Adjacent Buildings Caused by Deep Excavation Construction in Saturated Soft Loess Geology 421
Hanjuan Yao, Wenyao He, Man Wang, and Liudi Yang

Influence of Subgrade Excavation on Vertical Deformation of Collinear Metro Tunnels in Soft Soil Area 429
Jianzhao Li and Qingyuan Zeng

Study on the Influence of Pile Arrangement of Wet Sinking Loess Foundation on the Settlement of Outlet Tower Foundation

Haichen Zhang, Zihao Ren, Jiaming Li, Yahui Zi, and Xu Xu

Abstract Due to load variations and foundation disturbances, hydraulic buildings on self-weighting wet sinking loess foundations are vulnerable to uneven settlement, which compromises the project's structural safety. Understanding the settlement and settlement characteristics of a group pile foundation supporting a pumping station outlet tower structure under different load distributions and different pile spacing arrangements for a wet-submerged loess foundation is essential for future research. To simulate and analyze the effects of load and group pile spacing variations on the settlement characteristics of wet sinking loess foundations, the finite element simulation software ABAQUS was used in this study to build numerical simulation models of group pile foundations with pile spacing of 1.2, 1.6, and 2.0 m under variable load conditions. The research results indicate a positive correlation between the soil settlement of collapsible loess foundations and the magnitude of load pressure. Under various pile spacing configurations, the center of the bearing plate is where the foundation's maximum deformation settlement occurs, and 18 m below the surface is where the inflection point of the settlement change at the pile-soil contact region emerges. Uneven settlement will result in the distortion of the soil mass due to the pile group impact of the pile foundation. The foundation with 1.2 m pile spacing has the highest uneven settlement deformation, followed by 1.6 m pile spacing, and the settlement with 2.0 m pile spacing has the most consistent deformation. It is possible to analyze the uneven settlement of collapsible loess foundations and improve building design by using the research techniques and simulation results presented in this article as suggestions and references.

Keywords Wet sinking loess · ABAQUS · Group pile foundation · Foundation settlement

---

H. Zhang
Water Resources Engineering Construction Center of Ningxia, Yinchuan, China

Z. Ren (✉) · J. Li · Y. Zi · X. Xu
School of Water Conservancy, North China University of Water Resources and Electric Power, Zhengzhou, China
e-mail: 1132968657@qq.com

1 Introduction

The foundation is essential in the construction and long-term operation of water conservancy projects [1–3]. The gene of collapsible loess has the characteristics of uneven settlement, large deformation, and complex stress. In long-term operation, it is highly susceptible to changes in external adverse factors such as climate conditions and exhibits complex mechanical properties, resulting in deformation and instability failure [4–6]. In order to ensure the structural safety of hydraulic structures located on collapsible loess foundations, engineers and technicians often use Pile foundations to support the upper buildings to maintain the structural stability of buildings. Engineering practice has proved that the layout types of the pile foundation and the settlement characteristics of the wet sinking loess have a specific coupling linkage effect; in the building bearing and foundation soil stress process, the stress of the pile foundation and soil settlement due to different layout types and present a different characterization. Therefore, for important buildings, it is necessary to conduct in-depth research on the stress characteristics of pile foundations to further optimize the pile foundation design scheme in such projects based on the evaluation of engineering safety [7, 8].

With the rapid development of computer technology, numerical analysis methods for the relationship between pile foundations have emerged, with finite element analysis being the mainstream technical means [9, 10]. Jie et al. [11] studied the settlement of the foundation during the construction period of collapsible loess at Luliang Airport in Shanxi, China. They used the plane strain finite element method, considering linear changes in modulus to simulate the stress and strain changes caused by the layered filling of four typical sections and calculated and analyzed the settlement results of the high-fill airport. The results indicate that the settlement caused by relatively "soft" loess parameters is more significant than that caused by relatively "hard" parameters; the thicker the filling body, the greater the settlement. Li et al. [12] focused on the relationship between the lateral earth pressure of the pile and the soil and found that the bond around the pile has a significant effect on the lateral ultimate earth pressure of the pile and the relative displacement of the soil that reaches the lateral ultimate stress state of the pile decreases along with the increase of E/Cu. Hu et al. [13] used the finite element analysis software ANSYS to establish a three-dimensional numerical model of the building's structure, foundation and foundation. It was found that the simulation model established using the numerical model is more reliable and intuitive than the theoretical calculation method, and its calculation results can better reflect the uneven settlement pattern of the building. Sheng et al. [14] used ABAQUS finite element translation software to establish a three-dimensional finite element calculation model of pile-soil to analyze the pile-soil action under vertical load. The analysis results show that: under the same load, increasing the pile distance or pile length can reduce the settlement; with the increase in the number of piles, the settlement of the pile foundation is also increasing. Although the finite element analysis method has been successfully applied to the simulation of pile foundation and foundation, there is no unified standard and mature theoretical system for the settlement study of collapsible loess foundation group Pile foundation [15, 16], which is still the direction of active exploration in geotechnical engineering [17, 18].

The study of the deformation and settlement patterns caused by wet-submerged loess when the building is load-bearing and revealing its effect on the stresses in the pile foundations of the building is essential for the safety of the outlet tower. This paper establishes a numerical model of multiple piles with variable pile spacing, and ABAQUS finite element analysis software is used to carry out force analysis on the bearing capacity of wet-submerged loess foundations under different loading conditions. This paper establishes a numerical model of multi-pile variable pile spacing. ABAQUS finite element analysis software is used to conduct force analysis on the bearing capacity of the foundation of wet-submerged loess under different loading conditions. Under the condition that the Mohr–Coulomb criterion is satisfied, the settlement characteristics of wet-submerged loess foundations carried out by studying the distribution of foundation stress and the settlement pattern of the soil, which is of great practical significance in guiding the structural optimization design of self-weighted wet-submerged loess foundations.

2 Project Overview

2.1 *Engineering Background*

Ningxia solid sea expansion irrigation district is located in the southern part of the Ningxia Loess Plateau area, and it is a typical high-head gradient water-raising irrigation project. It is also the Ningxia Hui Autonomous Region's major poverty alleviation migration resettlement project, and the poverty alleviation Yang Huang irrigation project is an important part. The irrigation area is built with 12-step pumping stations, and the total length of the project water transmission trunk canal is 83 km. Class IV self-weight collapsible sites account for about 1/3, with the most profound thickness of collapsible soil layers reaching 25–36 m and loess collapsibility reaching up to 200 cm. Most of the project backbone buildings are built on wet sink loess foundations. With the construction of an irrigation area and significant area of irrigation infiltration, Water outlet towers, pumping stations, ducts, reservoirs, water pipelines, and other buildings on wet sink loess foundations have different degrees of wet sink settlement safety problems. The regular use of buildings has been seriously affected, bringing substantial potential risks to the safety of irrigation district projects and the production and life of people in the irrigation area.

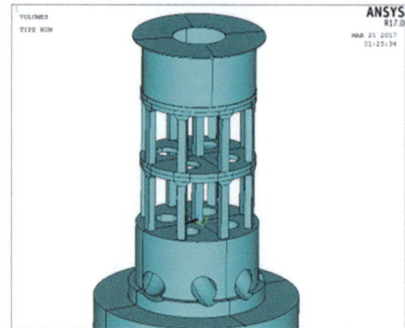

Fig. 1 Out of the water tower structure and foundation support type schematic diagram

2.2 Characteristics of Wet-Submerged Loess Foundation and Model Parameter Setting

This paper takes the foundation of the outlet tower of the total dry four pumping station of the solid sea irrigation expansion project as the object of study, which is a typical self-weight wet sinking loess foundation in this area. The foundation soil has low Compressibility and high strength under natural humidity; When soaked in water, it exhibits characteristics such as collapse deformation, large settlement amount, fast settlement speed, and a significant reduction in bearing capacity and strength. The water tower foundation uses the end-bearing pile foundation support type, the water tower structure and the foundation support type, as shown in Fig. 1.

3 Finite Element Model Construction

3.1 Finite Element Model for Variable Load Multi-pile Foundations

Due to the complex and variable force on the primary body of the wet sinking yellow ground, the traditional formula calculation is difficult to reflect the specific action relationship between the soil and the pile foundation concisely and efficiently. Therefore, ABAQUS non-linear finite element software was chosen for modeling simulation, and analysis to obtain higher accuracy.

Following the actual arrangement scheme, Hypermesh 14.0 software was used to construct a nine-pile model with a pile length of 14 m, a pile diameter of $D = 0.8$ m, and pile spacing of $1.5\,D = 1.2$ m, $2.0\,D = 1.6$ m, and $2.5\,D = 2\,0.0$ m in that order. The total dimensions of the soil model around the pile were set at 30 m × 30 m × 30 m. Specific material performance parameters are shown in Tables 1 and 2. When the model is constructed, the different materials, such as bedding, bearing plates, piles, and their regional meshes, are encrypted to reduce their calculation process while ensuring the accuracy of the calculation. The constructed finite element mesh model is shown in Fig. 2.

Table 1 Table of soil material parameters for the foundation of the water outlet tower

Soil layer	Thickness (m)	internal friction angle (°)	Elastic modulus (Mpa)	Poisson's ratio	Porosity ratio	Permeability coefficient (cm/s)
Plain fill soil	2	23	16.1	0.3	0.726	6.29E−03
Loamy soil	5	16	13.6	0.32	0.758	8.47E−03
Powdered sand	2	32	12	0.15	0.705	0.227
Sandy loam soil	4	20	10	0.23	0.8	0.034
Angular gravel	17	26	26	0.18	0.708	0.227

Table 2 Table of material parameters for the support base of the water tower

Material name	Thickness (m)	Density (kg/m³)	Elastic modulus (MPa)	Internal friction angle (°)	Poisson's ratio
Bedding	0.5	1500	35	38	0.3
Pressure-bearing plates	0.5	7900	2.06e5	–	0.3
Piles	14	2500	160	–	0.15

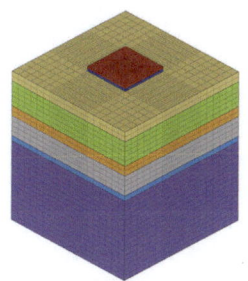 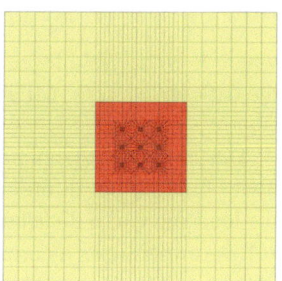

Fig. 2 Schematic diagram of the finite element model of the pile foundation structure

3.2 Calculation Model Selection

The ABAQUS software contains a variety of soil calculation models, such as the wireline elastic model, the porous media elastic model, and the Mohr–Coulomb principal structure model. The Mohr–Coulomb model conforms to the stress situation of collapsible loess material and the supporting structure of the water outlet tower. Therefore, in the model selection, the Mohr–Coulomb yielding criterion in ABAQUS software was chosen for the unit properties of the frictional contact surface between each soil layer to accurately calculate the pile foundation's force conditions and displacement state. The shear strength of the τ_f soil is shown in Eq. 1.

$$\tau_f = c + \sigma \tan \varphi \tag{1}$$

where σ is the normal stress of the action (KPa); φ is the angle of friction within the soil (°), c is the cohesive force (KPa) within the soil, and the principal stresses in their ultimate equilibrium state are related as in Eqs. 2 and 3.

$$\sigma_1 = \sigma_3 \tan\left(45° + \frac{\varphi}{2}\right) + 2c \tan\left(45° + \frac{\varphi}{2}\right) \tag{2}$$

$$\sigma_3 = \sigma_1 \tan\left(45° - \frac{\varphi}{2}\right) + 2c \tan\left(45° - \frac{\varphi}{2}\right) \tag{3}$$

where σ_1 and σ_3 represent the state of the principal stress at a fixed point, respectively. The yield surface function for the principal stress space is shown in Eq. 4.

$$F = R_{mc}q - p \tan \varphi - c = 0 \tag{4}$$

$$R_{mc}(\Theta, \varphi) = \frac{1}{\sqrt{3} \cos \varphi} \sin\left(\Theta + \frac{\pi}{3}\right) + \frac{1}{3} \cos\left(\Theta + \frac{\pi}{3}\right) \tan \varphi \tag{5}$$

where F is the yield force (MPa), R_{mc} is the radius of the molar stress circle (m), q is the deflective stress (MPa), p is the pressure (Pa), Θ is the angle of polar declination (°).

Due to the complexity of the upper bearing capacity and load changes of the outlet tower foundation, in order to simplify the calculation process, this study simplified the load force as a distributed load perpendicular to the cushion direction. ABAQUS software manifests as creating different distributed load levels with additional 200, 250, and 300 kPa based on load steps. The settlement characteristics and responses to wet-submerged loess foundations at three pile spacings are simulated.

4 Finite Element Model Simulation Results and Analysis

4.1 Simulation Results for Variable Load Foundations

The resulting displacement clouds and their settlement values, calculated using ABAQUS software, are shown in Fig. 3.

From the displacement clouds of the lateral profiles for the nine cases, we can obtain that under distributed loads of 200, 250, and 300 kPa, the wet-submerged soil layers of the foundations all show different degrees of downward displacement, and their settlement tends to spread from the middle to the sides. The maximum displacements of the soil in the bearing layer at 1.2 m pile spacing are 18.24, 23.47 and 32.09 cm; the maximum displacements of the soil in the bearing layer at 1.6 m pile spacing are 19.59 cm, 25.29 and 32.75 cm; the maximum displacements at 2.0 m pile spacing are 20.09 cm, 26.89 cm and 34.31 cm, respectively. The overall tendency of the soil to fall increases with the load, and the maximum value of displacement in all scenarios occurs in the middle of the soil in contact with the pile foundation, which is in line with the natural force conditions.

4.2 Analysis of Settlement Impact at Different Depths of Soil Centerline

In order to explore the specific numerical changes and trends of pile settlement, the analysis field extraction function in ABAQUS software was used to quantitatively plot the settlement curves of three different pile spacing soil masses at different depths of the pile foundation under distributed loads of 200, 250, and 300 kPa, based on the center soil position before the two piles. The details are shown in Fig. 4.

From the curves of total soil settlement at different depths in Fig. 4d, we can see that the variation in soil settlement will show a negative correlation in settlement characteristics with increasing depth in three cases. The settlement displacement curve of the same load with different pile spacing shows a positive correlation trend of larger pile spacing and a more significant slope. The inflection point of settlement for all nine curves occurs at around 18 m, with the overall settlement of the soil converging to 0 m when the depth reaches below 21 m.

Meanwhile, the settlement displacement of the soil around the 1.2 m pile spacing bearing piles under the three loads showed the same trend as the 1.6 m pile spacing. However, the soil at the 2.0 m pile spacing showed a stagnant settlement in the 200 kPa load pressure, especially in the 5-10 m depth range from the surface, which showed a stagnant decline at 10 cm and 13 cm, respectively.

Fig. 3 Displacement clouds of the lateral profile of the soil under different loads

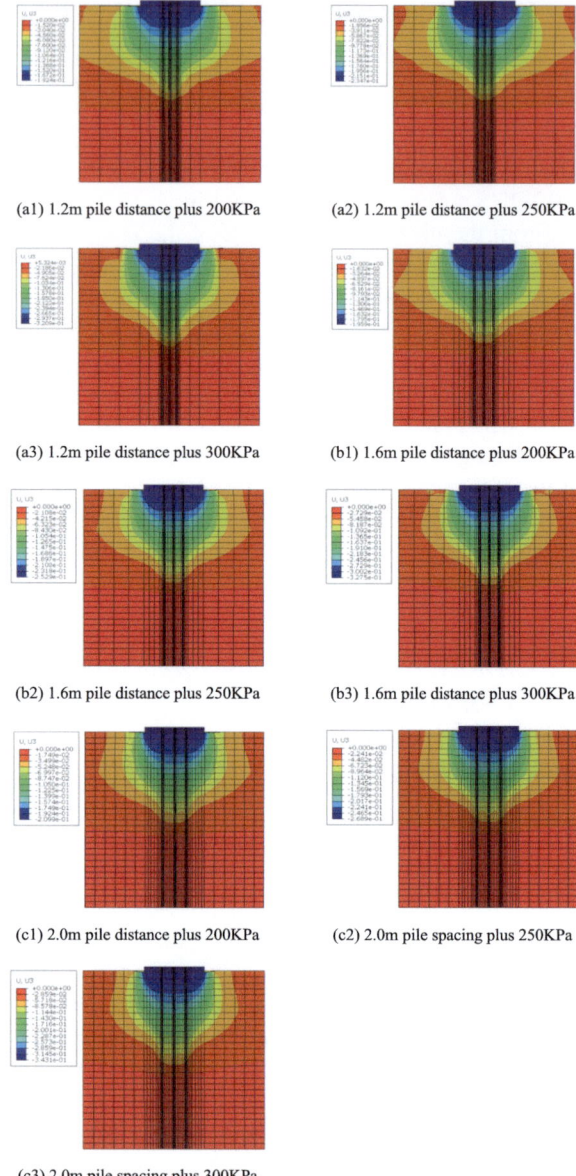

(a1) 1.2m pile distance plus 200KPa

(a2) 1.2m pile distance plus 250KPa

(a3) 1.2m pile distance plus 300KPa

(b1) 1.6m pile distance plus 200KPa

(b2) 1.6m pile distance plus 250KPa

(b3) 1.6m pile distance plus 300KPa

(c1) 2.0m pile distance plus 200KPa

(c2) 2.0m pile spacing plus 250KPa

(c3) 2.0m pile spacing plus 300KPa

4.3 Research on Settlement Deformation of Pile Foundation

In order to investigate in-depth the total pile settlement deformation under the three different pile spacing arrangement types, pile vertical displacement clouds were extracted for the nine scenarios, as shown in Fig. 5.

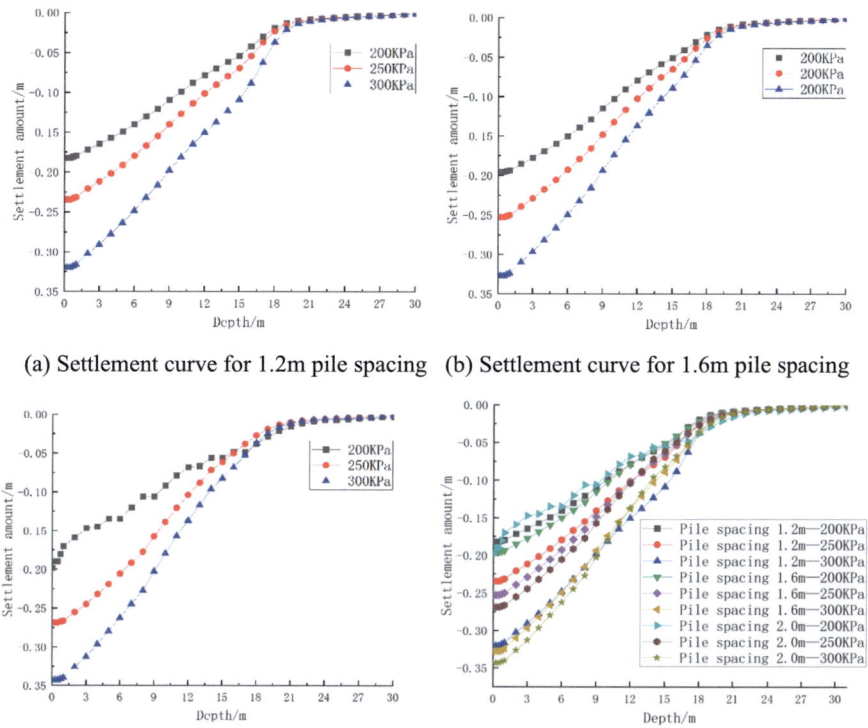

Fig. 4 Soil settlement displacement curve

As can be seen from Fig. 5, the displacements under the additional distributed loads of 200–300 kPa increase with increasing load for the three pile spacing arrangements, with the maximum vertical displacement of the pile occurring at the 300 kPa load case. The settlements were 31.32 cm, 32.03 cm and 33.57 cm for the pile spacing of 1.2 m, 1.6 m and 2.0 m, respectively, with a slight difference of 6.7% relative displacement. The difference between the maximum settlement of displacement under 200 kPa load and 300 kPa is significant, 18.24 cm, 19.16 cm, and 20.54 cm, respectively, with a settlement difference of 41.69, 40.18, and 38.81%, showing an overall decreasing trend. For the piles, the load effect at the time is the primary influence on the variation of the total settlement. The location of the maximum displacement settlement is at the top of the pile body for the different arrangement scenarios, and there is also a somewhat linear increase in the settlement at the top of the group piles concerning the pile body compression.

As can be seen from the settlement displacement clouds, the settlement of each pile part of the nine piles also differs to a certain extent under a fixed distributed load. The central pile displacement decreases faster than the surrounding pile settlement,

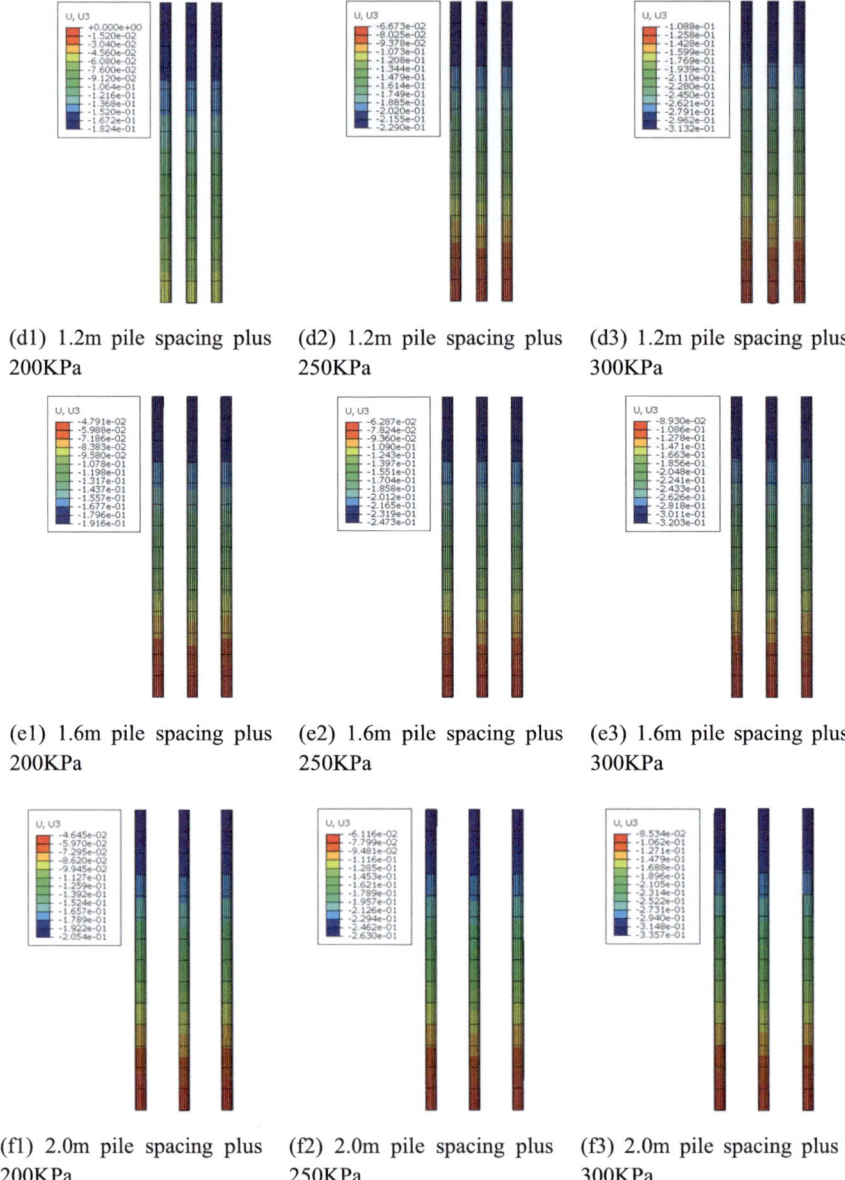

(d1) 1.2m pile spacing plus 200KPa (d2) 1.2m pile spacing plus 250KPa (d3) 1.2m pile spacing plus 300KPa

(e1) 1.6m pile spacing plus 200KPa (e2) 1.6m pile spacing plus 250KPa (e3) 1.6m pile spacing plus 300KPa

(f1) 2.0m pile spacing plus 200KPa (f2) 2.0m pile spacing plus 250KPa (f3) 2.0m pile spacing plus 300KPa

Fig. 5 Pile displacement clouds under 200–300 kPa load

and the upper part settles more than the lower part, reflecting the characteristics of the pile foundation's group pile effect [19] in the wet-submerged loess foundation.

4.4 Settlement Deformation Analysis of Bearing Slabs and Bedding Layers

In order to analyze the effect of group pile spacing arrangement on the bearing layer of a wet sinking foundation, the displacement clouds of the bearing layer settlement for three pile spacings with a distributed load of 300 kPa were extracted for analysis. The results are shown in Fig. 6.

It can be seen from Fig. 6 that the overall settlement of the foundation mat and the pressure plate of the outlet tower increases with increasing pile spacing. The group pile foundation with 1.2 m pile spacing shows a decreasing settlement trend from the center to the ends, which is more evident than the settlement clouds of 1.6 and 2.0 m pile spacing. The soil settlement around the nine piles is small, and there is an uneven settlement between the various pile spacings. In this study, a diagonal path is set up in the plane, which can extract specific settlement displacement variation values for a portion of the soil mass. The settlement extraction method for the diagonal path on the plane is shown in Fig. 7.

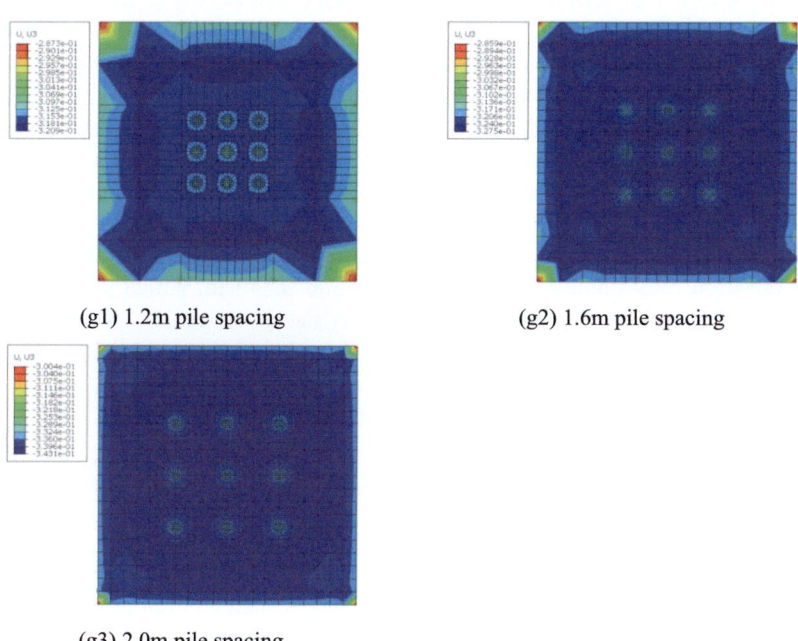

(g1) 1.2m pile spacing (g2) 1.6m pile spacing

(g3) 2.0m pile spacing

Fig. 6 Settlement clouds of the bearing layer under 300 kPa load

Fig. 7 Schematic diagram of the soil settlement extraction path

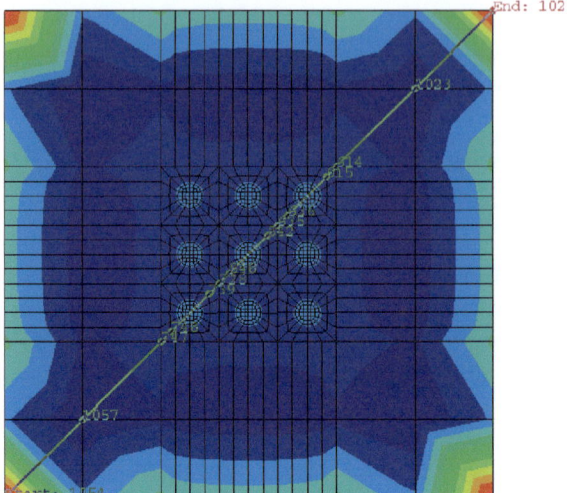

Figure 7 shows a schematic diagram of the path extraction for a 1.2 m pile spacing under a 300 kPa load. This overall path extraction method covers all areas of the pile arrangement and is representative. The settlement curves under the three scenarios are shown in Fig. 8.

Figure 8 shows the settlement displacement curves of the soil in the bearing layer for different pile spacing schemes. The results in the figure show that the central part of the bearing layer in the 1.2 m pile spacing arrangement scheme shows significant uneven settlement, with the maximum fluctuation occurring at the 11.5 m position and fluctuating between 0.318 and 0.324 m. The 1.6 m pile spacing arrangement has the highest frequency of fluctuating changes in an uneven settlement at its intermediate position. However, the change in an uneven settlement is smaller than in the 1.2 m pile spacing case. The 2.0 m pile spacing arrangement has the least uneven settlement fluctuations in the bearing layer, with only four locations, and the settlement variation only fluctuates between 0.338 and 0.343 m, which shows that the settlement phenomenon in the bearing layer is the most stable. Therefore, when considering the arrangement of pile spacing on a wet-submerged loess foundation, attention needs to be focused on the center of the soil where the group pile effect gathers to avoid uneven settlement of the foundation.

5 Conclusions

Self-weighting wet sinking loess has characteristics such as high porosity, uneven settlement, and high deformation. Traditional in-situ testing methods make it difficult to visually and accurately analyze the settlement changes and fluctuations under load-bearing forces. This paper uses the non-linear finite element software ABAQUS

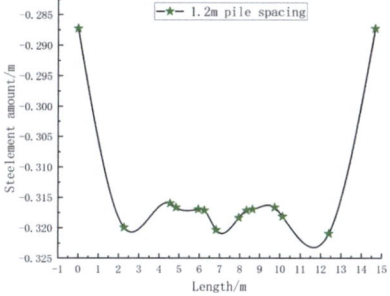 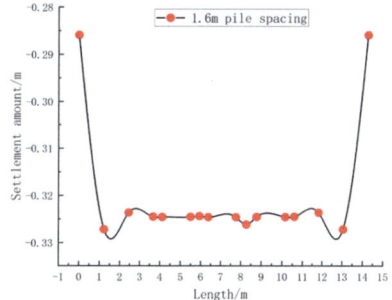

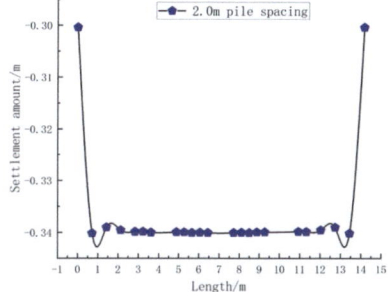

Fig. 8 Settlement displacement curves for 3 pile spacings under 300 kPa load

as a means of calculation by constructing a finite element model of the foundation of the pumping station outlet tower supported by nine piles and simulates the settlement deformation characteristics of the wet sinking loess foundation and its supporting structure under the action of pile spacing of 1.2 m, 1.6 m, 2.0 m, and load distribution of 200 Kpa, 250 Kpa, and 300 Kpa, respectively, and obtains the following conclusions.

(1) The maximum settlement of the foundation soil was 31.32, 32.03, and 33.57 cm when the load reached 300 kPa for a 0.8 m diameter water outlet tower nine pile foundation with pile spacing of 1.5 D, 2.0 D, and 2.5 D, respectively, with the maximum displacement occurring at the center of the soil bearing layer. The relative displacement of the piles for the three pile spacing arrangements is small at 6.7% for the same load. The difference in pile displacement for variable loads is more extensive, with a maximum variation of 41.69%. The maximum load and the pile foundation layout are the most critical elements in determining the foundation settlement for buildings located on wet sinking loess foundations and need to be studied in a targeted manner during design.

(2) By extracting the settlement in the path of the bearing layer for the three different pile spacings, it was found that the settlement at the center of the bearing plate for all three pile spacings fluctuated between 0.31 and 0.34 m, with the most

significant fluctuation in the center of the bearing plate for the pile spacing of 1.2 m, indicating that the risk of uneven settlement is much greater than for the 1.6 and 2.0 m pile spacings. It can be seen that the settlement distribution in the path of the bearing layer will be affected by the group pile effect. When the bearing pile distance of the bearing plate is between 2.0 D and 2.5 D, the uneven settlement variation of the foundation soil is minor, and when optimizing the pile arrangement in the central part of the bearing plate, reference to this value can reduce the number of pile foundations based on ensuring the uneven settlement is minimized.

It is worth noting that, as there is an inevitable variability in the characteristics and foundation conditions of the wet-submerged loess in different areas, it is necessary to do an excellent job of testing the characteristics of the soil and in-situ tests when modeling in order to ensure the reliability of the modeling data, to establish a more accurate finite element simulation model and realize the simulation calculation of the settlement of the foundation bearing process in order to provide numerical assurance for engineering construction safety.

Acknowledgements This work was supported by the Ningxia Water Conservancy Science and Technology Tackling Project (GKGX-KY-01), Henan Province Science and Technology Research Project (212102310273), and Henan Province Higher Education Key Research Project Plan (20A570006). Sincere gratitude is extended to the editor and anonymous reviewers for their professional comments, which greatly improved the presentation of the paper.

References

1. Zhang YJ, Wang X, Liang QG et al (2021) Model tests on bearing behavior of pile groups in collapsible loess ground under water immersion. Chin J Geotech Eng 43(S1):219–223
2. Wu HY, Zhai KJ, Fang HY et al (2022) Bell-and-spigot joints mechanical properties study of PCCP under the uneven settlement of foundation: simulation and full-scale test. Structures 43
3. Zhao ZF, Ye SH, Zhu YP et al (2021) Scale model test study on negative skin friction of piles considering the collapsibility of loess. Acta Geotechnica
4. Shan Y, Huang A J, Qin XG et al (2022) Long-term in-situ monitoring on foundation settlement and service performance of a novel pile-plank-supported ballastless tram track in soft soil regions. Transport Geotech 36
5. Weng XL, Hu JB, Mu XH et al (2022) Model test study on the influence of the collapsibility of loess stratum on an urban utility tunnel. Environ Earth Sci 82(1)
6. Zhakulin AS, Zhakulina AA, Zhusupbekov AZ et al (2022) Prediction of the settlement of foundations by an elastic-plastic model of clay soils. Soil Mech Found Eng 59(3)
7. Nguyen QV, Fatahi B, Hokmabadi AS (2016) The effects of foundation size on the seismic performance of buildings considering the soil-foundation-structure interaction. Struct Eng Mech 58:1045–1075
8. Deng JH, Zeng T, Yuan S et al (2022) Interval prediction of building foundation settlement using kernel extreme learning machine. Front Earth Sci
9. Peng WZ, Zhao MH, Zhao H et al (2020) A two-pile foundation model in sloping ground by finite beam element method. Comput Geotech 122(C)
10. Han XL, Li YK, Ji J et al (2016) Numerical simulation on the seismic absorption effect of the cushion in rigid-pile composite foundation. Earthquake Eng Eng Vibr 15(2)

11. Jie YX, Wei YJ, Wang DL et al (2021) Numerical study on settlement of high-fill airports in collapsible loess geomaterials: a case study of Lüliang Airport in Shanxi Province, China. J Central South Univ 28(03):939–953
12. Li L, Ma R, Li JM et al (2022) The group effect on the lateral earth pressure of piles subjected to lateral soil movement. J Disaster Prevent Mitigation Eng 42(03):561–570
13. Hu Q, Jiang J, Yan XS (2005) Three-dimensional numerical simulation analysis of settlements of buildings on soft soil. Rock Soil Mech 12:2015–2018
14. Sheng ZQ, Shi YC, Sun JJ et al (2013) Three-dimensional FEM analysis of settlement and deformation of pile-soil system under vertical load using ABAQUS. Chin J Geotech Eng 35(S1):366–371
15. Xu YR, Leung CF, Yu J et al (2018) Numerical modelling of hydro-mechanical behaviour of ground settlement due to rising water table in loess. Nat Hazards 94(1)
16. Mehdi M, Ali S, Mojtaba H et al (2012) Evaluation of soil collapse potential in regional scale. Nat Hazards 64(1)
17. Xing HF, Liu LL (2018) Field tests on influencing factors of negative skin friction for pile foundations in collapsible loess regions. Int J Civil Eng 16(10)
18. Charles W, Hamed S, Fardin J (2016) Compression and shear strength characteristics of compacted loess at high suctions. Canad Geotech J 54(5)
19. Ong DEL, Leung CF, Chow YK et al (2015) Severe damage of a pile group due to slope failure. J Geotech Geoenviron Eng 141(5)

Open Access This chapter is licensed under the terms of the Creative Commons Attribution 4.0 International License (http://creativecommons.org/licenses/by/4.0/), which permits use, sharing, adaptation, distribution and reproduction in any medium or format, as long as you give appropriate credit to the original author(s) and the source, provide a link to the Creative Commons license and indicate if changes were made.

The images or other third party material in this chapter are included in the chapter's Creative Commons license, unless indicated otherwise in a credit line to the material. If material is not included in the chapter's Creative Commons license and your intended use is not permitted by statutory regulation or exceeds the permitted use, you will need to obtain permission directly from the copyright holder.

Research on Intelligent Mining of Super-High Core Wall Dam Complex Soil Stockyard

Chao Hu, Jialin Yu, Guike Zhang, and Zhongyao Wang

Abstract This research takes the mining planning of complex soil yards and the dam construction process as research objects. Based on the spatial morphological characteristics of complex soil yards, the three-dimensional models of the soil yards are established. Aiming at the multi-source and multi-dimensional data structure and type characteristics of soil material characteristics, the integration and unified management of soil material multi-source data are realized based on the data warehouse. The spatial interpolation algorithm was used to analyze and query the physical characteristics of soil material at any position. A virtual analysis model of soil material information system is constructed. Aiming at the characteristics of core wall filling construction process, a construction progress simulation model considering the impact of low temperature and rainy conditions on core wall filling was established, and the dynamic simulation of core wall construction progress with the filling layer as the basic unit, and its three-dimensional visualization display were realized. Based on the simulation data of the core wall filling process and considering the quality requirements of the core wall filling material, a dynamic planning model for the soil stockyard mining system was established, and the dynamic planning for the multi-material field earth material mining was realized. The above models and methods were verified on a practical project.

Keywords Super-high core wall dam · Complex soil yard · Intelligent mining · Construction simulation

---

C. Hu · J. Yu (✉) · Z. Wang
China Renewable Energy Engineering Institute, Beijing, China
e-mail: yujialin_creei@163.com

C. Hu
e-mail: huchao0621@vip.163.com

Z. Wang
e-mail: wangzhongyao@creei.cn

G. Zhang
Yalong River Hydropower Development Company, Ltd., Chengdu, China
e-mail: zhangguike@ylhdc.com.cn

© The Author(s) 2026
G-F. Zhao (ed.), *Geotechnical Modeling and Intelligent Systems*,
https://doi.org/10.1007/978-981-96-6925-7_2

1 Introduction

Southwest China has abundant hydropower resources. Hydropower projects in these areas are usually located in high mountain canyons, with complex geological conditions, inconvenient external transportation, and harsh construction environments. The rockfill dam has been widely used in the construction of hydropower projects because of its good adaptability to basic conditions, soil and rock dams can obtain materials locally, and can make full use of excavation materials of structures, with low cost and good structural stability [1–5]. In recent years, China has built a number of super-high core wall dams in the southwestern region.

The core wall dam has high control standards for filling materials and large filling volume [3, 6]. Due to the special construction conditions, the control of materials mainly has the following problems: (1) the current exploration information of the stock yard is limited, and it is impossible to fully grasp the material quality distribution and reserve information before excavation [7, 8]. (2) The quality distribution of soil material in the soil yard is quite different, and the proportion of soil mixed with gravel and the adjustment method of the water content in each yard are different [9]. (3) The demand for soil material at different elevations at different times is quite different [11, 12]. Therefore, it is important to figure out the accurate quality and reserve distribution the soil material and then design a reasonable mining plan for dam construction.

This paper introduces an intelligent analysis for multi-source information fusion simulation system, and establishes an analysis model of intelligent mining system for soil material field. The practice results show it can help planning soil excavation.

2 Multidimensional Information Modeling of Complex Soil Stockyard

2.1 Morphological Scanning of Soil Field Based on 3D Laser Scanning

Three-dimensional laser scanning technology is a measurement technology developed in recent years. It has the characteristics of real-time, initiative, non-contact, high precision, and so on. The scanning process is mainly divided into field operations and internal operations. The field operations mainly refer to the collection of data. In field scanning, the scanner and RTK are used together to establish a common target system to complete the data positioning operation. The scanning process is as shown in Fig. 1, and the point cloud is as shown in Fig. 2.

Fig. 1 Scanning process (original image)

Fig. 2 Point cloud data of soil stockyard

2.2 Point Cloud Denoising

Because the data collection during the excavation process is susceptible to unavoidable factors such as vibration, dust, and construction equipment obstruction, there will be "offset points" in the collected data. These "offset points" can also be called noise [13]. In the point cloud data, the noise can be divided into large-scale noise and small-scale noise according to its distance from the target. The large-scale noise is an isolated point or a cluster of points that clearly deviate from the target, and small-scale noise is points or clusters of points scattered near the target surface. Large-scale denoising uses the hypergeometric sphere denoising method, and small-scale denoising uses the improved bilateral filtering method. The large-scale and small-scale noise reduction effects are shown in Figs. 3 and 4, respectively.

3 Three-Dimensional Modeling

The multi-dimensional information model of the stockyard and the three-dimensional geometric information are the basic carriers of various types of information[10]. In addition, it also includes other information such as the geological information, structure information, material characteristic information, physical and mechanical parameters of the soil stockyard. Various types of information can be divided into static information and dynamic information according to their data structure characteristics. Static information is mainly three-dimensional spatial information and its

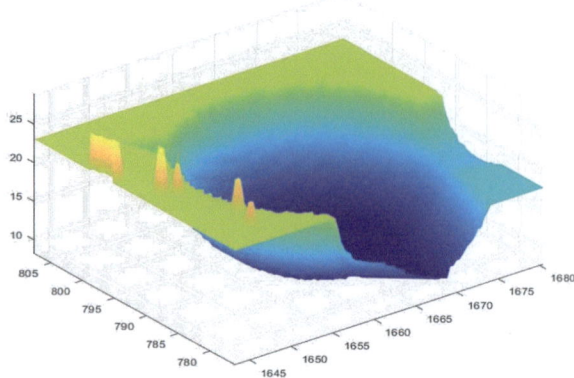

Fig. 3 3D topographic map before denoising

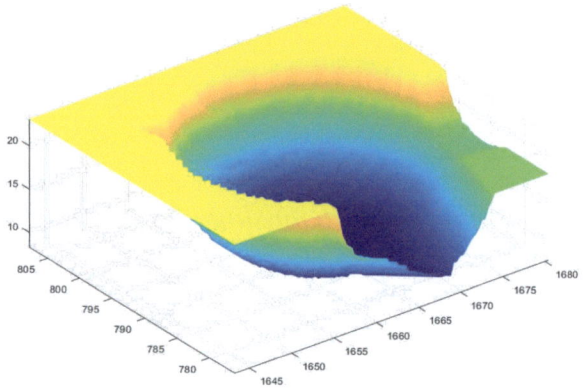

Fig. 4 3D topographic map after denoising

related characteristic attributes. Dynamic information is information that needs to be continuously updated with the construction process, such as quality and construction attributes. According to different information characteristics, different collection methods and modeling methods are selected. In order to realize the rapid update of the stock yard shape, this paper introduces the 3D laser scanning technology to obtain the real-time image of the stock yard excavation. The points obtained with the 3D laser scanner are all irregularly distributed 3D data points. Thus, the Delaunay triangle formation method based on adaptive meshing is used to model the 3D solid terrain. A schematic diagram of the point-by-point interpolation method is shown in Fig. 5, and a three-dimensional model of the material yard established by this method is shown in Fig. 6.

Fig. 5 Schematic diagram of triangle network generation by point insertion

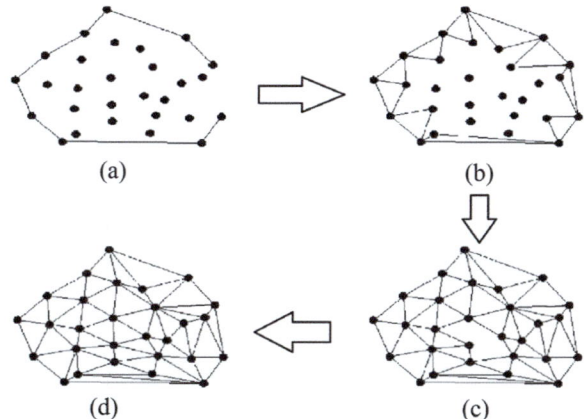

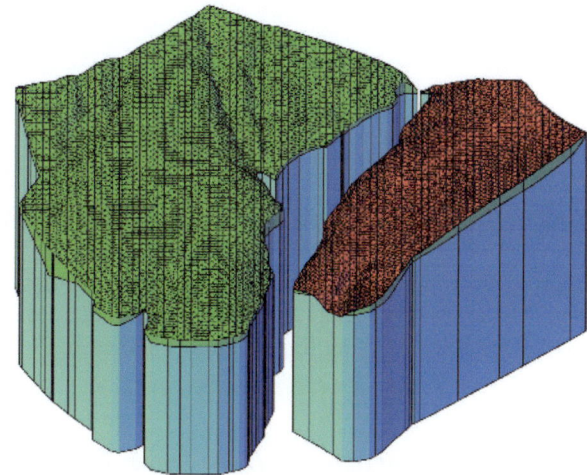

Fig. 6 Three-dimensional solid model

3.1 Spatial Feature Interpolation of Soil

Spatial interpolation is a key step in the analysis of soil material quality characteristics. Through spatial interpolation, point information of unknown locations can be estimated based on limited soil detection data. From the position relationship between the interpolation point and the observation point, interpolation can be divided into two types: interpolation and extrapolation [14, 15]. Data is encrypted within the known data as interpolation, and according to the current data segment, it is judged as outside by trend analysis. In this study, the inverse distance weighting method and kriging method were used to perform spatial fitting on the soil material quality. After multiple experiments, it was found that the interpolation accuracy and the number of pits directly affect the interpolation method and accuracy. When smaller, the inverse

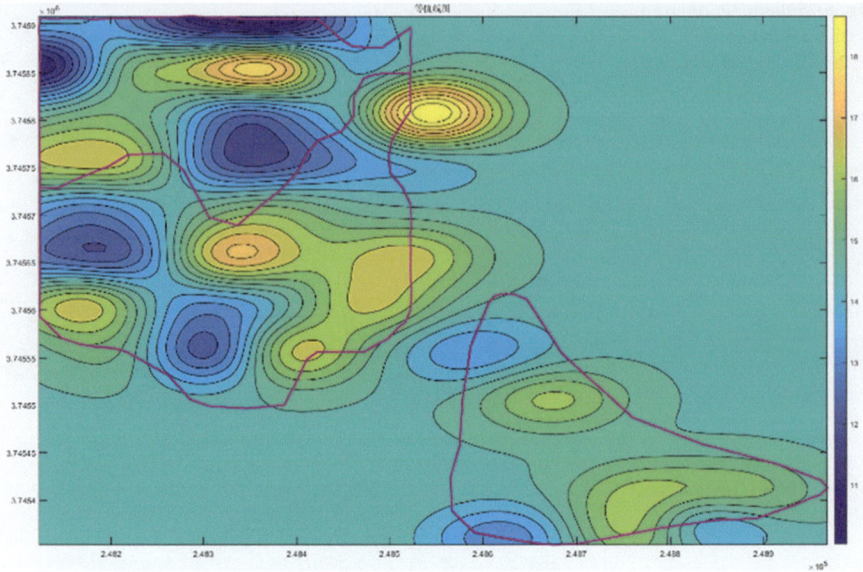

Fig. 7 Contour map of water content distribution

distance weighted interpolation has better accuracy. When the original number is larger, the kriging interpolation result is obviously better than the weighted average interpolation method. The main method used in this research is kriging interpolation. The analysis of water content and P5 content distribution at a certain depth in a certain stockyard using this method is shown in Figs. 7, 8, and 9.

3.2 Multi-dimensional Data Integration for Soil Yard

According to the analysis of the characteristics of the soil material field data, a multi-dimensional information model of the soil material field can be established, including the three-dimensional terrain model of the soil material field, geological exploration information, characteristic data of the material field, and physical parameters of the material field. Expressed as:

$$\Omega = \{\Omega_1, \Omega_2, \cdots \Omega_n\}$$

n is the information dimension included.

Among them: Ω_1 represents a collection of three-dimensional spatial models of soil yards, $\Omega_1 = \{V_1, V_2, \ldots, V_m\}$ (m represents the number of space divisions), Ω_2 represents the set of soil P5 content in each subspace of the soil yard, $\Omega_2 = \{P_1, P_2, \ldots, P_m\}$, Ω_3 represents the collection of soil moisture content in each

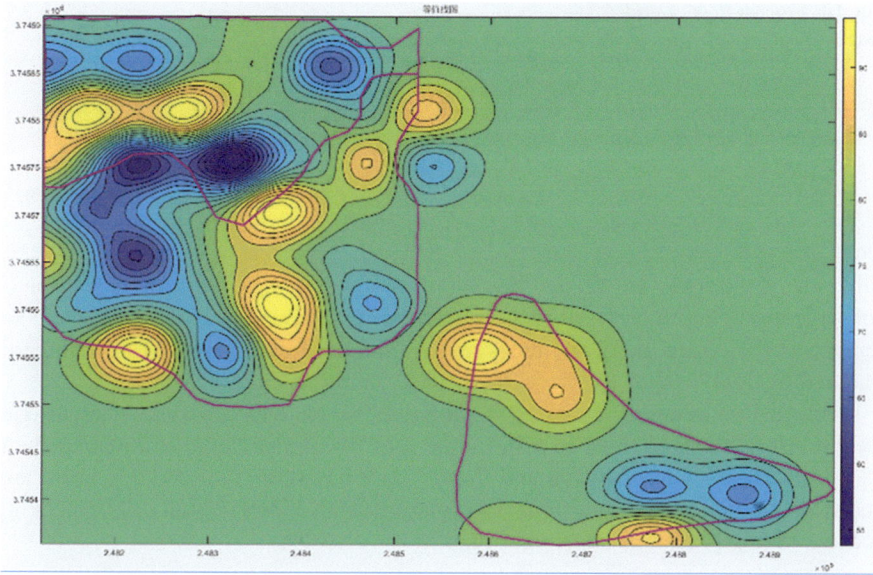

Fig. 8 Contour map of particle 5mm soil

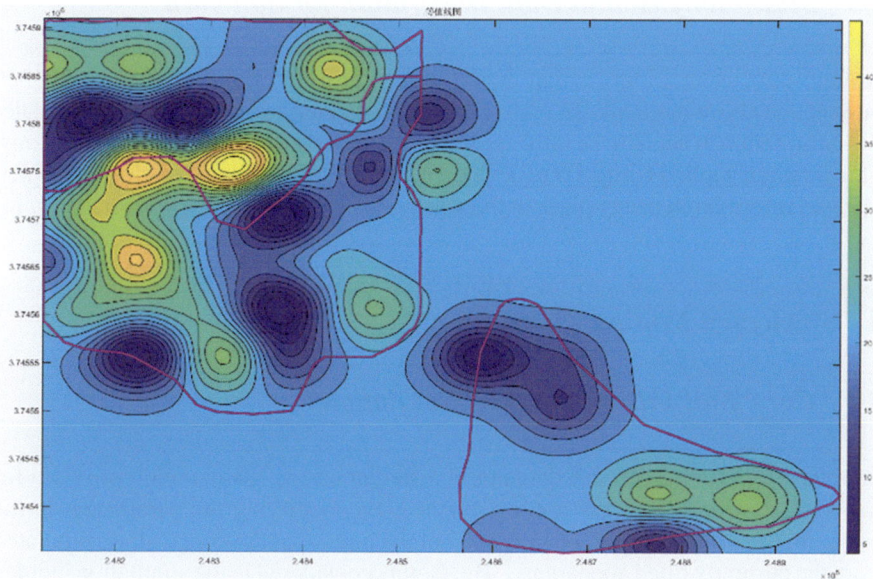

Fig. 9 Contour map of P5 content distribution

Fig. 10 Process flow of grid space

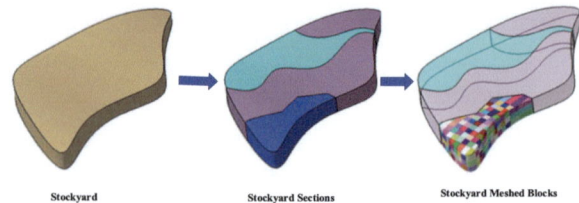

Stockyard   Stockyard Sections   Stockyard Meshed Blocks

subspace of the soil yard, $\Omega_3 = \{W_1, W_2, \ldots, W_m\}$, Ω_4 represents the collection of soil gradation data in each subspace of the soil yard. $\Omega_4 = \{G_1, G_2, \ldots, G_m\}$, Ω_5 represents the collection of the recoverable amount of earth material in each earth material field, $\Omega_5 = \{v_1, v_2, \ldots, v_m\}$.

Each feature contains a variety of data dimensions and contains different data format information. In order to achieve the effective organization and management of various types of information and to provide a basis for engineering control and decision-making, multidimensional information needs to be integrated. The stocking space gridding process is shown in Fig. 10.

3.3 Multi-dimensional Data Integration for Soil Yard

A multi-dimensional information model of the earth material field was established through the above methods. To realize the intuitive query and analysis of earth material information, it needs to be implemented with the help of virtual reality technology. This research applied Unity 3D to develop a virtual query analysis platform. The soil material information in any place in the stockyard can be queried as shown in Fig. 11.

4 Stockyard Mining Planning

4.1 Core Wall Filling Process Simulation

In order to determine the soil demand in different stages, a simulation program for the core wall filling process was established. By modifying the parameters in the program, the simulation can be started from the bottom of the core wall or simulated from the currently filled appearance. The factors considered in the simulation include the size of the soil surface, the type of soil material, the mechanical configuration, the transportation conditions, and meteorological and climatic factors.

Fig. 11 Query of soil material information

4.2 Stockyard Mining Planning Based on Simulation Result

According to the dam body filling progress plan, the simulation analysis can obtain the actual demand for different soil materials in different stages. Decompose according to the demand and the actual reserves, remaining reserves, and difficulty of mining of each soil yard, and simulate the mining methods and mining processes of each soil yard to configure appropriate mining machinery and equipment. Dynamic simulation state changes are shown in Fig. 12.

Suppose that the state s_k at stage k, after executing the selected decision x_k, the state becomes $s_{k+1} = T_k(s_k, x_k)$. According to the optimal principle, after taking the optimal sub-strategy for the sub-process of $k + 1$, The optimal index function of the process behind k is:

$$f_k(s_k) = \underset{x_k \in D_k(s_k)}{opt} \{r_k(s_k, x_k) \Theta f_{k+1}(s_{k+1})\} = \underset{x_k \in D_k(s_k)}{opt} \{r_k(s_k, x_k) \Theta f_{k+1}(T_k(s_k, x_k))\} \quad (1)$$

$$k = n, n-1, \ldots, 1$$

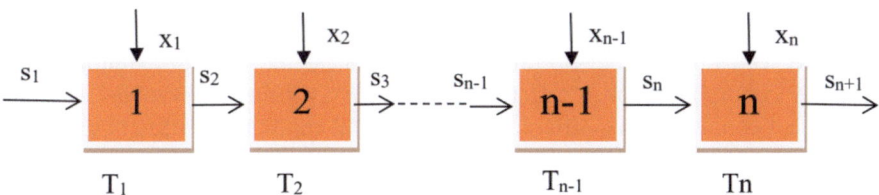

Fig. 12 State change diagram

The process of solution optimization is to minimize the total cost of the entire process, so the following objective functions can be established:

$$F = \min\left(\sum_y \sum_z \sum_{St} C_{yz}^{St} X_{yz}^{St}\right) \quad (2)$$

5 Engineering Applications

A hydropower station in the southwest China, the volume of core wall is 4.4 million m3. The material sources of the gravelly soil core wall are XDi(Area A, B), PGY (A, B), YZ (A, B, C), GL (A, B1, B2), and PBR (A, B) have a total of 5 materials yards and 12 warehouses. During the implementation stage, increase the bulk yard of PBR C area.

5.1 System Module Design

According to the demand analysis of the intelligent mining management system for complex soil yards, in view of the integrated terrain and geological model of the system and its huge multi-dimensional data, a C/S implementation method can be used to ensure operational performance and user experience. The entire system includes three modules: full information model management, mining plan simulation, and system management. It involves the dynamic management of the three-dimensional soil material field geological model, the spatial grid processing of the three-dimensional soil material field model, and the addition of multi-information and multi-information data management, intelligent simulation of the stockyard mining program simulation, interactive query of full information model data show 5 main functions.

5.2 Core Wall Filling Process Simulation

The core wall filling process simulation can realize the progress simulation of boundary conditions such as different time periods, different elevations, different mechanical configurations, different meteorological conditions, different operating parameters, etc. Figure 13 show the simulation interface and results from the bottom of the core wall. Figure 14 shows the relationship between rising height and filling volume.

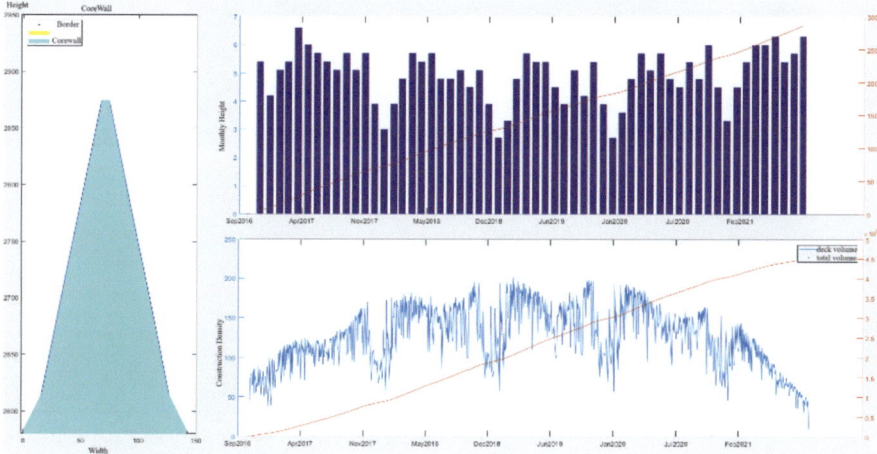

Fig. 13 System simulation process

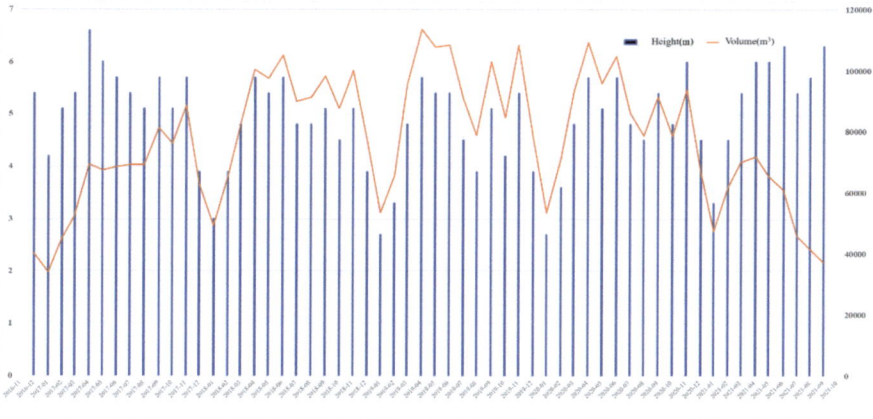

Fig. 14 Relationship between rising height and filling volume

5.3 Stockyard Mining Planning

According to the simulation results, mining planning is performed for the excavation process of the earth material yard. The planning data comes from the core wall filling simulation result. The planning can be carried out from boundary conditions such as different periods, different elevations, and different materials. The calculation interface as shown in Fig. 15 and the planning result as shown in Fig. 16.

The results of a certain plan are shown in Table 1. As can be seen from the table, the planning results fully take into account the principle of good materials and full use.

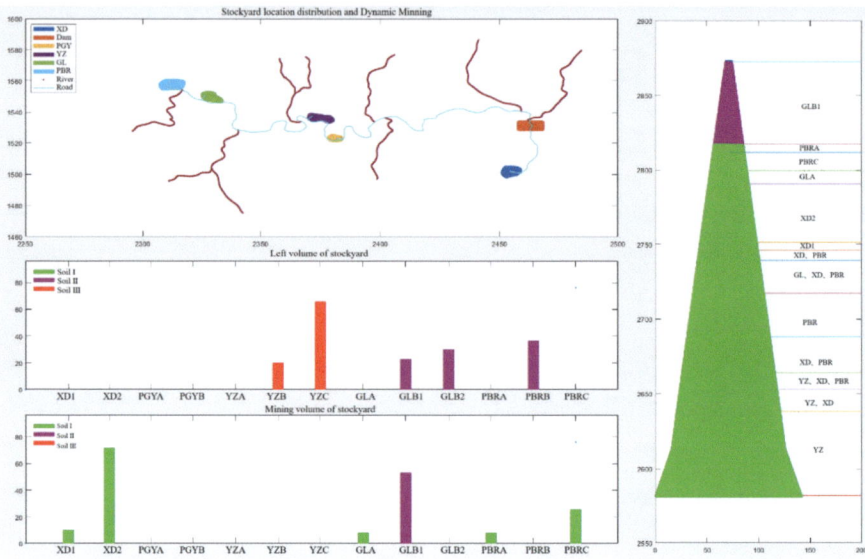

Fig. 15 Stockyard mining planning interface

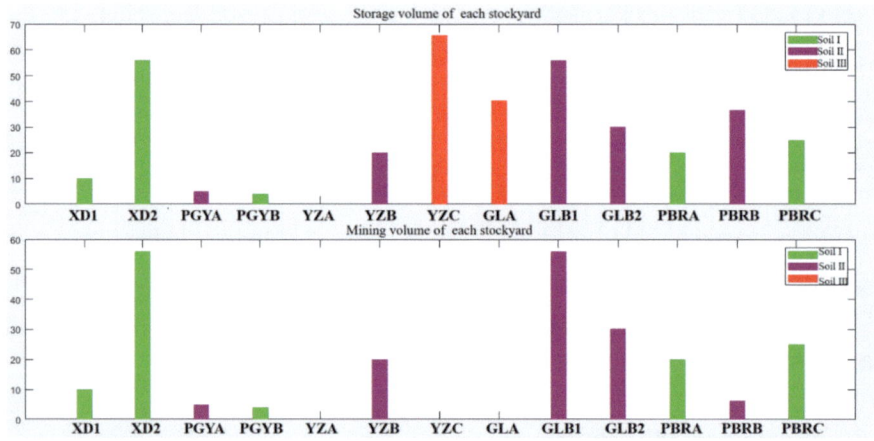

Fig. 16 Planning results

This paper takes complex earth yard mining planning as the research object. By establishing a multi-dimensional information analysis model of earth yard and a soil yard mining planning model based on core wall filling simulation, it provides scientific basis of information management and planning for earth mining. The main research contents are as follows.

Table 1 Result of the entire mining yard planning (10,000 m^3)

Stockyard	Soil type I			Soil type II			Soil type III		
	Original	Mined	Left	Original	Mined	Left	Original	Mined	Left
XD1	10	10	0	0	0	0	0	0	0
XD2	145.5	145.5	0	0	0	0	0	0	0
PGYA	0	0	0	0	0	0	0	0	0
PGYB	0	0	0	0	0	0	0	0	0
YZA	65.75	65.75	0	0	0	0	0	0	0
YZB	0	0	0	0	0	0	20	0	20
YZC	0	0	0	0	0	0	65.8	0	65.8
GLA	27	27	0	0	0	0	0	0	0
GLB1	0	0	0	76.12	45	31.12	0	0	0
GLB2	0	0	0	30.2	0	30.2	0	0	0
PBRA	96.3	96.3	0	0	0	0	0	0	0
PBRB	0	0	0	36.6	0	36.6	0	0	0
PBRC	38	38	0	0	0	0	0	0	0

(1) A method of systematically constructing a three-dimensional model of the soil stockyard, and the integration and unified management of soil material multi-source data are realized based on the data warehouse; the spatial interpolation algorithm is used to analyze and query the physical characteristics of soil material at any location. A virtual analysis model of soil material information was constructed based on the three-dimensional model of the soil material field and the multi-dimensional characteristic data.

(2) According to the characteristics of core wall filling construction, a construction progress simulation model considering inner and out parameters was established, and the dynamic simulation of core wall construction progress with the filling layer as the basic unit was realized and its three-dimensional visualization was displayed.

(3) Based on the simulation data of the core wall filling process, considering the quality requirements of the core wall, a dynamic planning model for the earth material field mining was established, and the dynamic planning for the multi-material field earth material mining was realized. Integrate the above theories and methods to build an intelligent mining system, and realize some functions such as integrated management of stockyard information, virtual analysis and query of earth material information, simulation of core wall construction process, and so on. And the above models and methods were verified on a practical project.

Acknowledgements This work was supported by Yalong River Hydropower Development Company, Ltd, and Open Foundation of Hubei Key Laboratory of Construction and Management in Hydropower Engineering (2020KSD07).

References

1. Zhou Y, Zhou H, Chen T et al (2023) Detection of rockfill gradation based on video image recognition. Autom Construct 154:104965
2. Zhang S, Wu G, Yang X et al (2018) Digital image-based identification method for the determination of the particle size distribution of dam granular material. KSCE J Civil Eng 22(8):2820–2833
3. Hassan M, Saeid S, Hadi A (2021) A new approach to improve the assessment of rock mass discontinuity spacing using image analysis technique. Int J Rock Mech Min Sci 143:104760
4. Hu C, Chen Y, Chen T et al (2019) Research on fast calculation method of complex soil yard excavation volume based on 3D laser scanning. IOP Conf Ser Earth Environ Sci 304(3):32076
5. Yuan S, Zhao W, Gao S et al (2022) An adaptive threshold-based quantum image segmentation algorithm and its simulation. Quantum Inf Process 21(10)
6. Haitao G, Carlos QJ, Le HV et al (2021) Three-dimensional simulation of asphalt mixture incorporating aggregate size and morphology distribution based on contact dynamics method. Construct Build Mater 302
7. Olde Scholtenhuis LL, Hartmann T, Dorée AG (2016) 4D CAD based method for supporting coordination of urban subsurface utility projects. Autom Constr 62:66–77
8. Kassem M, Dawood N, Chavada R (2015) Construction workspace management within an industry foundation class-compliant 4D tool. Autom Constr 52:42–58
9. Fuchs S, Scherer RJ (2017) Multimodels—instant nD-modeling using original data. Autom Constr 75:22–32
10. Johansson M, Roupé M, Bosch-Sijtsema P (2015) Real-time visualization of building information models (BIM). Autom Constr 54:69–82
11. Liang S, Liu W, Wang N et al (2020) Application of consumer drone in the earthwork estimation of artificial deposit in construction project. Sci Soil Water Conservation, 2020, 18132–138.
12. Chao H, Zhou Y, Zhao C et al (2015) Slope excavation quality assessment and excavated volume calculation in hydraulic projects based on laser scanning technology. Water Sci Eng 8(2):164–173
13. Marinoni O (2003) Improving geological models using a combined ordinary–indicator kriging approach. Eng. Geol 69(1)
14. Yang Q, Wei J, Zheng J et al (2019) Research on interpolation method of DEM based on discrete point cloud. Sci Survey Map 44(07)
15. Yang Y, Li S, He F et al (2019) Influence of semivariogram and sampling interval on Kriging method in marine stratigraphic analysis. J Eng Geol 27(04)

Open Access This chapter is licensed under the terms of the Creative Commons Attribution 4.0 International License (http://creativecommons.org/licenses/by/4.0/), which permits use, sharing, adaptation, distribution and reproduction in any medium or format, as long as you give appropriate credit to the original author(s) and the source, provide a link to the Creative Commons license and indicate if changes were made.

The images or other third party material in this chapter are included in the chapter's Creative Commons license, unless indicated otherwise in a credit line to the material. If material is not included in the chapter's Creative Commons license and your intended use is not permitted by statutory regulation or exceeds the permitted use, you will need to obtain permission directly from the copyright holder.

Three-Dimensional Numerical Simulation Study on the Influence of Reservoir Water Level Variation on the Stability of Multiple Taluses

Zelin Zhou, Jie Zhang, Chao Feng, and Maoyi Liu

Abstract In order to investigate in-depth the impact of water level fluctuation on the stability of slopes with multiple accumulated deposits, this article takes the example of the deep accumulated deposits in Longwangxi. Based on the exploration of engineering geological conditions and the surrounding areas affected by historical landslides, a large-scale 3D numerical model is established to calculate the stability, deformation, and plastic zone variations of different accumulated deposits on the slope under different water level conditions through fluid-soil coupling analysis. The influence of water level changes on the stability of accumulated deposits is analyzed based on the numerical results. The analysis shows that when the water level is below the range of accumulated deposits, the impact of water level fluctuation on their stability can be neglected. However, as the water level rises above the range of accumulated deposits, significant deformations occur at the front edge of the deposits due to the softening and suspension-induced reduction in weight of the soil-rock mass, leading to even larger overall deformations of the accumulated deposits. In extreme cases, such as when the upstream reservoir gates of tributaries are opened, resulting in a sudden drop in water level, the entire slope of the accumulated deposits may become unstable. Furthermore, accumulated deposits located above the highest water level line may experience significant displacement due to disturbances caused by the instability of deposits below, thus indirectly affected by water level fluctuations.

Keywords Three Gorges Reservoir · Accumulated deposits · Water level variation · Fluid-soil coupling

---

Z. Zhou · J. Zhang · C. Feng
China 19Th Metallurgical Corporation, Chengdu, Sichuan Province, China

M. Liu (✉)
Chongqing Construction Investment (Group) Co., Ltd., Chongqin, China
e-mail: 41824251@qq.com

1 Introduction

The Three Gorges Reservoir Area, located in the southwestern mountainous region of China, is a key area for landslide prevention and control. The density of geological hazards in this area is 3.3 times higher than the national average. Among the geological disasters occurring in this reservoir area, landslides of loose accumulated deposits from the Quaternary period account for 42.85% of the total. These loose accumulated slopes may experience landslide disasters during changes in reservoir water levels.

Numerous scholars have conducted research on the stability of bank slopes in the Three Gorges Reservoir Area with respect to water level fluctuations. Ding et al. [1] analyzed the seepage field and stress field of the Nanqiaotou landslide under different water level conditions, and studied the deformation trends and failure characteristics of the landslide under the coupling effect of stress and seepage. Zhang and Cheng [2] simulated the transient seepage field of the landslide under combined rainfall infiltration and reservoir water level conditions, and then used the distribution of transient pore water pressure for limit equilibrium analysis of the landslide. Tao et al. [3], based on existing monitoring data, analyzed the variation patterns of deformation with time and deformation with water level, and analyzed the stress–strain characteristics inside the landslide under reservoir water level change conditions using two-dimensional numerical simulation. Fang [4] studied the stability of the Hujia slope landslide in the Three Gorges Reservoir Area under the combined action of reservoir water level and rainfall, and predicted the deformation and failure of the slope during subsequent reservoir water level changes. Zhang et al. [5] analyzed the influence of reservoir water level changes on the stability and sliding mode of bank slopes in the Three Gorges Reservoir Area through three-dimensional numerical modeling considering fluid-soil coupling. Liu et al. [6] conducted numerical simulations on the 1# landslide deposit of a loess slope near the river, analyzing the variations of stress and strain within the landslide under conditions of reservoir water level fluctuations, and proposed a strength reduction method considering cyclic loading. Lei and Lei [7] studied the deformation patterns and stability of accumulated deposit landslides in the reservoir area under different reservoir water level conditions through analysis of monitoring data, limit equilibrium analysis, and numerical simulations. Their research concluded that the safety factor of accumulated deposit landslides increases during the process of water level rise and decreases when the water level rapidly drops.

Currently, there have been numerous studies on the impact of reservoir water level fluctuations on bank slope stability. However, research on bank slopes in the Three Gorges Reservoir Area beyond the conventional water level range of 145–175 m is relatively limited. Additionally, there is a lack of research on the interaction between different landslides in different sections with the help of three-dimensional numerical analysis. Therefore, it is necessary to conduct research in this area.

This study focuses on the talus in the Longwangxi section, which is located on the side of the Caijiaba landslide at the Caijiaba Dam. The south abutment approach bridge of the Fengming-Chongqing Expressway Project passes through the slope angle of this talus. Instability in this slope area could cause damage to the approach bridge and even lead to further instability of the Caijiaba landslide above it. The source of Longwangxi upstream is the Nongjian Reservoir in Fengming Town, with significant differences in flow and water levels between the rainy season and dry season. During the rainy season, the maximum floodwater level released from the reservoir can reach around 185 m, while during normal times, it remains at the same level as the Yangtze River. In order to investigate the stability of the talus in the Longwangxi section under different water level conditions, this study, based on the engineering geology and geotechnical characteristics of the talus, combines Rhino modeling and Flac 3D finite element software to establish a three-dimensional model for stability analysis under different water level fluctuation conditions.

2 Engineering Geological Overview

2.1 Topography and Landforms

The deep talus of Longwangxi belongs to the landform of structural erosion and valley erosion. The slope terrain is steep, with a slope angle generally ranging from 25° to 45°. There are local steep cliffs with elevations ranging from 192.0 to 284.0 m. The overall slope presents a stepped topography with steep lower sections, gentle middle sections, and steep upper sections. The slope is covered with abundant vegetation, and the bottom of the slope has a gentle terrain with a gradient of 3°–10°. The lowest elevation of the Longwangxi riverbed is approximately 180.0 to 186.0 m. Three accumulation bodies, G1, G2, G3, and G4, are distributed on the studied slope (see Fig. 1). There is no obvious geological boundary between accumulation bodies G3 and G4, so this study considers them as the same accumulation body for modeling purposes. G1 is located below G2.

2.2 Geological Structure

The slope is located on the northern flank of the Wanzhou syncline. The attitude of the rock layers is 120°–151° with a dip angle of 8°–13°. Two sets of fractures were observed in the bedrock: (1) strike direction of 2°–5° with a dip angle of 80°–85°, the fracture surface is relatively flat and smooth, with an opening of 4–10 mm, extending 2–3 m, and spaced at 3–5 m apart. There is a small amount of muddy filling, and the bond is relatively poor, indicating rigid structural planes; (2) strike direction of 283°–295° with a dip angle of 70°–78°, the fracture surface is relatively flat and smooth, with an opening of 5–100 mm, extending 1.5–5.5 m, and spaced at 2–3 m

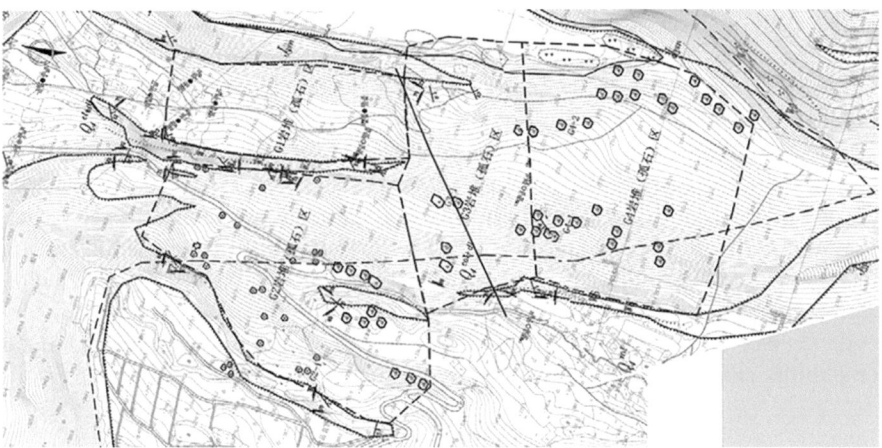

Fig. 1 Geological plan of the Longwangxi talus

apart. There is muddy filling, and the bond is weak, also indicating rigid structural planes.

2.3 Lithology

According to the field drilling data, the interface between the rock and soil is steep, generally at 8°–12°, with the rear portion exceeding 55°. The length of the landslide deposit is approximately 640 m, with a width of 100–180 m and a thickness of 8–32 m. The landslide deposit mainly consists of sand, shale fragments, and silty clay, with a content of about 20–70% gravel. The underlying bedrock is composed of Upper Jurassic Penglai Formation (J3p) sandstone and shale. The bottom of the deposit is mainly composed of purple-red shale detrital fragments.

2.4 Hydrogeological Conditions

The region falls under a subtropical monsoon climate zone with mild temperatures and abundant rainfall. The average annual rainfall ranges from 1149.3 to 1213.5 mm, with the maximum annual precipitation recorded at 1614.8 mm. The distribution of rainfall throughout the year is uneven, with a concentration of rainfall occurring between May and September, accounting for 70% of the total annual rainfall. The maximum daily rainfall reached 190.3 mm in June 2017. The normal operating water level of the Three Gorges Reservoir fluctuates between 145 and 175 m. There are multiple small tributaries flowing into the Longwang Creek, primarily relying on rainfall as the water source. During the rainy season, the water flow can reach a maximum of approximately 30 m³/s, while during the dry season, it decreases to a minimum of around 0.3 m³/s. The water level is normally 20 m higher than that of the Three Gorges Reservoir, but during flooding situations when upstream reservoirs discharge water, it can rise up to a maximum flood level of 183.5 m. The 175 m water level reaches the lower edge of the G3 and G4 deposits. Based on on-site pumping tests, the permeability coefficient of the deposit was measured at 8.28 m/d, indicating a moderate permeability.

2.5 Geotechnical Properties

The investigated slope is located on the side wing of Caijiaba. Very limited research has been conducted on the Caijiaba landslide. Zhang Yan et al. extensively discussed the mechanism of the Caijiaba landslide based on multiple sources of data and determined the saturated shear strength parameters of the sliding soil as $C = 14.16$ kPa and $\varphi = 8.26°$ through laboratory and field experiments. Our team compared the content of rocks in the rock-soil interface of the Longwangxi thick deposit with that of the sliding soil in the Caijiaba landslide through field surveys and laboratory tests, and found that the former had a higher rock content. Research conducted by Luis and Roger [8], Li et al. [9], and Li et al. [10] concluded that the cohesive strength of a deposit decreases as the rock content increases, while the internal friction angle shows the opposite trend. Both cohesive strength and internal friction angle decrease with increasing saturation. Taking into account the research results of Zhang et al. [5] and the shear strength parameters obtained from indoor tests on the Longwangxi deposit under natural and saturated conditions, our team determined the shear strength indicators as $C = 8.0$ kPa and $\varphi = 27°$ in its natural state, and $C = 5.0$ kPa and $\varphi = 24°$ in a saturated state. The mechanical model of the deposit follows the Mohr–Coulomb model, and Fig. 2 provides a comparison of the saturated shear strength between the sliding soil in the Caijiaba landslide and the rock-soil interface of the Longwangxi deposit in its natural and saturated states.

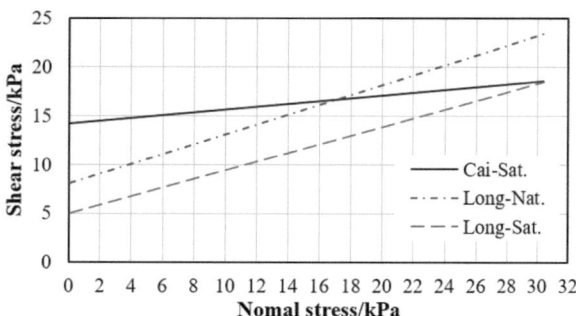

Fig. 2 Relationship between the slip zone soil of the Caijiaba landslide and the rock-soil boundary soil of the Longwangxi talus (σ–τ diagram)

3 Three-Dimensional Numerical Simulation Analysis of Taluses Under Reservoir Water Level Variations

3.1 Calculation Model

In order to analyze the stability of the accumulated slope under changing reservoir water levels, this study adopts FLAC3D, a fluid–solid coupling three-dimensional simulation calculation. The calculation is carried out using a fluid–solid coupling method, and then the overall safety factor of the model is determined using the strength reduction method.

The three-dimensional accumulated slope model in this paper is built and divided into grids using Rhino software and the Griddle plugin. In order to calculate the stability of the accumulated mass, the model is divided into three parts: accumulated mass, soil-rock interface, and underlying bedrock, based on their physical and mechanical characteristics. To obtain a more realistic spatial distribution of the soil-rock interface, we use the method of interpolating and draping based on 107 geological borehole information in Rhino software. After dividing the grids, they are imported into FLAC3D for calculation using the Mohr–Coulomb criterion, as shown in Fig. 3. The model extends 1064 m in the direction of the Longwang River (X-direction), 605 m in the direction of the Yangtze River (Y-direction), with a maximum height of 425 m (Z-direction). The model grid consists of tetrahedra, totaling 60,460 nodes and 322,059 elements. The boundary conditions of the model include lateral normal constraints, full constraints at the bottom, and free surface at the top.

According to the data on water level regulation of the Three Gorges Reservoir collected by Fang [4], during the impoundment phase, the fastest rising rate of the water level in the Three Gorges Reservoir is 1 m/d, and during extreme discharge, the fastest receding rate of the water level is 3 m/d. The highest water level during this discharge phase is 162 m.

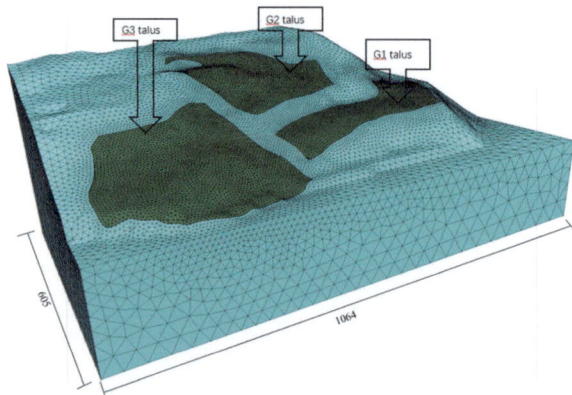

Fig. 3 Three-dimensional numerical calculation mesh

Table 1 Numerical calculation parameters

Formation	γ (kN/m)	E (GPa)	v	φ (°)		C (MPa)		K (m²/Pa s)	n
				Nat	Sat	Nat	Sat		
Talus	22.0	0.85	0.29	33	30	0	0	9.58 × 10^{-9}	0.35
Interface	–	–	–	24	22	8	5	9.58 × 10^{-9}	0.4
Bedrock	26.8	4	0.15	18	–	50	–	5 × 10^{-13}	0.2

Based on historical hydrological data and field investigations, it is known that the water level of the Longwang Creek section studied in this paper is affected by both the downstream water level of the Three Gorges Reservoir and the opening of the gates of the upstream Nongjian Reservoir. During the rainy season, when heavy rain coincides with the opening of the gates of the Nongjian Reservoir, the water level in the Longwang Creek section can reach a maximum of 185 m. When the upstream gates are closed, the water level rapidly drops to match the water level of the Three Gorges Reservoir, with a fastest recession rate of 10 m/d. Considering the actual conditions during rising water levels, four scenarios are calculated: Scenario 1: reservoir water level of 145 m; Scenario 2: reservoir water level of 160 m; Scenario 3: reservoir water level of 175 m; Scenario 4: water level in Longwang Creek of 185 m; Scenario 5: water level in Longwang Creek dropping from 185 to 175 m. The values for the geotechnical parameters used in this study are shown in Table 1, based on indoor testing data and survey reports.

The constitutive model used for the calculation of the rock-soil mass in the analysis is the Mohr–Coulomb criterion, and it is calculated based on the ideal elastic–plastic model.

3.2 Analysis of Calculation Results

Stability Analysis of the Accumulated Slope Under Different Scenarios

Figure 4 shows the safety factors of the Longwang Creek accumulated slope obtained through the strength reduction method for each scenario.

From Scenario 1 to Scenario 4, as the water level rises, the safety factor of the slope decreases. At the critical highest water level of 185 m, the safety factor reaches 1.12, approaching the instability limit. When encountering the extreme condition of Scenario 5, where the water level in Longwang Creek drops abruptly from 185 m to the maximum impoundment level of 175 m in the Three Gorges Reservoir, it was found through fluid–structure interaction analysis that the safety factor is less than 1.0 after the fourth hour of water level decline, indicating overall instability of the accumulated slope. Therefore, under normal impoundment conditions of the

Fig. 4 Variation of safety factor of the talus under different working conditions

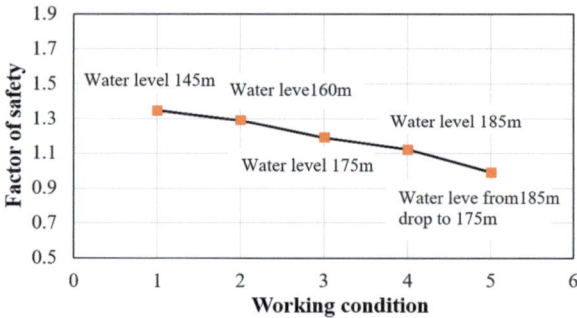

Three Gorges Reservoir, the accumulated slope remains generally stable. However, it is relatively sensitive to sudden drops in water level in Longwang Creek, which can lead to unstable landslides in such situations. Considering the stability of the accumulated slope under different water level scenarios, at high water levels, the mechanical properties of the accumulated slope decrease due to being located below the water surface, resulting in increased deformation and decreased safety factors due to reduced effective stress within the soil and rock mass. Even under extreme water level conditions alone, the safety factor still remains above 1.0, indicating that the accumulated slope remains basically stable. However, when high water levels are combined with sudden drops in water level, the overall safety factor of the accumulated slope is less than 1.0, indicating that the internal seepage force caused by the abrupt water level decline disrupts the balance state of the lower part of the accumulated slope, leading to overall instability and failure.

Deformation Analysis Under Different Scenarios

Figure 5 illustrates the contour maps of displacement in the X-direction (downslope) of the slope under different scenarios. From the figure, it can be observed that the deformation of the slope gradually decreases from the slope surface downwards, with the maximum deformation occurring at the toe of the slope, indicating reasonable simulation results. From Fig. 5a, b, it can be observed that when the water level displacement is below the range of the deposit, the change in water level has minimal impact on the deformation and stability of the deposit. From Fig. 5c, d, it is observed that as the water level rises within the range of the accumulated slope, there is a downslope deformation appearing first at the forefront of the G3 accumulation, which increases in magnitude with the rise in water level. In the process of water level rise mentioned above, the G1 and G2 accumulations remain stable without significant deformation. Figure 5e shows that in Scenario 5, where the water level drops abruptly from 185 to 175 m, a large area of significant deformation occurs at the forefront of the G3 accumulation, which also triggers small-scale significant deformation within the range of the G1 and G2 accumulations. This indicates that if the G3 accumulation experiences deformation instability due to the disturbance caused by the water level drop, it will induce instability phenomena in the G1 and G2 accumulations as well.

Fig. 5 Contour lines of X-direction displacement of the talus under different working conditions

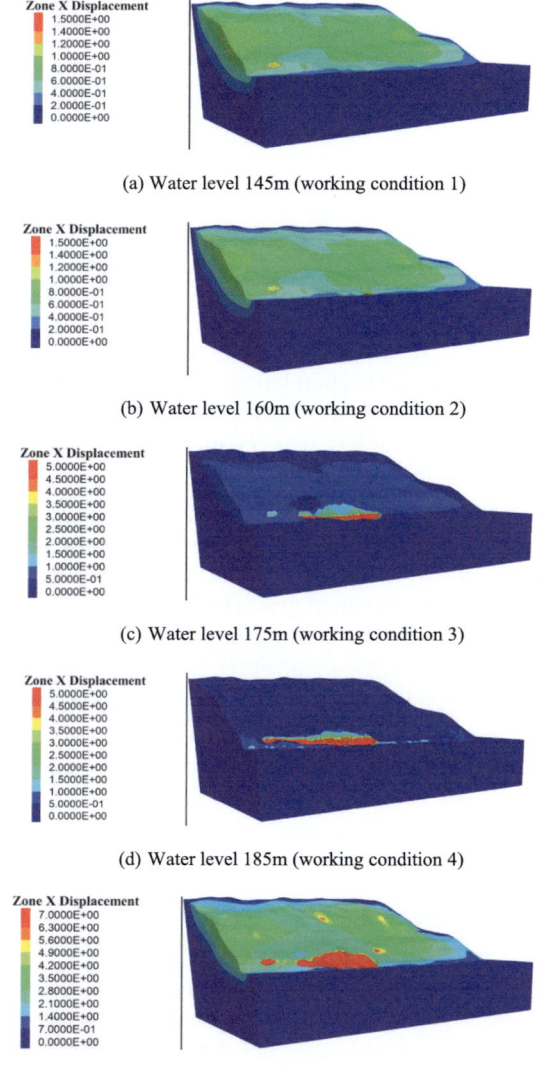

(a) Water level 145m (working condition 1)

(b) Water level 160m (working condition 2)

(c) Water level 175m (working condition 3)

(d) Water level 185m (working condition 4)

(e) Water level drops abruptly from 185m to 175m (working condition 5)

Figure 6 shows the variation of maximum displacement in the X-direction under different scenarios. It can be observed that when the water level does not reach the lower edge of the accumulation, the displacement of the accumulation is very small, around 1 cm. Although it appears at the prominent part of the forefront of the accumulation, the overall displacement at the rear of the accumulation remains relatively consistent. When the water level continues to rise to the height of the accumulation, the maximum displacement in the X-direction generally increases linearly with the rise in water level, and the maximum displacement occurs at the submerged section at

Fig. 6 Variation of maximum displacement in the X-direction under different working conditions

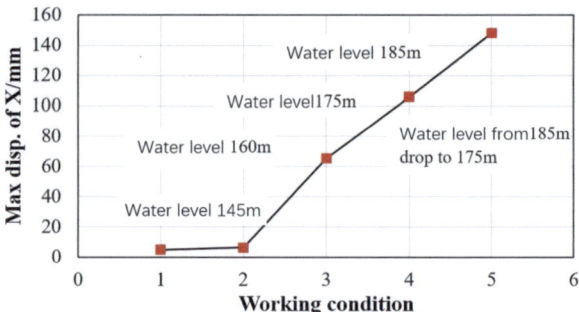

the forefront of the accumulation. Under the scenario of a sudden drop in water level when the accumulation is at a high water level, an extreme X-direction displacement value of 14.8 cm occurs, which represents a relative increase of 39.6% in X-direction displacement compared to the water level at 185 m.

Combining the analysis of the X-direction displacement under different scenarios of the accumulation, we can observe that when the water level is below the accumulation, the deformation of the accumulation occurs primarily through translation. Due to the high content of gravel in the accumulation and its relatively good mechanical properties, the overall stability of the talus is relatively controllable. When the water level rises to the range of the accumulation, the submerged section of the accumulation experiences buoyancy reduction, leading to initial large displacements. Subsequently, the upper part of the accumulation slides and deforms due to the loss of support at the forefront, forming a traction landslide. This deformation exhibits a characteristic of propagation from the forefront to the rear.

3.3 Analysis of Plastic Zone in the Accumulation.

Figure 7 shows the positions of the plastic zone within the accumulation. (a) represents the positions of the plastic zone in G1 and G2 accumulations; (b) represents the positions of the plastic zone in G3 accumulation. From the figure, it can be observed that the plastic zone is primarily distributed along the interface between rock and soil. The plastic zone inside the G1 accumulation has not yet fully connected, but the distribution of the accumulation is more pronounced below the range of 185 m compared to above. This indicates that even under extreme water level conditions, the G1 accumulation may also become unstable. The upper edge of the plastic zone in the G2 accumulation is sporadically distributed, while the plastic zone in the G3 accumulation is essentially connected and has become unstable.

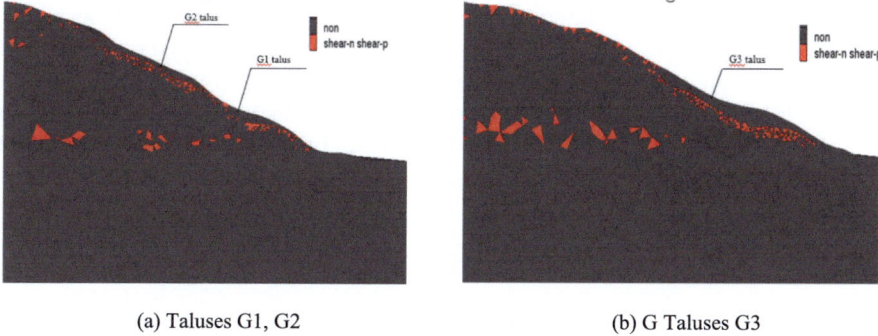

(a) Taluses G1, G2   (b) G Taluses G3

Fig. 7 Position of the plastic zone under working condition 5

4 Discussion and Conclusion

The stability of the talus in the reservoir bank is greatly influenced by water level fluctuations. This study considers the water level fluctuations of both the reservoir and tributaries. A three-dimensional realistic model is constructed, and the simulation calculation method of fluid–solid coupling and strength reduction of rock-soil materials is applied to analyze the stability of the talus under various water level conditions. Additionally, there are few studies on the impact of water level fluctuations on accumulations above the maximum water level line. In this study, a larger-scale three-dimensional model is established, considering three accumulations located at the inflow position of the Three Gorges Reservoir's tributaries, indirectly studying the stability of accumulations above the maximum water level line affected by water level fluctuations. The conclusions obtained in this study are as follows:

(1) The slope of the Longwangxi deep accumulation is very close to the location of the Caijiaba landslide. However, due to the steeper slope and higher stone content in the Longwangxi deep accumulation area, there are differences in the mechanical properties of the accumulations at these two locations.

(2) During the reservoir filling process, when the water level does not reach the range of the accumulations, the impact on the stability of the accumulations can be ignored. When the water level reaches the forefront of the accumulations, significant deformation occurs due to the softening and buoyancy reduction effects on the rock-soil mass. Subsequently, the deformation mode of the accumulations changes from translation to traction.

(3) Compared to the weakening effect of the reservoir water level rise on the stability of the taluses, the abrupt drop in the water level caused by the opening and closing of gates in the upstream reservoir of the tributary during extreme conditions has a destructive effect on the stability of the taluses, which can directly lead to instability and landslide disasters in a short period.

(4) Although the accumulations in the area above the maximum water level line are not affected by the softening and buoyancy reduction caused by water level changes themselves, they can still experience larger deformation due to disturbances from other accumulations outside this area that are influenced by water level changes. In extreme cases, instability and destruction of the taluses can also occur.

References

1. Ding X, Fu J, Zhang Q (2004) Stability analysis of the Fengjie Nanqiao Tou landslide under the fluctuation of water level in the Three Gorges Reservoir. Chin J Rock Mech Eng 17:2913–2919
2. Zhang G, Cheng W (2011) Stability prediction of the Zigui Bazimen landslide under the joint action of rainfall and reservoir water level. Chin J Geotech Eng 32(S1):476–482
3. Tao HG, Fan SK, Xu GL et al (2014) Deformation and Stability of Collapse and Gliding Landslide with Changes of Water Level. Water Resources and Power 32(05):96–100
4. Fang S (2016) Three-dimensional fluid-solid coupling analysis of landslides under the joint action of reservoir water level changes and rainfall. J Geotech Found Eng 30(06):662–666+671
5. Zhang Y, Chen GQ, Zhang GF et al (2016) Numerical analysis of influence of water level variation on Guanyinping landslide stability. J Eng Geol 24(04):501–509
6. Liu LL, Song L, Jiao YY, et al (2017) Study of stability of Huangtupo riverside slumping mass #1 under reservoir water level fluctuations. Rock and Soil Mechanics 38(Suppl. 1):359–366
7. Lei J, Lei J (2018) Research on the stability of accumulation landslides on the reservoir bank under the fluctuation of reservoir water level. Water Resour Hydropower Eng 39(04):25–28
8. Vallejo LE, Mawby R (2000) Porosity influence on the shear strength of granular material-clay mixtures. Eng Geol 58(2):125–136
9. Li X, Liao QL, He JM (2004) In-situ tests and a stochastic structural model of rock and soil aggregate in the Three Gorges Reservoir area, China. Int J Rock Mech Min Sci 41(3):702–707
10. Li W, Wu A, Ding X (2006) Study on the influencing factors of shear strength parameters of slip zone soil in the Three Gorges Reservoir Area. Chin J Rock Mech Eng 01:56–60

Open Access This chapter is licensed under the terms of the Creative Commons Attribution 4.0 International License (http://creativecommons.org/licenses/by/4.0/), which permits use, sharing, adaptation, distribution and reproduction in any medium or format, as long as you give appropriate credit to the original author(s) and the source, provide a link to the Creative Commons license and indicate if changes were made.

The images or other third party material in this chapter are included in the chapter's Creative Commons license, unless indicated otherwise in a credit line to the material. If material is not included in the chapter's Creative Commons license and your intended use is not permitted by statutory regulation or exceeds the permitted use, you will need to obtain permission directly from the copyright holder.

Compaction Quality Control of Soil-Rock Mixture Subgrade Based on DEM

Zhuoling He, Junyun Zhang, Xiaolong Luo, Tao Yang, Xiaofei Wu, and Le Zhang

Abstract Over-compaction or under-compaction of soil-rock mixture subgrade could lead to diseases after the expressway is put into service. Selecting the appropriate compaction process and compaction quality testing method is the key to controlling the compaction quality of soil-rock mixture subgrade. Based on continuous compaction control technology, the subgrade porosity under rolling was simulated by two-dimensional discrete element method, considering lay thickness, roller mass, and roller passes. The suitable filler type and recommended compaction process for the soil-rock mixture subgrade in Xingtai Section of Taihangshan Expressway were given: the lay thickness of soil-rock mixture subgrade is 0.6–0.7 m; the mass of the roller is 26 t, and the number of roller passes is rolled 5–6 passes. The evolution law of the subgrade porosity considering the particle crushing was analyzed. A method to control the compaction quality of the soil-rock mixture subgrade was proposed: whether the average difference between the in-site measurements from the last two roller passes exceeded 1% of the average value of the last pass.

Keywords Soil-rock mixture subgrade · Discrete element method · Rolling process · Porosity · Particle crushing behavior

1 Introduction

Continuous compaction control technology has great advantages in subgrade compaction measurement. The subgrade compaction state is the physical and mechanical properties of the subgrade under the roller rolling. During the roller rolling, the roller passes affect the mechanical properties of the subgrade [1, 2]. The decrease in the number of roller passes could cause under-compaction of the

---

Z. He · J. Zhang (✉) · X. Luo · T. Yang · X. Wu
School of Civil Engineering, Southwest Jiaotong University, Chengdu, China
e-mail: zjywxfbb@swjtu.edu.cn

L. Zhang
Sichuan Highway Planning Survey and Design Institute Ltd., Chengdu, China

subgrade, while the increase in the number of roller passes may lead to over-compaction of the subgrade [3]. Under-compaction or over-compaction leads to deterioration of the subgrade after the expressway is put into service [4–6]. Accordingly, reasonable compaction process and subgrade compaction quality testing methods are important.

The particle size of the filler excavated near the Xingtai Section of the Taihangshan Expressway varies greatly, the maximum particle sizes of some blocks can reach 80 mm, and the particle sizes of some soil particles do not exceed 0.25 mm. Moreover, during the rolling process of soil-rock mixture (S-RM) subgrade, the spectral composition of the vibration wheel response signal is complex. It is difficult to establish a general regression relationship between continuous measurements and point measurements in the S-RM subgrade, and using linear regression relationships to determine the compaction quality of the subgrade is not reliable [7–11].

And only in-situ monitoring tests cannot effectively provide the meso-behavior of the internal system of materials and the compaction mechanism of S-RM subgrade. The discrete element method (DEM) can reflect the composition characteristics of heterogeneous materials and can simulate the meso-behavior of granular materials [12–15].

In this study, through two-dimensional (2D) DEM simulation experiments, the recommended compaction process of Xingtai Section of Taihangshan Expressway is given, considering lay thickness, roller mass, and roller passes. We investigated the evolution laws of porosity (n) and particle crushing in subgrade compaction, to provide theoretical support for compaction quality control of S-RM subgrade.

2 Materials

Located in the Beijing-Tianjin-Hebei regions, the Xingtai section of the Taihangshan Expressway spans 64.2 km. As shown in Fig. 1a, the micro-geomorphic units mainly consist of low mountain terrain and low mountain–hilly terrain. The monitoring site is located at K8+105-K8+341 in the Xingtai section. The center of the monitoring site has a height of 19.2 m, and the maximum height of the section is 19.8 m, as shown Fig. 1b. The filler for this section was excavated from the adjacent road graben with a lithology of granitic gneiss: deep-extreme weathering (Filler A) and incipient-intermediate weathering (Filler B), as shown in Fig. 1c.

3 Roller Rolling Process Research

In the subgrade compaction quality control process, the control of filler is fundamental. And the control mainly includes two aspects: (1) Lay thickness. Smaller lay thickness causes over-pressure of the subgrade and larger lay thickness results in under-pressure of the subgrade. (2) Roller rolling. Roller rolling is also a key part of

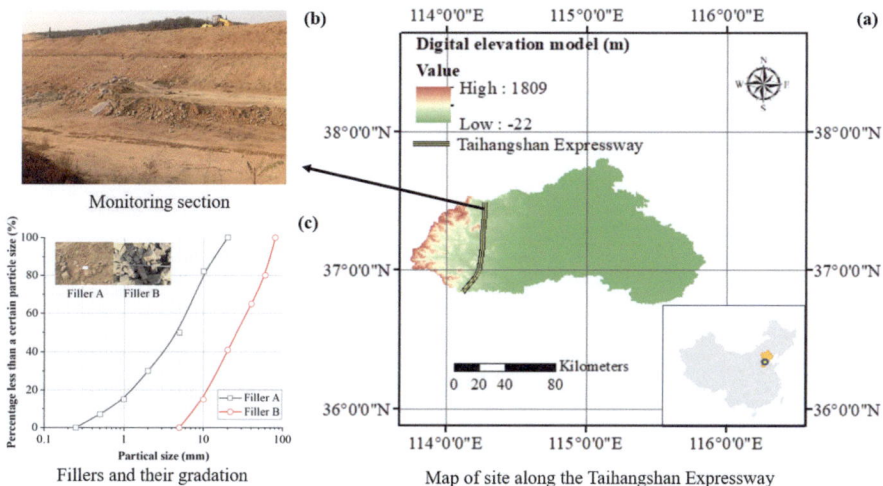

Fig. 1 Site location and fillers

the formation of the subgrade structure. The relevant parameters such as the mass, passes, and the velocity of the roller are crucial.

In this study, the driving velocity of the roller is nearly uniform, approximately 3 km/h. Considering the lay thickness, the mass of the roller, and roller passes, PFC2D5.0 was used to realize the DEM simulation of the roller rolling process, and the *n* of the S-RM subgrade was analyzed.

3.1 2D DEM Modeling

The modeling process is shown in Fig. 2. Firstly, a 2D outline of the block in PFC was generated based on a random polygon, and randomly distributed in the box according to the gradation. Then the clumps were filled in geometric shapes, and each clump is numbered. After recording the size and location of each clump, each pebble in the clump was replaced with a ball until all the pebbles were replaced by the ball. And the unbreakable clump turned into a breakable cluster. The balls of soil particles with size of 2–5 mm were generated, and the balls that coincide with the cluster were deleted. Then the boundary conditions were given: the left, right and lower borders were closed, and the upper border was free, as shown in Fig. 3a. A circular wall was used that applied downward eccentric force and horizontal reciprocating motion, as shown in Fig. 3b. The roller traveled from the left to the right at an average speed of 3 km/h. When the roller arrived at the right boundary, it rolled in the opposite velocity. A 2D DEM model of the subgrade compaction process was established. Four measurement regions were monitored, the n of subgrade was calculated by the average value of the n of the four regions.

Fig. 2 2D DEM model modeling process

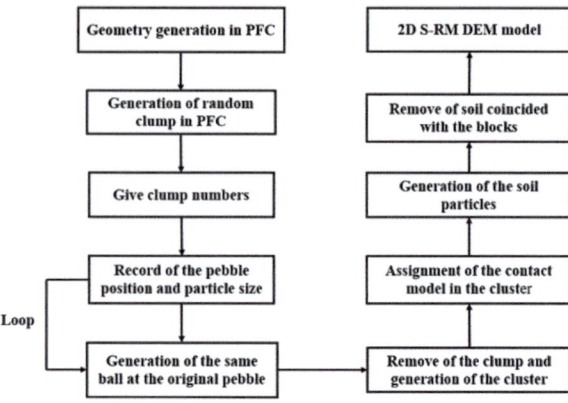

Fig. 3 2D DEM simulation of roller rolling

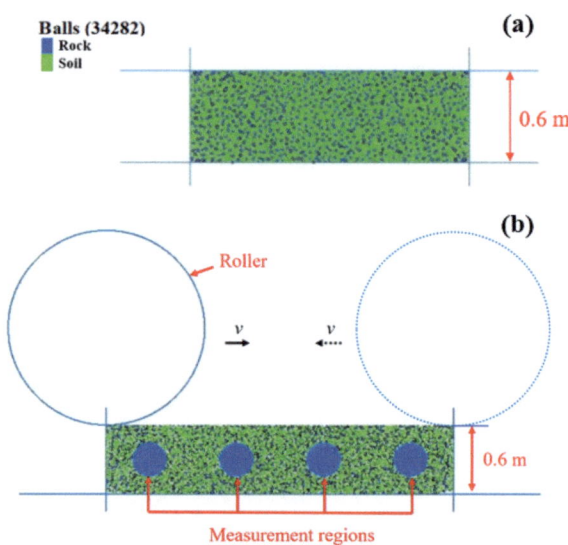

According to the results of indoor tests, the parameters required for the parallel bonding model and rolling resistance model were selected by trial-and-error method, as shown in Table 1.

3.2 Lay Thickness

Figure 4 is the section view model of the S-RM subgrade with different lay thickness. The gradation of the filler in Fig. 4 is $A:B = 2:5$, which is mostly used on site. The length of the four models is 3 m, and the thickness is 0.5 m, 0.6 m, 0.7 m, and 0.8 m,

Table 1 Selection of 2D DEM parameters for S-RMs

Material	Parameter	Value
Filler A	Density of particles (kg·m^{-3})	1800
	Elastic modulus of particle contacts (MPa)	40
	Poisson's ratio of particle contacts	0.5
	Friction coefficient	0.45
	Rolling resistance coefficient	0.05
Filler B	Density of particles (kg·m^{-3})	2400
	Elastic modulus of particle contacts (MPa)	100
	Poisson's ratio of particle contacts	0.5
	Friction coefficient	0.65
	Bond shear strength (MPa)	1.5
	Bond tensile strength (MPa)	1.5
Contacts between Filler A and Filler B	Elastic modulus of contacts (MPa)	100
	Poisson's ratio of particle contacts	0.5
	Friction coefficient	0.65
	Rolling resistance coefficient	0.1

respectively. The number of balls of the blocks in subgrade models of different thicknesses was 18,453, 22,150, 25,761, and 29,634, respectively. The number of balls of these subgrade models was 28,569, 34,282, 39,925, and 45,856, respectively.

As shown in Fig. 5, during the compaction process, the n decreases rapidly during pass 1, and with the increase of the lay thickness, the n gradually decreases. When the lay thickness is 0.5–0.7 m, the n shows a more obvious step shape with the compaction process. With the progress of the compaction process, the subgrade with the lay thickness of 0.5–0.7 m is gradually compacted, and the n reduction rate is basically stable.

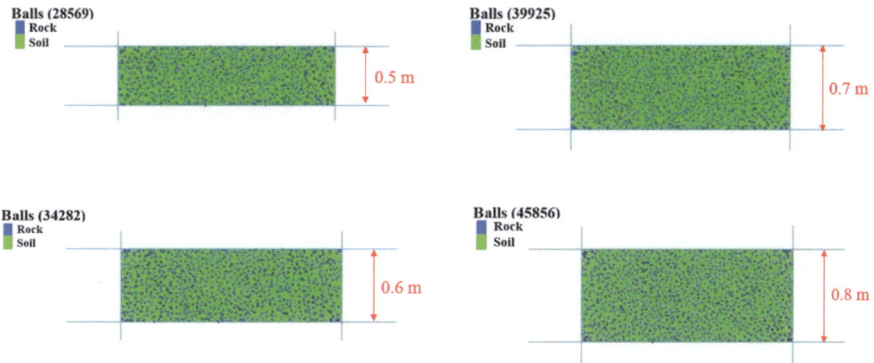

Fig. 4 2D DEM model with different lay thickness

Fig. 5 The n of S-RM subgrade with different lay thickness

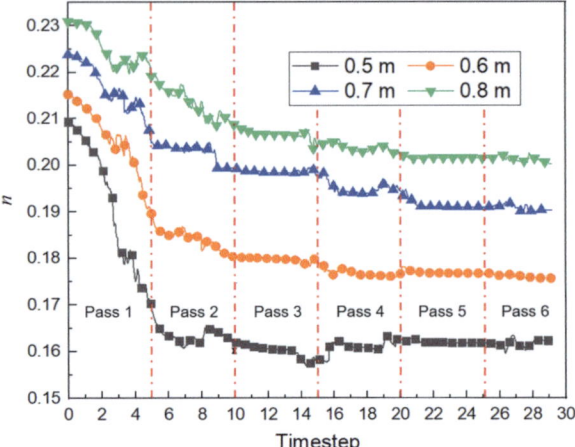

However, when the lay thickness is 0.5 m, the n of the subgrade at the time step of 14 is already at a relatively low level. And then with the rolling progress, the block particles are further broken, and the n rises. The curve is prone to over-pressure phenomenon. When the lay thickness is 0.8 m, the n change form of the subgrade is parabolic and still slowly decreasing, therefore, when the lay thickness is 0.8 m, the compaction quality of the subgrade is also difficult to control, and the rolling of six passes cannot compaction the subgrade, and the subgrade presents a state of under-pressure. Therefore, lay thickness between 0.6 and 0.7 m is a more suitable thickness.

3.3 Roller Mass

The vertical eccentric force has a relatively large influence on the compaction quality of the S-RM subgrade, considering the tonnage of commonly used vibrating rollers: 14, 22, 26, and 33 t. According to the relevant parameters of the roller, the maximum eccentric forces of these four tonnage rollers are 185 kN, 400 kN, 430 kN, and 725 kN, respectively. The n of the subgrade slowly decreases when the mass of the roller is 14 and 22 t, as shown in Fig. 6.

The curve does not appear to be stepped down, but rather a parabolic descent. The subgrade shows the characteristics of "under-pressure", and the compaction of 6 passes cannot compact the subgrade. However, for the roller with the tonnage 26 t, the particles appear less breakage and the n of the subgrade shows a stepwise reduction, as far as the numerical simulation results show that after the six passes of the 26 t vibrating roller, the n of the subgrade gradually flattens and the subgrade is significantly compacted. When the tonnage is 33 t, the n of the subgrade is reduced to a lower level, and the subgrade eventually flattens. The curve is also oscillating,

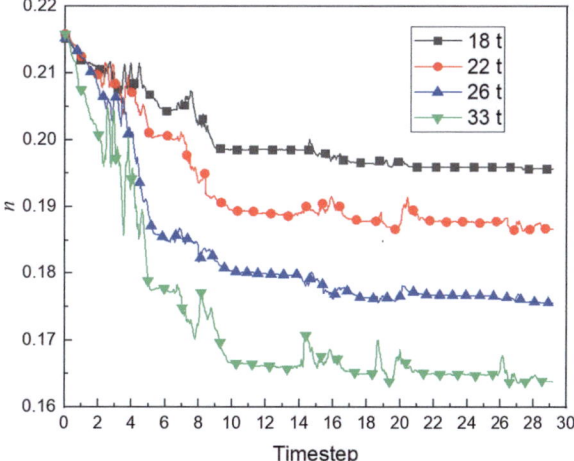

Fig. 6 Effect of different tonnage rollers on compaction quality

and there is a clear fragmentation behavior, the subgrade shows the characteristics of "over-pressure".

3.4 Recommended Compression Process

According to the above analysis, this study proposes that the lay thickness of 0.6–0.7 m is proposed as the single-layer filling thickness of the S-RM subgrade; the roller mass is 26 t; the rolling passes are suggested as 5–6 passes.

3.5 The Evolution Law of Subgrade State During Roller Rolling

As shown in Fig. 7, in the process of vibrating roller rolling, the mechanical state of S-RM subgrade can be divided into four sections: severe particle crushing stage (section *OA*), serious particle crushing stage (section *AB*), weak particle crushing stage (section *BC*), and compaction stage (section *CD*).

Section OA: The subgrade at this stage is continuously compacted, and the numerical simulation test results show that the curve oscillates more intensely. The blocks with relatively large particle size at this stage produce more serious crushing in the process of compaction.

Section AB: The n of the second section first decreases rapidly and then slow down to show a "step" reduction, the compaction quality of the S-RM subgrade at this stage is relatively close to the final stage, and the particle crushing behavior is weakened.

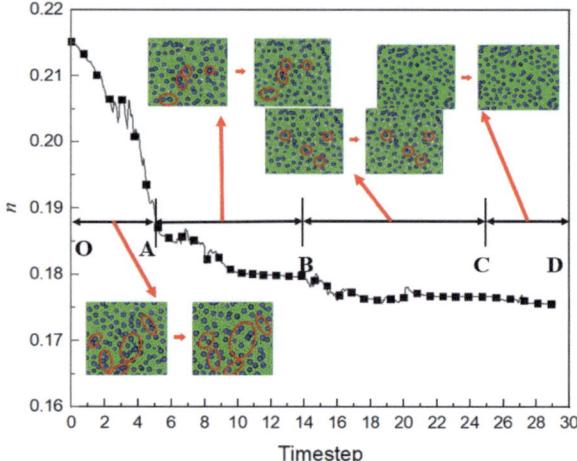

Fig. 7 Discrete element simulation test results

Section BC: The n of this stage is close to the development law of section AB, and the evolution of n is also similar to the n of section AB. The compaction quality of the subgrade is further improved, and the particle crushing behavior is further weakened.

Section CD: The n in this stage is basically flattened, the amplitude of the n curve oscillation is already at a relatively low level, and the particle fragmentation behavior is significantly lower than that of the section BC.

In this study, Fig. 8 show that the difference between the n of the last two passes is 0.001, which is approximately 1% of the mean of the last pass. Based on this result, this study proposes that the compaction quality control method of S-RM subgrade is: whether the difference between the average value of the continuous index after the last two rolling operations is less than 1% is used to assess whether the compaction quality of the S-RM subgrade meets the requirements.

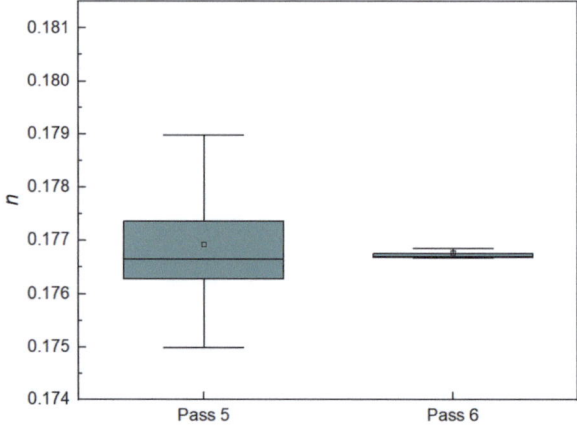

Fig. 8 Analysis of subgrade structure in the last two passes

4 Conclusion

The DEM can accurately simulate the subgrade rolling process, and the crushing behavior of block particles can be considered. Based on the DEM simulation of soil-rock mixture subgrade rolling in the Xingtai section of Taihangshan Expressway, the lay thickness is 0.6–0.7 m; the mass of the road roller is 26 t; the number of roller passes is 5–6 passes. After the construction of this process, the subgrade filler can be fully compacted, and a certain skeleton can be formed between the soil and the block, and the crushing behavior of the block particles is relatively slight. The results of 2D DEM simulation show that the porosity of the subgrade structure changes very weakly in the average of the porosity of the last two passes. The difference between the average value of the continuous index after the last two rolling operations is less than 1% of the mean of the last pass. In this study, 2D DEM simulation is considered for the subgrade analysis under roller rolling, some shortcomings exist: the variation of the porosity is a three-dimensional problem. Using the three-dimensional discrete element method in subsequent research to study the compaction of soil-rock mixture subgrade would be more convincing.

Acknowledgements This work was supported by the National Natural Science Foundation of China (Grants Nos. 42072313), Sichuan Province Science and Technology Support Program (2021-A-2).

References

1. Beainy F, Singh D, Commuri S et al (2014) Laboratory and field study on compaction quality of an asphalt pavement. Int J Pavement Res Technol 7:317–323
2. Liu P, Hu J, Falla GC et al (2019) Primary investigation on the relationship between microstructural characteristics and the mechanical performance of asphalt mixtures with different compaction degrees. Constr Build Mater 223:784–793
3. Zeinali A, Blankenship PB (2020) Employment of mechanical testing to evaluate the effect of density on asphalt pavement performance. J Test Eval 48:1014–1030
4. Gao Y, Wang R, Yu W (2011) Effective compactible time of asphalt pavement. Adv Mater Res Trans Tech Publ 255–260:3156–3160
5. Williams KL, Cox BC, Howard IL et al (2015) Models of asphalt concrete field compactibility with focus on lift thickness. Transp Res Rec 2504:135–147
6. Fan L (2011) Study on the rapid detection method of compaction quality with earth-rock subgrade based on PFWD. Chang'an University
7. Mooney M, Adam D (2007) Vibration roller integrated measurement of earthwork compaction: an overview. In: Myrvoll F (ed) Field measurements in geomechanics. CRC Press, Boca Raton
8. Chang G, Xu Q, Rutledge J et al (2008) Accelerated implementation of intelligent compaction technology for embankment subgrade soils, aggregate base, and asphalt pavement materials. In: Federal Highway Administration/Transportation Pooled Fund (TPF). Federal Highway Administration, Washington, DC
9. Barman M, Nazari M, Asif SI et al (2014) Application of intelligent compaction technique in real-time evaluation of compaction level during construction of subgrade. In: Proceedings of Transportation Research Board, 93rd Annual Meeting, No. 14-5183. Transportation Research Board, Washington, DC

10. Kumar SA, Mazari M, Garibay J (2016) Compaction quality monitoring of lime-stabilized clayey subgrade using intelligent compaction technology. In: Proceedings of International conference on transportation and development. ASCE, Reston, pp 778–790
11. Wróbel M, Woszuk A, Franus W (2020) Laboratory methods for assessing the influence of improper asphalt mix compaction on its performance. Materials 13(11):2476
12. Xu W, Hu R (2009) Conception, classification and significations of soil-rock mixture. Hydrogeol Eng Geol 36(4):50–56
13. Liu Y, You Z (2009) Visualization and simulation of asphalt concrete with randomly generated three dimensional models. J Comput Civ Eng 23(6):340–347
14. Liu Y, You Z (2011) Discrete element modeling: impacts of aggregate sphericity, orientation, and angularity on creep stiffness of idealized asphalt mixtures. J Eng Mech 137(4):294–303
15. Liu Y, You Z (2013) Fundamental study on pavement wheel interaction forces through discrete element simulation. J Int J Pavement Res Technol 6(6):689–695

Open Access This chapter is licensed under the terms of the Creative Commons Attribution 4.0 International License (http://creativecommons.org/licenses/by/4.0/), which permits use, sharing, adaptation, distribution and reproduction in any medium or format, as long as you give appropriate credit to the original author(s) and the source, provide a link to the Creative Commons license and indicate if changes were made.

The images or other third party material in this chapter are included in the chapter's Creative Commons license, unless indicated otherwise in a credit line to the material. If material is not included in the chapter's Creative Commons license and your intended use is not permitted by statutory regulation or exceeds the permitted use, you will need to obtain permission directly from the copyright holder.

Analysis of the Influence of Deep Soft Soil Foundation Cofferdam Filling on Lateral Compression of Bridge Pile Foundations

Ruiqi Zhang, Aimin Liu, Guoliang Ye, and Xu Liu

Abstract In order to prevent seawater from flowing back into cities, road construction in Indonesia's coastal areas often requires the construction of breakwaters on the outside of bridges. The area is widely distributed on deep soft foundations, and the foundation soil at the breakwater is reinforced using the preloading method. The road is built on friction prefabricated pipe piles as the bridge foundation. Using PLAXIS 3D to analyze the beneficial effects of bamboo piles and bamboo rafts as working cushion layers on the filling of deep soft soil foundation and the settlement and deformation characteristics of the foundation, and simulating the adverse effects of lateral compression on the bridge pile foundation caused by the soft soil foundation under different loading conditions of the sea embankment. The results show that the flexible working cushion layer of bamboo piles and bamboo rafts is conducive to improving the loading stability of the soft soil foundation, when the second layer of pile is consolidated and stabilized before carrying out bridge pile foundation construction, the lateral compression generated by the soil will be smaller.

Keywords Deep soft foundation · Lateral compression · Cofferdam · Bridge pile foundation · Adverse effects

---

R. Zhang (✉) · A. Liu · X. Liu
Tianjin Port Engineering Institute Co., Ltd., CCCC First Harbor Engineering Co., Ltd., Tianjin, China
e-mail: 804105324@qq.com

R. Zhang · A. Liu · G. Ye · X. Liu
CCCC First Harbor Engineering Company Ltd., Tianjin, China

R. Zhang · A. Liu · X. Liu
Key Laboratory of Port Geotechnical Engineering of the Ministry of Communication, Tianjin, China

Key Laboratory of Port Geotechnical Engineering of Tianjin, Tianjin, China

1 Introduction

When constructing bridges on deep soft soil foundations, engineering problems caused by geological reasons are not uncommon. Scholars at home and abroad have proposed corresponding research methods based on lateral displacement of soil. According to Poulos [1] discusses the pile-soil interaction relationship within the elastic range, believing that soil pressure is directly proportional to lateral displacement of soil. The pile foundation is treated as an elastic foundation beam for solution, which is only suitable for situations with small horizontal deformation and not suitable for soft soil with high porosity. Springman [2] used model experiments to simulate the stress shape of adjacent pile foundations under lateral pile loading, and based on this, proposed a calculation method for bridge abutment pile foundations under passive diagram action using a triangular assumed soil displacement mode on the pile side. Li et al. [3] studied the effect of soft soil drainage on the lateral soil pressure of passive piles, and obtained that in soft soil foundation pile loading construction, the maximum bending moment of the pile body under slow loading will be 20–50% smaller than that under fast loading, and the lateral soil pressure of the pile will also be smaller. Carrying out pre-loading construction first and then carrying out pile driving operations is beneficial for controlling the displacement of the pile body. Simultaneously controlling the loading rate will also reduce the force on the pile foundation. Huang [4] used three-dimensional finite element method to simulate engineering examples, aiming to improve engineering design through numerical analysis methods. The results have good reference significance for actual construction, but the analysis results are still far from previous research conclusions, and cannot effectively solve on-site problems. Hou [5] used a combination of theoretical calculations, numerical simulations, and on-site monitoring to analyze the maximum horizontal displacement and maximum pile body cracks of pile foundations under original differential loading and optimized differential loading conditions. Jiang et al. [6] used finite element numerical calculation methods to simulate construction and analyze the effects of additional loads on the negative friction resistance, internal force of the pile body, and deformation of the pile foundation. Che [7] conducted numerical analysis on the adjacent area of the existing pile foundation for loading operations, resulting in significant long-term displacement of the pile foundation. Xu [8] calculated through numerical analysis and theoretical calculations that as the road load increases, the neutral point of pile foundation friction gradually moves downwards, and the negative friction and horizontal displacement of the pile foundation significantly increase.

At present, there are few reports on the risks and determination of the loading time for both pile foundation treatment and bridge pile foundation construction in coastal areas with deep soft foundations. Based on a certain highway in Indonesia, studying the lateral compression of bridge pile foundations on deep soft foundations under different loading conditions of cofferdams has certain reference value for similar projects.

2 Engineering Background

The design scheme of a certain highway project in Indonesia intends to adopt a "bridge embankment separation" scheme, which is a road formed by a sea side water retaining seawall and a landside bridge; Based on the on-site water depth situation, some bridge sections are planned to adopt the "island building and road pile foundation construction" process, with a standard section width of 45.1 m for island building and road backfilling; The seawall adopts the construction technology of bamboo rafts, bamboo piles, and drainage boards for preloading. The total height of the stacking is 9.5–10.0 m, the spacing between drainage boards is 1 m, and the depth of drainage board installation is 23.0–33.0 m. Bamboo stakes are made up of seven bamboo sticks with a length of 8 m tied together into a bundle as one bamboo stake. The bamboo row consists of three bamboo pieces grouped together, with a spacing of 1 m. The upper and lower layers of bamboo are crisscrossed and tied together, totaling nine layers of bamboo; the distance between the seawall and the bridge island is 35–42 m; The top elevation of the backfill for the access road and island construction is 1.6 m. A typical cross-sectional view of the design scheme is shown in Fig. 1.

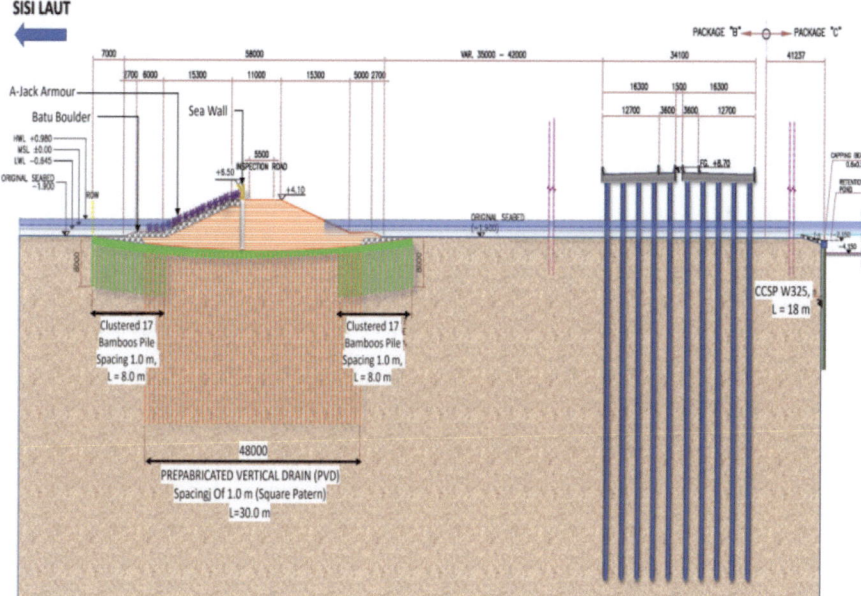

Fig. 1 Typical cross-sectional view of design scheme (the image is sourced from the design drawings and can be publicly used)

3 Numerical Analysis Calculation

3.1 Modeling

Using large-scale finite element PLAXIS 3D [9, 10] to conduct computational analysis on the stability of seawall loading and the impact of seawall loading construction on the safety of bridge pile foundations. The soft soil foundation adopts the soil hardening HS model, which is based on the hyperbolic relationship between deviatoric stress and axial strain in triaxial drainage tests. Bamboo piles and bamboo rafters are rigidly connected, with bamboo piles equivalent to pile units and bamboo rafters equivalent to board units.

The model calculates a horizontal (vertical to the axis of the seawall) length of 250 m in the area. The foundation treatment form of the cofferdam is preloading: using 9-layer bamboo rows + 8 m long bamboo piles (16 rows on each side, with a spacing of 1.0 m), plastic drainage boards are installed at a depth of 33.0 m, with a spacing of 1.0 m. The seawall is filled in three levels, with the first layer being 3.0 m high, the second layer being 2.0 m high, and the third layer being 3.7 m high.

The total length of the bridge pile foundation is about 62.6 mm, of which the pile length above the surface is 8 m, and the pile length below the surface is 54.6 mm, with a pile diameter of 0.8 m. The spacing between the seawall and the nearby seawall side pile foundation is 35 m. The calculation model is shown in Fig. 2.

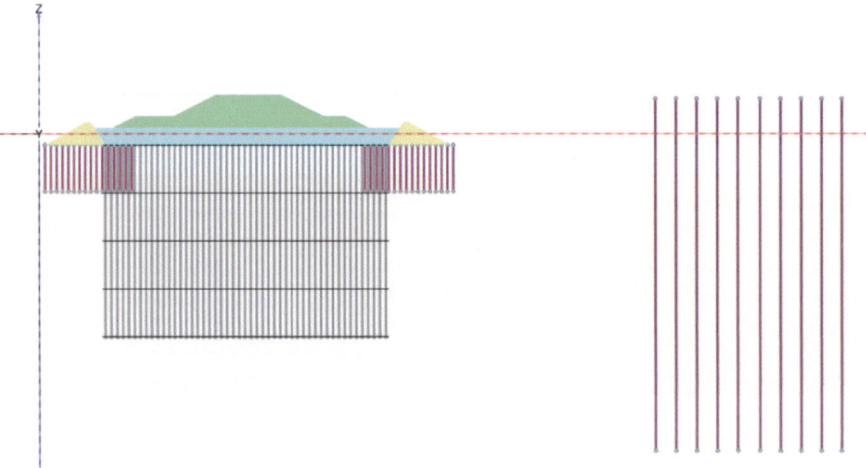

Fig. 2 Layout of test area (the image is sourced from our company's genuine PLAXIS 3D software and can be publicly released)

Table 1 Bamboo pile and bamboo raft parameters

Material name	Stiffness (MPa)
Bamboo raft	1810
Bamboo pile	240

Table 2 Pipe pile parameters

Material name	Stiffness (GPa)	Diameter (m)	Wall thickness (m)	Severe (kN/m^3)
Pipe pile	33.8	0.8	0.12	10

3.2 Selection of Model Parameters

The parameters of the finite element model are obtained from the survey report, and some parameters are obtained from the inversion of measured settlement in the field test section. According to literature research [11], The calculation parameters of bamboo piles and bamboo rafts are shown in Table 1, the design requirements of pipe piles are shown in Table 2, and the soil properties of the model foundation are shown in Table 3.

4 Numerical Analysis Result

Due to time constraints, the construction of bridge pile foundations needs to be carried out simultaneously during the embankment filling process. During the process of seawall filling, the soil will undergo vertical and lateral deformation, which may cause deformation or instability of the bridge pile foundation due to lateral soil compression. The determination of the optimal construction time for bridge pile foundation is a key factor in measuring the safety and progress of this project. According to the characteristics of soil consolidation deformation, the impact of layered embankment filling on bridge pile foundations is analyzed in two working conditions:

Working condition 1: After the first layer of the seawall is filled and consolidated, the construction of the bridge pile foundation begins.
Working condition 2: After the second layer of the seawall is filled and consolidated, the construction of the bridge pile foundation begins.

4.1 Total Settlement of Seawall

The consolidation compression caused by the filling of seawall on deep soft foundation is only related to the height of the pile, the thickness of the soft soil layer, and

Table 3 Foundation soil properties parameters

Depth (m)	Soil name	Natural bulk density (kN/m³)	Modulus (MPa)	Stiffness stress power exponent	C (kPa)	φ (°)	Permeability coefficient $K_x = K_y$ (m/d)	Permeability coefficient K_v (m/d)
0.0–5.0	Oc very soft clay	15.5	0.6	0.5	13	19	6.15×10^{-6}	3.07×10^{-6}
5.0–12.5	Very soft clay	15.5	0.7	0.5	15	19	6.15×10^{-6}	3.07×10^{-6}
12.5–27.5	Very soft to medium stiff clay	16.2	1.9	0.5	15	19	6.15×10^{-6}	3.07×10^{-6}
27.5–53	Very soft to medium stiff clay	15.8	3.8	0.5	15	19	6.15×10^{-6}	3.07×10^{-6}
53–60	Stiff to very stiff clay	17.5	15.2	0.5	15	25	4.75×10^{-3}	4.75×10^{-3}
60–80	Very stiff to hard clay	18.0	21.9	0.5	15	30	4.75×10^{-3}	4.75×10^{-3}

the depth of the drainage plate installation. The construction process of bridge pile foundation will not affect the final settlement of the seawall foundation soil.

According to the calculation results, the final total settlement of the seawall is 2.85 m. The specific settlement calculation cloud map is shown in Fig. 3.

4.2 Horizontal Displacement of the Pile Foundation Closest to the Seawall

The filling of seawall will cause lateral deformation of the soft soil on both sides of the seawall, and the lateral displacement of the soil near the pile will compress the pile foundation and cause horizontal displacement. Different loading processes have different effects on the lateral compression of the pile foundation. Through numerical analysis, the maximum horizontal displacement of the pile foundation closest to the seawall in working conditions one and two is 116 mm and 53 mm, respectively, in

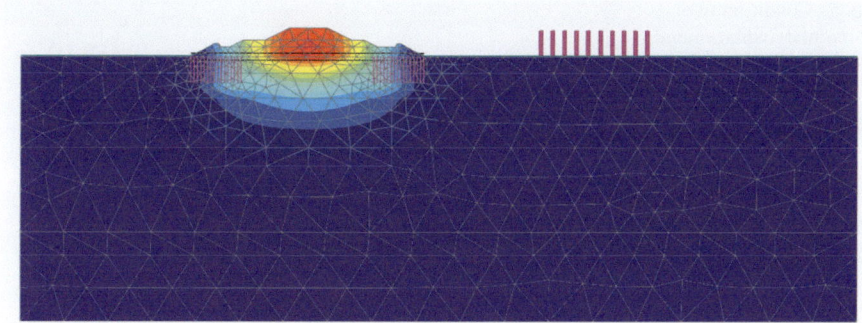

Fig. 3 Cloud map of total settlement calculation for seawall (the image is sourced from our company's genuine PLAXIS 3D software and can be publicly released)

the direction away from the seawall side. The horizontal calculation cloud maps of the pile foundation are shown in Figs. 4 and 5.

Fig. 4 Condition two: Horizontal displacement of pile top

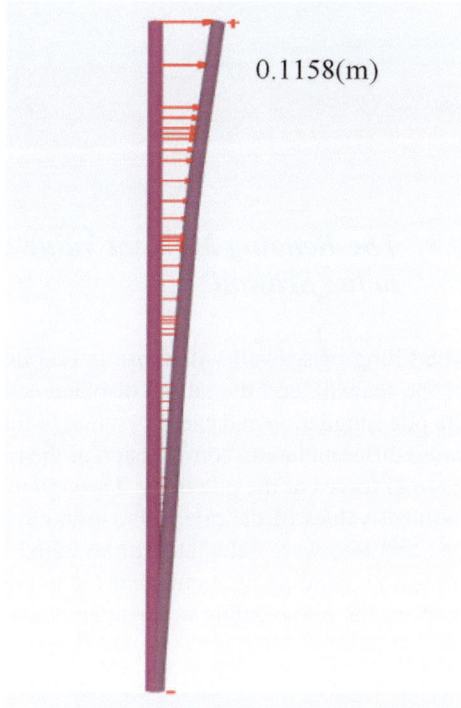

Fig. 5 Condition two: Horizontal displacement of pile top

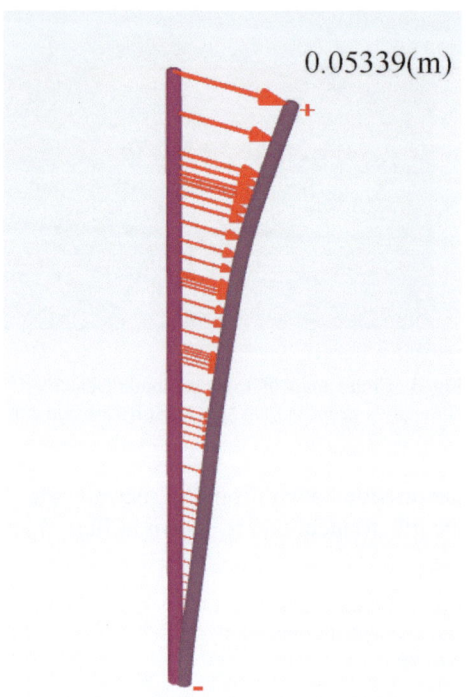

4.3 The Bending Moment Value of the Pile Body Closest to the Seawall

The filling of seawall will cause lateral deformation of the soft soil on both sides of the seawall, and the lateral displacement of the soil near the pile will compress the pile foundation and cause bending deformation. Different loading processes will cause different lateral compression of the pile foundation, and also result in different internal forces of the pile body. Through numerical analysis, the maximum bending moment values of the pile foundation closest to the seawall in working conditions one and two were calculated to be 220.2 kN·m and 88.6 kN·m, respectively. The maximum bending moment occurred below the mud surface. The calculation cloud maps of the pile bending moment are shown in Figs. 6 and 7.

Fig. 6 Condition one: Pile bending moment

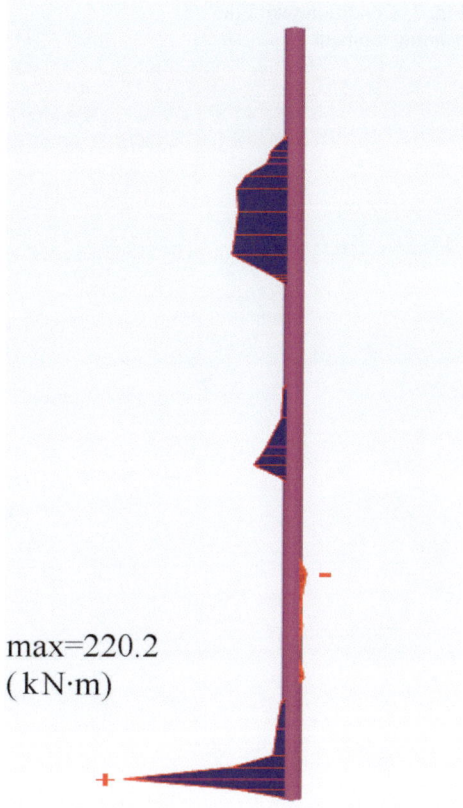

4.4 The Shear Force Value of the Pile Closest to the Seawall

The filling of seawall will cause lateral deformation of the soft soil on both sides of the seawall, and the lateral displacement of the soil near the pile will compress the pile foundation, generating a certain shear force on the pile body. Different loading processes will cause different lateral compression of the pile foundation, and also cause different internal forces in the pile body. Through numerical analysis, the maximum shear force values of the pile foundation closest to the seawall in working conditions one and two were calculated to be 101.7 and 40.1 kN, respectively. The maximum shear force occurred at the lower part of the pile body, and the shear force calculation cloud maps of the pile body are shown in Figs. 8 and 9.

Fig. 7 Condition two: Pile bending moment

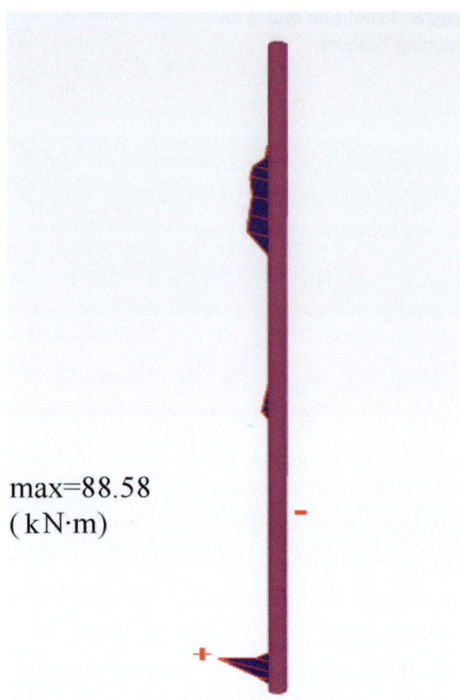

Fig. 8 Condition one: pile body shear force

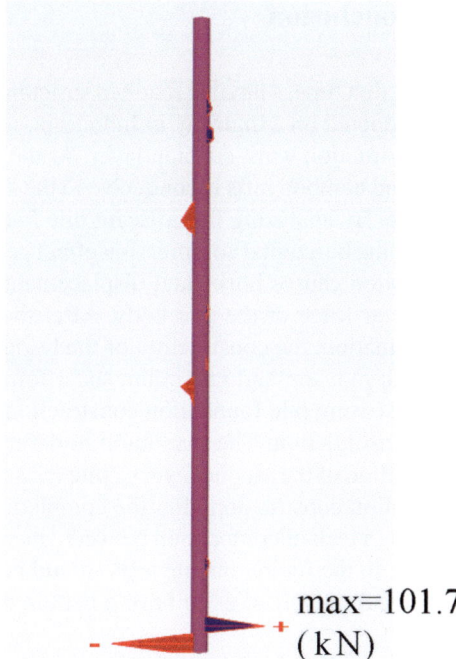

Fig. 9 Condition two: pile body shear force

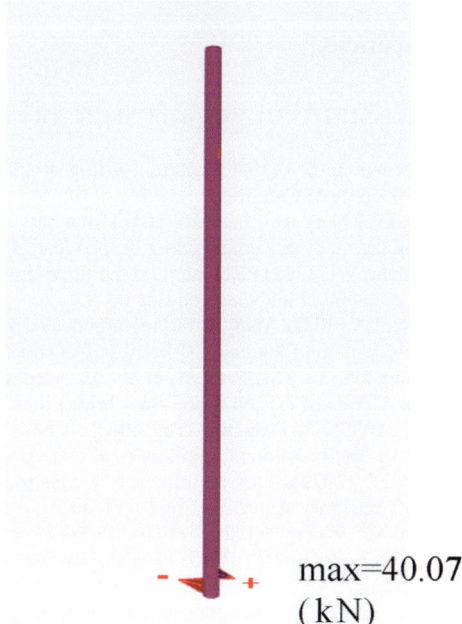

5 Conclusion

For the first time, a flexible roadbed structure with bamboo piles and bamboo rafts has been adopted on a highway in Indonesia, which is conducive to the rapid formation of construction work cushion layer. At the same time, the overall effect of bamboo piles and bamboo rafts is conducive to the stability of soft soil foundation under later loading. By analyzing the adjacent pile foundation construction at different loading stages, the horizontal compression effect generated by the loading process on the pile foundation causes horizontal displacement, bending deformation of the pile body, and shear force of the pile body. After the completion of the second layer of pile consolidation, the construction of the bridge pile foundation begins. The maximum bending moment and maximum shear force experienced by the pile body are 40% of the starting pile foundation construction after the completion of the first layer of pile consolidation. The maximum horizontal displacement of the pile top after the completion of the second layer of pile consolidation is 50% of the starting bridge pile foundation construction after the completion of the first layer of pile consolidation. There is a lack of comparison between on-site measured data and numerical analysis results. In the future, on-site tests should be carried out to verify the accuracy of the numerical model, so as to have a certain reference value in similar projects in the future.

References

1. Poulos HG (1973) Analysis of piles in soil undergoing lateral movement. J Soil Mech Found Div 99:391–406
2. Springman SM (1989) Lateral loading on piles due to simulated embankment construction. University of Cambridge
3. Li XF, Zhang JH, Zheng JL (2011) Summary of the influence on piled bridge abutment by soft soil lateral displacement under heaped load. J Sub Grade Eng 1:12–15
4. Huang WD (2021) Research on the three-dimensional behavior of passive pile groups under pile loading. Field Geotech Eng 09
5. Hou LX (2022) Analysis of the influence of preloading on adjacent bridge pile foundations in deep soft soil foundation. J Railway Construct Technol (06):140–144
6. Jiang JW, Yu YM, Ning SL et al (2020) Influence of sea embankment underpass on the pile foundation of existed expressway bridge in deep soft soil area. J Adv Eng Sci (04):109–116
7. Che PY (2020) Research on the effect on the stability of the main bridge pile foundation brought by the construction of auxiliary road in deep soft foundation area. J Southeast Univ
8. Xu YL (2022) Study on influence of construction road on the bridge pile foundation in deep soft clay area. Adv Civil Eng 8:111–115
9. Liu XX, Zhang HQ (2014) PLAXIS 3D basic tutorial. Machinery Industry Press, Beijing
10. Liu XX, Zhang HQ (2015) PLAXIS advanced application tutorial. Machinery Industry Press, Beijing
11. Liu XZ (2020) Research on the response of bamboo characteristics to the performance of laminated timber and its environmental benefits. J Northeast Forest Univ

Open Access This chapter is licensed under the terms of the Creative Commons Attribution 4.0 International License (http://creativecommons.org/licenses/by/4.0/), which permits use, sharing, adaptation, distribution and reproduction in any medium or format, as long as you give appropriate credit to the original author(s) and the source, provide a link to the Creative Commons license and indicate if changes were made.

The images or other third party material in this chapter are included in the chapter's Creative Commons license, unless indicated otherwise in a credit line to the material. If material is not included in the chapter's Creative Commons license and your intended use is not permitted by statutory regulation or exceeds the permitted use, you will need to obtain permission directly from the copyright holder.

Drilling Optimization Using Digital Image Stratigraphic Modeling

Zhengqiang Zeng and Yongchang Cai

Abstract How to use survey data to build digital stratigraphic models, assess the uncertainty of models, and provide reasonable survey optimization schemes is one of the important tasks in the digital analysis of geotechnical engineering. This article offers a digital modeling concept based on surface and borehole data, with digital images as the medium. It uses the standard deviation of Kriging prediction to establish the twin-error map of the digital strata, and evaluates the uncertainty of the stratum boundaries using this twin-error map as an indicator of the accuracy of the digital strata. In addition, based on the proposed model concept, a sequential borehole optimization method based on the column error rate indicator is proposed. Case analysis shows that the sequential borehole optimization method, based on the digital stratigraphic model, significantly improves the refinement of the geological model and more effectively reduces the uncertainty of the digital model compared to conventional distance-based borehole arrangements.

Keyword Digital image stratigraphy · Stratigraphic uncertainty · Borehole optimization · Column error rate

1 Introduction

Geological drilling is a commonly used tool in geological investigations and for obtaining stratigraphic information. However, the acquisition of borehole data is costly and generally sparse. Optimizing drilling to obtain as much accurate geological information as possible at the lowest cost is of great significance. Establishing

Z. Zeng (✉) · Y. Cai
College of Civil Engineering, Tongji University, Shanghai, P. R. China
e-mail: zzqiang@tongji.edu.cn

State Key Laboratory of Disaster Reduction in Civil Engineering, Tongji University, Shanghai, P. R. China

as accurate a geological model as possible based on limited borehole data, and evaluating its accuracy, holds substantial research value for improving the precision of geological and numerical models [1, 2].

Numerous scholars have investigated drilling optimization methods across various contexts. For instance, Drew [3] examined the selection of optimal drilling patterns when targeting elliptical objectives, establishing a functional relationship between optimal hole spacing and drilling costs, target value, target shape, and the probability of target occurrence. Soltani and Hezarkhani [4] explored different mineral exploration and drilling systems within the framework of information systems, proposing a model to assess the value of exploration drilling information. Hossein and Memarian [5] introduced a dynamic multi-level sampling algorithm capable of dynamically identifying the optimal drilling pattern with size and directional parameters. While existing drilling optimization methods have made certain advancements in exploring mineral deposit distributions, they rarely consider the impact on the accuracy of digital models. Therefore, further research and exploration into the effects of drilling optimization methods on digital modeling and geotechnical engineering analysis are warranted.

Recent research into theoretical inversion of high-fidelity stratigraphic models from survey data has seen broad interest and notable advancements. Random field theory [6] has been applied to analyze soil spatial variability. Studies employing random fields have explored this variability, encompassing soil classification, geotechnical parameters, and physical indices, and associated these with slope stability and reliability uncertainties. Reports have also detailed geological uncertainty analyses using Markov models [7], which integrate surface and borehole data, and stratigraphic orientation, even with limited raw data. Yet, challenges in defining horizontal transition probabilities and integrating all borehole data limit the use of coupled Markov chains, often resulting in discontinuous stratigraphic boundaries [8]. Geostatistical semivariogram simulations, notably more accessible than other methods, leverage the flexibility of semivariogram models in kriging variance [9]. These models facilitate the derivation of geological uncertainties—predictive standard deviation, quantiles, probabilities—during digital modeling, which are instrumental in refining borehole strategies.

This research proposes a borehole optimization approach for stratigraphic modeling based on digital imaging to obtain more precise numerical analysis models. The approach is structured around three elements: digital image stratigraphy, twin-error maps, and optimization of the survey scheme. Digital image stratigraphy renders soil units as pixels on images, derived from stratigraphic boundary interpolations via the Kriging method. Utilizing the Kriging prediction standard error, we introduce an accuracy metric for digital image predictions. Moreover, a clear and concise method for optimizing the sequence of additional boreholes is developed, grounded in the analysis of column prediction error rates.

2 Generation of Digital Image Stratigraphy

2.1 Kriging Method

For the development of precise stratigraphic models, digital modeling is required, coupled with real-time assessment of the digitized strata's level of detail. Kriging offers a flexible approach to spatial interpolation, allowing for the estimation at unsampled locations using available data points and the simultaneous evaluation of the interfaces within the digital stratum.

Kriging is a regression algorithm that models and predicts (interpolates) spatial processes or random fields based on the covariance function (or semivariogram) [10]. The Kriging interpolation method can be used to predict the elevation of any unknown point, with its formula being:

$$z = \sum_{i=1}^{n} \lambda_i z_i \tag{1}$$

where z_i represents the elevation of sample point i, and λ_i is the weight. The weight λ_i is derived from Eq. (2), based on the semivariance values between the given and unknown points.

$$\begin{bmatrix} \lambda_1 \\ \lambda_2 \\ \cdots \\ \lambda_n \\ \varphi \end{bmatrix} = \begin{bmatrix} \gamma_{11} & \gamma_{12} & \cdots & \gamma_{n1} & 1 \\ \gamma_{21} & \gamma_{22} & \cdots & \gamma_{2n} & 1 \\ \cdots & \cdots & \gamma_{ij} & \cdots & \cdots \\ \gamma_{n1} & \gamma_{n2} & \cdots & \gamma_{nn} & 1 \\ 1 & 1 & \cdots & 1 & 0 \end{bmatrix}^{-1} \begin{bmatrix} \gamma_{10} \\ \gamma_{20} \\ \cdots \\ \gamma_{n0} \\ 1 \end{bmatrix} \tag{2}$$

where γ_{ij} represents the semivariance value for the distance between points j and i; $j = 0$ denotes the unknown point; $[\,]^{-1}$ indicates the inverse of a matrix. In Eq. (2), the semivariance values are obtained through the optimal fitting of empirical semivariogram data points. In this study, the subsequent integration of all geostatistical data, used for the generation of digital stratigraphic interfaces and the estimation of prediction error distribution, is based on the Gaussian model. The semivariogram function expression for the Gaussian model [9, 10] is:

$$\gamma(h) = \begin{cases} 0, & h = 0; \\ C_0 + C\left[1 - \exp\left(-\frac{h^2}{a^2}\right)\right], & h > 0. \end{cases} \tag{3}$$

where h denotes the intersample distance; C_0 is the nugget, linked to measurement error [11]; C is the sill; a relates to the range, reflecting spatial autocorrelation scope. The semivariogram escalates with increasing h up to a and plateaus beyond a. This indicates robust spatial autocorrelation among sample points within the threshold a, which diminishes past this limit, suggesting independence.

2.2 Generating Digital Image Stratigraphy

For the digital stratigraphy algorithm, pre-generation stratigraphic classification of digital geounits is required. Stratigraphic IDs are typically tied to interface IDs in practice. To optimize algorithmic efficiency, stratigraphic and interface IDs can be numerically sequenced based on sediment deposition chronology (mean elevation). In our approach, the primary interface between surface sediment and air is labeled with ID "0", with descending IDs for lower interfaces (e.g., $-1, -2, \ldots$). The bottom-most boundary ID is set at -999, below all primary interface IDs, while the topmost boundary ID is 999, above all. Secondary interfaces receive intermediary IDs, like -1.5, between adjacent primary interface IDs.

Following this, discrete nodal points for principal and subsidiary stratigraphic interfaces are created in accordance with the scale of the geologic units. These nodes are then refined using information from borehole records. Given the intrinsic structural features of stratified formations, the core methodology for the generation of pixelated stratigraphic imagery is delineated as: (1) Trimming the primary stratigraphic interfaces that protrude above the earth's surface with reference to the surface itself; (2) Trimming excess segments of subsidiary stratigraphic interfaces, which are employed in the modeling of tapering and prismatic strata, using the main stratigraphic interfaces. Post-trimming, only those discrete nodes that can precisely delineate the geologic section are preserved. In the final stage, geologic units situated between neighboring stratigraphic margins are populated at a consistent size, with each unit's material composition being determined by a previously formulated index linking stratigraphic interfaces to their respective strata.

3 Twin-Error Map

3.1 Error Map Based on Kriging Method

To evaluate the refinement of digital stratigraphic images, this study employs the Kriging standard deviation as a measure of uncertainty for each stratigraphic interface based on the Kriging method. The expression for Kriging variance is shown in Eq. (4), and the predicted standard deviation of deviation for each node on the stratigraphic interface from its estimated value can be calculated using Eq. (5).

Fig. 1 Results of geological uncertainty analysis based on Kriging prediction error

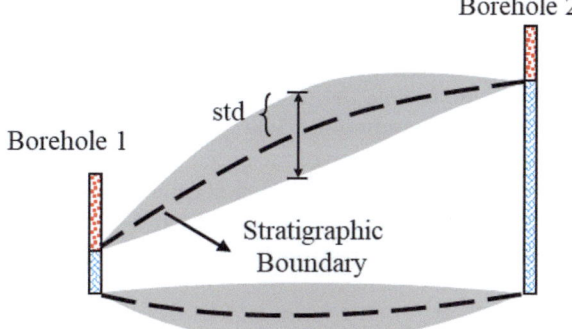

$$\text{var}\left(\widehat{Z}(x_0) - Z(x_0)\right) = \begin{pmatrix} \lambda_1 \\ \lambda_2 \\ \vdots \\ \lambda_n \\ \mu \end{pmatrix} \prime \begin{pmatrix} \gamma(x_1, x_0) \\ \gamma(x_2, x_0) \\ \vdots \\ \gamma(x_n, x_0) \\ 1 \end{pmatrix} \quad (4)$$

$$\text{std} = \sqrt{\text{var}\left(\widehat{Z}(x_0) - Z(x_0)\right)} \quad (5)$$

Hence, a twin-error map can be constructed for each stratigraphic boundary, as depicted in Fig. 1. Within this twin-error map, the gray regions denote the anticipated range of errors for the stratigraphic interfaces, which arise due to uncertainties in the stratigraphy, straddling the upper and lower sides of the forecasted stratigraphic boundaries. The greater the extent of the gray error range, the higher the expected magnitude of error in the prediction of stratigraphic layers within the digital image stratigraphy. The error margin typically expands with increasing distance from the closest borehole. It is evident that incorporating additional exploratory data can decrease the range of digital errors, thereby improving the granularity of the digitized stratigraphic layers. Importantly, the twin-error map shares the same scale as the digital image stratigraphy. Therefore, geological units (pixels) that fall within the gray area of the error map are considered to be uncertain; on the contrary, the status of the unit outside this area is considered to be certain.

3.2 Accuracy of Digital Image Stratigraphy

To evaluate the overall refinement of the generated digital image stratigraphy, an overall precision of the digital image stratigraphy can be defined based on the proportion of uncertain units:

$$R_F = 1 - \frac{M_u}{M} \tag{6}$$

where M represents the number of pixelated geological units, and M_u denotes the number of uncertain units. It is evident that the greater the value of R_F, the higher the simulation precision of the digital image stratigraphy model. Therefore, from the perspective of geological modeling and numerical analysis, the objective of borehole optimization is to optimize the location of boreholes to maximize the overall precision of the digitized strata as much as possible.

4 Borehole Optimization Based on Column Error Rate

In existing research, the entropy of information is commonly utilized for the optimization of geological boreholes. However, considering that in practical engineering, information entropy is not a very direct measure, it can be difficult for engineers to comprehend. Therefore, this paper proposes the use of the column error rate, a metric with direct significance, as an indicator for borehole optimization. The column error rate can be obtained by calculating the proportion of units identified as uncertain on each column in the error map. The column error rate is defined as:

$$\text{Err\_col}(k) = \frac{M_{ku}}{M_k}, \quad k = 1, 2, \ldots, N_x \tag{7}$$

where M_k represents the number of geological units (pixels) in the k-th column; M_{ku} denotes the number of uncertain units in the k-th column.

The column error rate records the average probability of geological unit prediction errors on each column and can serve as a supplementary part of the digital stratigraphy, used for analyzing the level of refinement in digital modeling. Viewing each geological unit column, k, as a potential borehole, optimization of additional boreholes can be based on this. From the perspective of geological modeling, the design for additional optimal boreholes should be located at potential boreholes with the highest column error. This is expressed as follows:

$$\text{col}_{\text{opti}} = \text{find}(\max(\text{Err\_col}) = \text{Err\_col}) \tag{8}$$

Drilling Optimization Using Digital Image Stratigraphic Modeling

5 Case Illustration

To demonstrate the applicability of the described method, this section presents a case study that establishes a two-dimensional digital stratigraphic model of a slope based on surface survey data and sparse borehole data. Based on this model, borehole optimization is conducted.

5.1 Initial Digital Stratigraphy Based on Boreholes

Figure 2a shows the borehole distribution for a survey project, including ten boreholes with only two near the target profile, marked by a red dashed line important for site engineering. For accurate stratigraphic modeling of this profile, additional boreholes are needed at strategic locations. The profile's orientation is along the x-axis, with the z-axis perpendicular to the plan view. Figure 2b presents the real profile and initial borehole locations, revealing four stratified soil layers: sandy soil, gravelly sand, sandstone, and bedrock, sequentially from top to bottom as Layers 1–4. The left side shows greater exposure depths for layers 2–4 than the right. However, the initial boreholes' uneven distribution complicates optimizing the number and placement of additional boreholes, a challenge this paper's proposed optimization method aims to solve.

In this case, surface point data with a sampling interval of 10 m were utilized. Given the existing borehole data, the Kriging method was applied to interpolate the strata, resulting in a digitized stratigraphic profile. This approach involves fitting the semivariogram function parameters using the available borehole and surface data. The stratigraphic boundaries were determined based on the semivariogram function fitting parameters listed in Table 1, without considering measurement errors.

In the recovered digital stratigraphy, each pixel (geological unit) represents a dimension of 1 m. Figure 3 displays the digital image stratigraphy and the twin-error map. In the twin-error map, black pixels represent the predicted stratigraphic boundary lines; gray pixels indicate that the geological unit is within the estimated error band range. The twin-error map reveals that the error in the stratigraphic boundaries converges at the borehole locations and increases further away from the boreholes, resulting in "error bars" forming between adjacent boreholes. The width of these bars signifies the magnitude of the prediction error. Figure 3a presents the initial digital image stratigraphy obtained after interpolating the combined surface and borehole data. The surface data enhanced the modeling accuracy of the surface, accurately depicting the undulating condition of the surface elevation. From the twin-error map in Fig. 3b, it can be observed that the shadows of prediction error by the Kriging method are primarily concentrated in the left area, indicating that the uncertainty of the stratigraphic results determined by the initial boreholes is greatest in the left area, which is also the focus of attention.

Fig. 2 Initial boreholes and real geological profile of an investigation project: **a** horizontal planar distribution of boreholes; **b** two boreholes on the reference line and the real geological cross section

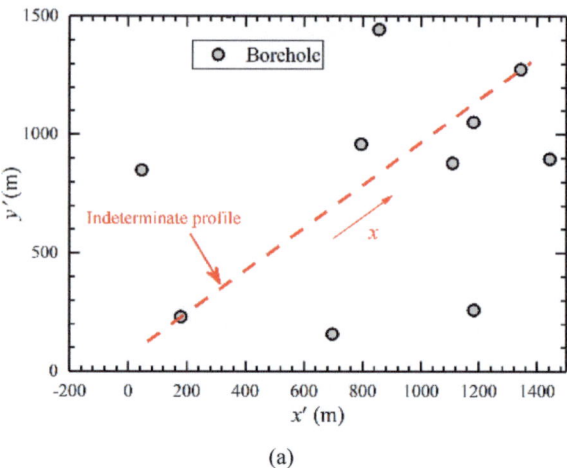

(a)

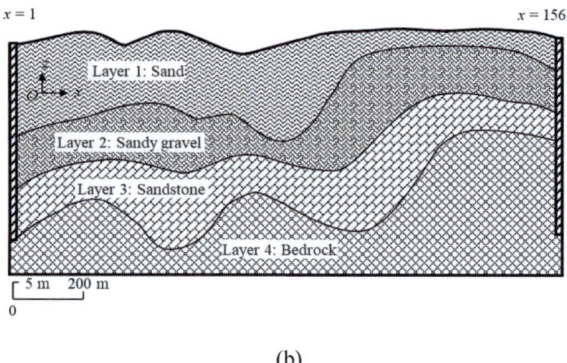

(b)

Table 1 Parameters of fitted semi-variance functions corresponding to stratigraphic boundaries (in pixels)

Stratigraphic boundary ID	Semivariogram model	Nugget	Sill	Parameter a
0	Gaussian	0	509.9	241.8
−1	Gaussian	0	7360.8	384.51
−2	Gaussian	0	5402.0	430.07
−3	Gaussian	0	10,586.4	363.68

5.2 Optimization of Additional Boreholes

In geological exploration, a common method to acquire stratigraphic data involves the addition of extra boreholes. When designing these additional boreholes, the aim is to collect the most useful data possible to create a more accurate digital stratigraphic

Fig. 3 Digital image stratigraphy model by combining surface and borehole data: **a** digital stratigraphic images obtained from borehole and surface data; **b** twin-error map

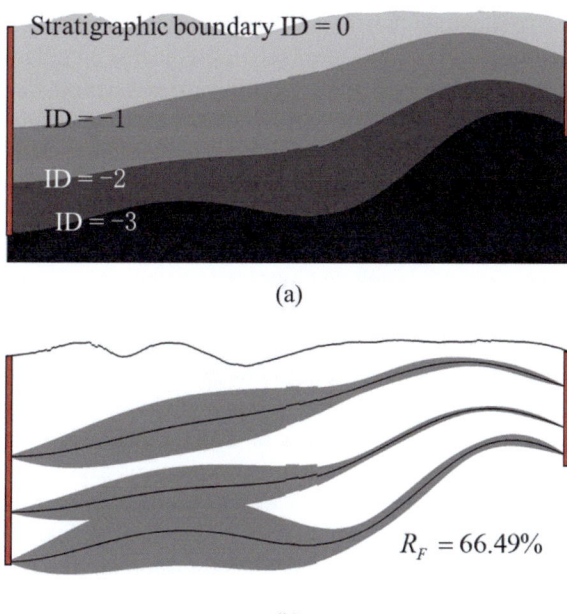

model. Before validating the effectiveness of the proposed borehole plan, this study first mapped the distribution of initial columnar error rates (refer to Fig. 4).

In Fig. 4, the column error rate is minimized at the borehole locations and gradually increases with the distance from the borehole positions. The point of maximum columnar error rate indicates the next potential optimal borehole location. Comparing the two curves in Fig. 4, it is evident that the introduction of surface elevation data has reduced the cumulative column error, with the maximum cumulative column error decreasing from 67.81 to 63.30%. This demonstrates that rich surface data can reduce the error in the model and enhance the efficiency of geological exploration. Moreover, with the inclusion of surface data, the position of the additional optimal borehole shifted from the original $x = 525$ to $x = 565$ m, indicating that surface data have a significant impact on the optimization of borehole placement.

Fig. 4 Column error rate distribution and predicted next additional optimal borehole location

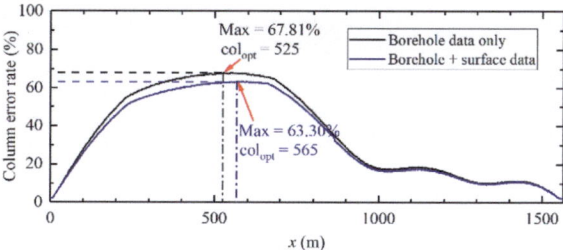

To illustrate the efficacy of the borehole optimization approach, the outcomes of additional borehole optimization predicated on columnar error rates are depicted in Fig. 5. Within Fig. 5, the red thick arrows denote the sites of each step's additional optimal boreholes, corresponding to the maximal cumulative columnar error rates. It is crucial to acknowledge that the semivariogram model parameters require an update to align with the best-fit conditions upon the addition of extra borehole data. Post-optimization, there is a notable decrement in the columnar error rates at the sites of the potential optimal boreholes, and the error rate curve undergoes truncation. Thus, the crux of this borehole optimization methodology lies in the progressive diminution of each step's peak curve, aiming to reduce the mean columnar error rate and curtail the cumulative area under the error rate curve. By iterating this process, the borehole arrangement is refined continuously until it satisfies the pre-established error criteria. Employing this sequential borehole optimization technique substantially bolsters the precision and dependability of digital stratigraphic modeling.

Figure 5 also illustrates the outcomes of stratigraphic uncertainty using a sequential optimization technique. Initial boreholes are denoted by red columns, while additional boreholes are indicated by blue ones. Additional boreholes enable a Kriging-derived digital stratigraphic model to capture finer details. For example, at the stratigraphic interface ID $= -3$, the initial boreholes only revealed a general left-to-right ascending trend with peaks on both ends. Introducing a third borehole transformed the boundary into a tri-peaked curve. Twin-error plot data show that adding boreholes truncates the thickest error bars, creating thinner error bars. Optimal additional borehole placement aims to reduce error bar thickness and minimize the count of uncertain elements (gray pixels in twin-error plot). The inclusion of four boreholes boosted the digital stratigraphy's overall accuracy from 66.49 to 94.21%, an increase

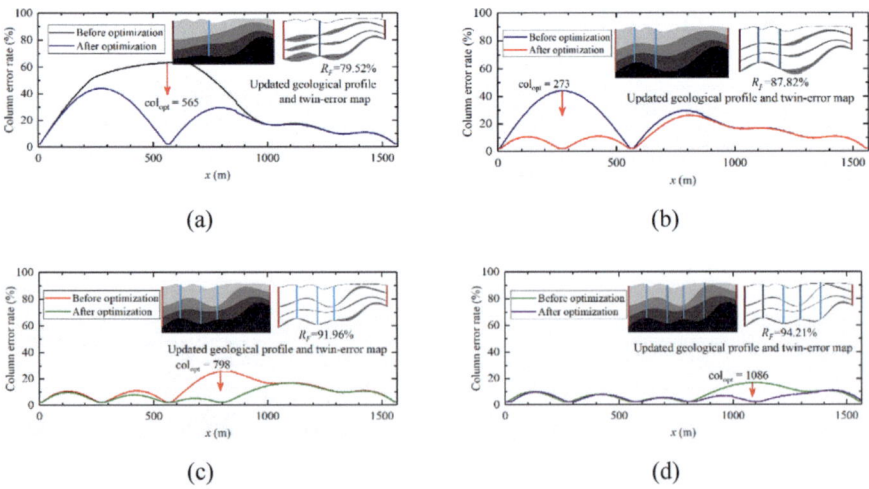

Fig. 5 Stepwise optimization of additional borehole locations based on column error rate: **a** initial model; **b** +1 borehole; **c** +2 boreholes and **d** +3 boreholes

of 27.72%. This underscores the significant enhancement in model detail achieved through new boreholes. Nonetheless, due to the original boreholes' uneven distribution, the first two additional boreholes on the left made the largest accuracy contribution to the profile model, aligning with the area's error distribution and underscoring the method's effectiveness in reducing geological uncertainty. The fourth borehole's contribution to error reduction was comparatively minor, attributed to the smaller error dispersion in that region.

5.3 *Efficacy Verification of Borehole Optimization Method*

To verify the effectiveness of the proposed optimization approach grounded on column error rates, this study compared it against a one-step equidistance-based optimization strategy. The study illustrates the method's efficacy by examining how adding the same number of extra boreholes enhances digital stratigraphic image accuracy and lowers column error rate curves. In the distance-based optimization, extra boreholes are evenly placed at equidistant points between the initial two boreholes. Conversely, in our optimization strategy, boreholes are incrementally added to achieve a finer stratigraphic model. Data from these additional boreholes are incorporated into updating the semivariogram model parameters.

Comparing both extra borehole strategies, this research illustrated the revised column error rate distribution curves, depicted in Fig. 6. Here, red arrows denote the additional borehole locations suggested by our method, and black arrows show those from the one-step optimization approach. The comparison of results from both strategies indicates that additional boreholes contribute to the establishment of a more accurate digital stratigraphic model. However, analyzing the column error rate curves reveals that the equidistant borehole optimization approach does not account for the significance of initial boreholes and the distribution of geological uncertainties. Consequently, it is significantly less effective in reducing the column error rate curve compared to the method proposed in this study. For instance, after adding the first additional borehole, the method presented in this paper achieved a peak reduction in the column error rate curve to 44.2% at $x = 273$; whereas, the distance-optimized method resulted in a peak value of 56.9% at $x = 303$, which is 12.7% higher than our method. With the introduction of two additional boreholes, the maximum column error rate dropped to 26.2% using our method, compared to a maximum of 47.1% with the distance-based optimization. This demonstrates that our borehole optimization strategy significantly reduces the uncertainty in geological distribution, especially in the left-side region.

Hence, the borehole optimization approach presented herein markedly outperforms traditional methods, reducing the reliance on additional boreholes and improving profile prediction precision. The comparison of schemes shows that borehole optimization, informed by twin-error mapping, adequately considers the effects of initial geological uncertainties. These insights provide a beneficial reference for digital stratigraphic modeling with extensive application potential.

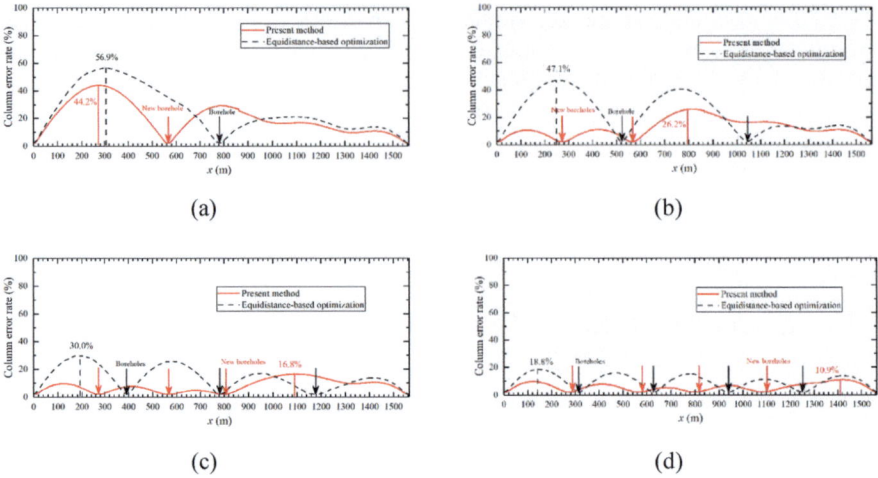

Fig. 6 Comparison of the column error rate curves obtained by this method with those obtained by the equidistance-based optimization method: **a** +1 additional borehole; **b** +2 additional boreholes; **c** +3 additional boreholes; **d** +4 additional boreholes

6 Conclusions

This paper introduces a digital image stratigraphy approach for the optimization design of geological survey boreholes. The Kriging method is employed for the generation of digital image stratigraphy, and a twin-error map is utilized to provide an assessment metric for the accuracy of the digital stratigraphy. A case study is presented to validate the effectiveness of the proposed borehole optimization method. In summary, the following conclusions can be drawn.

The borehole optimization method comprises three components: digital image stratigraphy, twin-error map, and column error rate map. Digital image stratigraphy employs a pixel discretization approach for geological modeling. The twin-error map assesses the uncertainty of stratigraphic boundaries, while the column error rate metric indicates the potential locations for subsequent optimal boreholes.

The Kriging method is employed for the generation of digital strata and the evaluation of stratigraphic uncertainty. By fitting the semivariogram, it is possible to obtain the stratigraphic boundaries at unknown locations. The standard deviation of Kriging prediction intuitively assesses the uncertainty of the stratigraphic boundaries.

A sequential borehole optimization approach based on column error rate has been proposed, which can predict potential borehole columns with the highest estimation errors. Comparative results with drilling schemes optimized based on distance suggest that the sequential optimization method guided by the column error rate metric yields superior outcomes.

References

1. Gong W, Tang H, Wang H et al (2019) Probabilistic analysis and design of stabilizing piles in slope considering stratigraphic uncertainty. Eng Geol 259:105162
2. Zhang W, Han L, Gu X et al (2022) Tunneling and deep excavations in spatially variable soil and rock masses: a short review. Undergr Space 7:380–407
3. Drew LJ (1979) Pattern drilling exploration: optimum pattern types and hole spacings when searching for elliptical shaped targets. J Int Assoc Mathe Geol 11:223–254
4. Soltani S, Hezarkhani A (2011) Determination of realistic and statistical value of the information gathered from exploratory drilling. Nat Resour Res 20:207–216
5. Hossein Morshedy A, Memarian H (2015) A novel algorithm for designing the layout of additional boreholes. Ore Geol Rev 67:34–42
6. Gong W, Zhao C, Juang CH et al (2021) Coupled characterization of stratigraphic and geo-properties uncertainties: a conditional random field approach. Eng Geol 294:106348
7. Cao W, Zhou A, Shen SL (2021) An analytical method for estimating horizontal transition probability matrix of coupled Markov chain for simulating geological uncertainty. Comput Geotech 129:103871
8. Li J, Cai Y, Li X et al (2019) Simulating realistic geological stratigraphy using direction-dependent coupled Markov chain model. Comput Geotech 115:103147
9. Kim M, Kim HS, Chung CK (2020) A three-dimensional geotechnical spatial modeling method for borehole dataset using optimization of geostatistical approaches. KSCE J Civ Eng 24:778–793
10. Pokhrel RM, Kuwano J, Tachibana S (2013) A kriging method of interpolation used to map liquefaction potential over alluvial ground. Eng Geol 152:26–37
11. Kumar P, Rao B, Burman A et al (2023) Spatial variation of permeability and consolidation behaviors of soil using ordinary kriging method. Groundwater Sustain Develop 20:100856

Open Access This chapter is licensed under the terms of the Creative Commons Attribution 4.0 International License (http://creativecommons.org/licenses/by/4.0/), which permits use, sharing, adaptation, distribution and reproduction in any medium or format, as long as you give appropriate credit to the original author(s) and the source, provide a link to the Creative Commons license and indicate if changes were made.

The images or other third party material in this chapter are included in the chapter's Creative Commons license, unless indicated otherwise in a credit line to the material. If material is not included in the chapter's Creative Commons license and your intended use is not permitted by statutory regulation or exceeds the permitted use, you will need to obtain permission directly from the copyright holder.

Research on the Influence of Ultra-Deep Foundation Pit Excavation on Adjacent Existing Tunnels and Deformation Control

Youpeng Wen, Mingxing Zhu, Kunpeng Wu, Xiaocong Liang, and Xiaozhou Yan

Abstract In view of the safety issues of subway tunnel proximity construction, taking a subway station pit project as an example, through the analysis of in situ measured data, the deformation impact of deep foundation pit excavation on the adjacent existing subway tunnel and the surrounding soil was studied, and numerical simulation to study the control effect of setting isolation piles, increasing the rigidity of the support in the pit, and unloading the soil above the tunnel on the deformation of the adjacent tunnel caused by the excavation of the foundation pit was used. The study shows that setting isolation piles near the tunnel side can effectively reduce the tunnel displacement, and the length of isolation piles should at least cover the bottom range of the tunnel. The support in the pit within the elevation of the tunnel section has a greater impact on the tunnel deformation, and enhancing the stiffness of the support or increasing the number of the supports in this range can effectively control the deformation of the tunnel as well. Unloading of the soil above the tunnel also has a significant effect on the vertical displacement of the tunnel, and it is recommended to adopt graded unloading in several layers to avoid one-time and massive-quantity unloading during the construction process.

Keywords Ultra-deep foundation pit · Existing tunnels · Proximity construction · Deformation control · Isolation piles

---

Y. Wen · M. Zhu (✉) · K. Wu · X. Liang
CCCC Fourth Harbor Engineering Institute Co., Ltd, Guangzhou, Guangdong, China
e-mail: zhumingxing@cccltd.cn

Key Laboratory of Environment and Safety Technology of Transportation Infrastructure Engineering, CCCC, Guangzhou, Guangdong, China

M. Zhu · K. Wu · X. Liang
Southern Marine Science and Engineering Guangdong Laboratory (Zhuhai), Zhuhai, Guangdong, China

X. Yan
The Second Engineering Company of CCCC Fourth Harbor Engineering CoLtd., Guangzhou, Guangdong, China

1 Introduction

With the deepening of the development and utilization of urban underground space, large-scale high-density and continuous development of housing and subways in prosperous areas, close-in construction problems such as deformation of existing tunnels caused by foundation pit excavation have become increasingly prominent. Existing structures under the influence of the surrounding environment have become increasingly prominent. Research on deformation control and safety assurance of tunnels has become one of the practical needs of the project.

In recent years, many scholars have conducted research on this problem from the aspects of theory, experiment, numerical, and actual measurement analysis [1–4]. Pu and Liu [5] considered the unloading stress of the bottom and side walls of the foundation pit and the foundation pit enclosure structure. The influence of the foundation pit excavation caused the calculation formula of the additional load adjacent to the existing tunnel was obtained; Liu et al. [6] conducted a centrifugal model test on the excavation plan of a foundation pit project adjacent to a subway tunnel in Shanghai and verified that large and small foundations of the block excavation method of pits can effectively protect the adjacent subway tunnels; Wang et al. [7], Li et al. [8] and others evaluated the protective effect of grouting on adjacent subway tunnels through model tests and field tests, respectively. Xu et al. [9] based on tunnel deformation monitoring data analyzed the entire process from the beginning of foundation pit retaining structure construction to the end of excavation; many scholars also proposed the effect of foundation pit excavation on the deformation of existing tunnels through numerical methods. The influence zone is divided [10], and the influence of measures such as soil unloading, isolation piles, increasing the stiffness of the struts, and optimizing the struts layout on the tunnel displacement and the evaluation of the control effect are analyzed [11–15].

This paper takes the foundation pit project of a new subway station adjacent to an existing subway tunnel as the background, combined with the analysis of on-site measured data, and uses the finite element method to study the control effect of foundation pit excavation on the deformation of the adjacent existing subway tunnel under different measures, which can be used for similar purposes providing reference for comparison and selection of engineering deformation control solutions.

2 Project Overview

The project and surrounding environment are shown in Fig. 1, including foundation pit A, existing line tunnels, and stations. The long axis of the foundation pit is parallel to the existing subway tunnel and runs northwest-southeast. The distance between the tunnel and the foundation pit enclosure is 0.50–16.60 m; at the standard typical section (2–2′), the tunnel depth is 6.45 m, foundation pit width 23.5 m, the standard section foundation pit depth reaches 30.08 m, the horizontal clear distance between

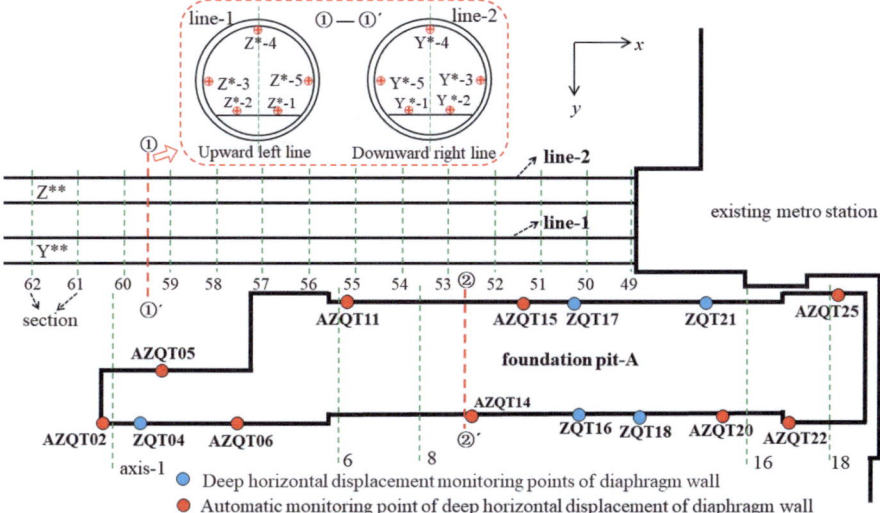

Fig. 1 Plane relationship between foundation pit and existing subway tunnel and layout of monitoring points

the tunnel and the outside of the foundation pit enclosure is 8.35 m, the main body of the enclosure structure adopts 1000 mm thick underground continuous wall plus internal supports.

The project site belongs to the landform of denuded residual hills and interplatform valleys caused by erosion and accumulation. After many constructions and transformations in the later period, the site is generally relatively gentle. The rock and soil layers exposed within the scope of the project are mainly fill layers, silt soil, silty clay, mixed granite residual soil, and weathered rock zones. The groundwater at the site is relatively shallow and mainly includes quaternary loose layer pore water and massive bedrock fissure water. Bedrock fissure water develops in strongly to moderately weathered zones. Combining on-site investigation and indoor test results, mixed granite residual soil and weathered rock belts have good mechanical properties in their natural state, but they are prone to disintegration and softening when exposed to water, and their strength decreases sharply. They have the characteristics of sandy soil and clayey soil feature.

3 Impact of Deep Foundation Pit Excavation on Adjacent Existing Tunnels

According to the monitoring layout of the foundation pit and adjacent tunnels at the project site (Fig. 1), taking a certain construction node as an example, at this time, the base plates of axes 1–6 of the foundation pit have been completed, axes

6–8 are under construction, and most of axes 8–16 are under construction. The construction of cushion layer and 16–18 axis base plate has been completed. At this time, the deep horizontal displacement of the foundation pit diaphragm wall is shown in Fig. 2. It can be seen from the figure that due to the unloading of the foundation pit excavation and the action of soil earth pressure, the diaphragm walls on both sides show obvious deformation in the direction of the foundation pit excavation. Trend, the overall horizontal displacement, shows a shape of small at both ends and large in the middle (bulging belly). Along the depth direction, the maximum horizontal displacement of the ground connecting walls on both sides is located at 1/2 of the depth of the foundation pit excavation. The displacement of the diaphragm walls on the side adjacent to the tunnel is larger than that of the connecting walls away from the tunnel. Tunnel side: The maximum horizontal displacement of the connecting wall on the side adjacent to the tunnel is about 40 mm, located in the middle of the foundation pit; the maximum horizontal displacement on the side away from the tunnel is about 30 mm.

Based on the automated monitoring data of the existing line tunnels, along the different monitoring section directions of the tunnel (Fig. 1), the lateral displacement of the left and right line tunnels is shown in Fig. 3. At the same monitoring section, the overall difference in the displacements of different monitoring points of the tunnel (vault, left and right apses, and track bed) is small, showing a relatively consistent deformation trend; from the side far away from the station to the side near the station, the overall lateral displacement gradually increases. According to the trend, part of

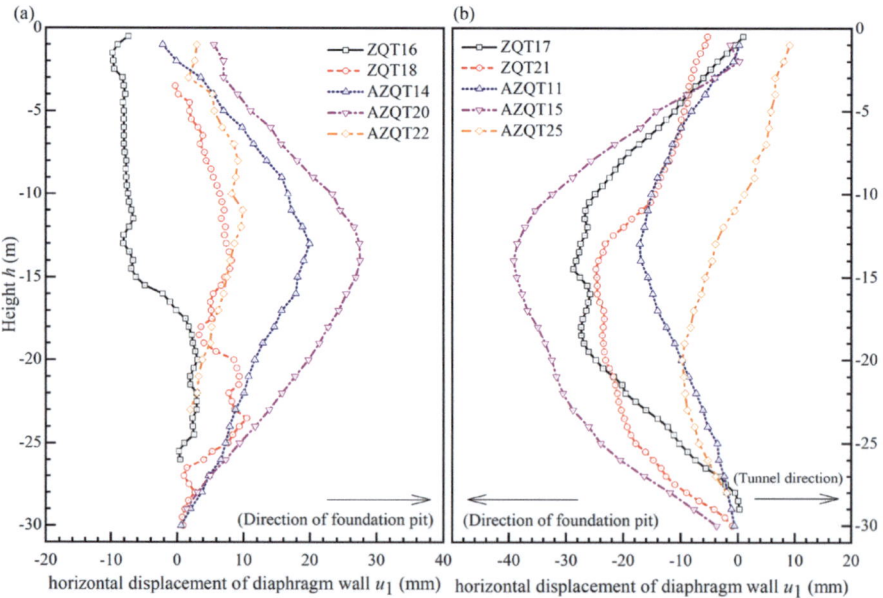

Fig. 2 Monitoring data of deep horizontal displacement of foundation pit diaphragm wall

the soil on the side away from the station deforms in the direction away from the foundation pit excavation.

The same as the lateral displacement, the vertical displacement of the existing line tunnel at this time is shown in Fig. 4. The overall change trend with the section is consistent with the lateral displacement, that is, the vertical displacement and deformation trend of different monitoring points on the same section remains consistent, from the side far away from the station to the side close to the station. On the station side, the vertical displacement of the tunnel changes from uplift to settlement and gradually increases.

Furthermore, the characteristic points of the tunnel section are selected, that is, the lateral displacement at the left and right arch waists of the tunnel, and the vertical displacement at the tunnel vault and track bed surface. The changes with different monitoring sections are shown in Fig. 5. The deformation of the right tunnel closer to

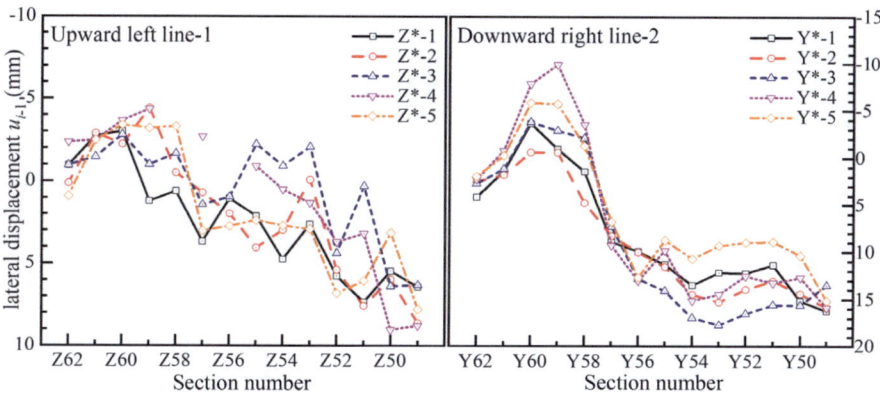

Fig. 3 Monitoring lateral displacement of existing line tunnels (positive values indicate deformation toward the foundation pit)

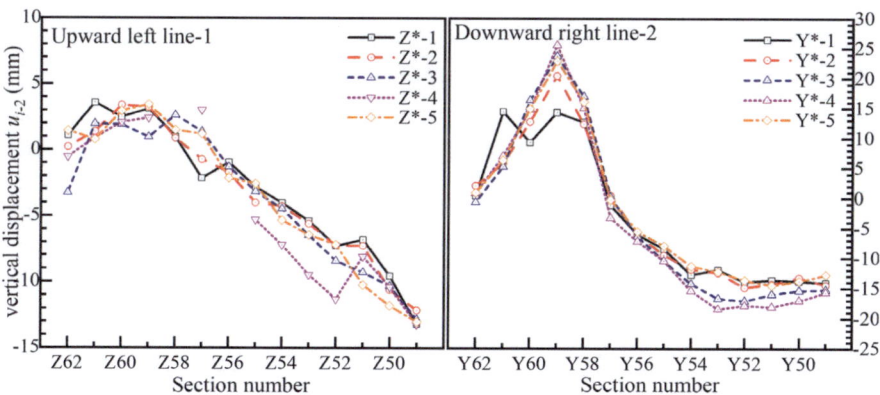

Fig. 4 Monitor vertical displacement of existing line tunnels (negative values indicate subsidence)

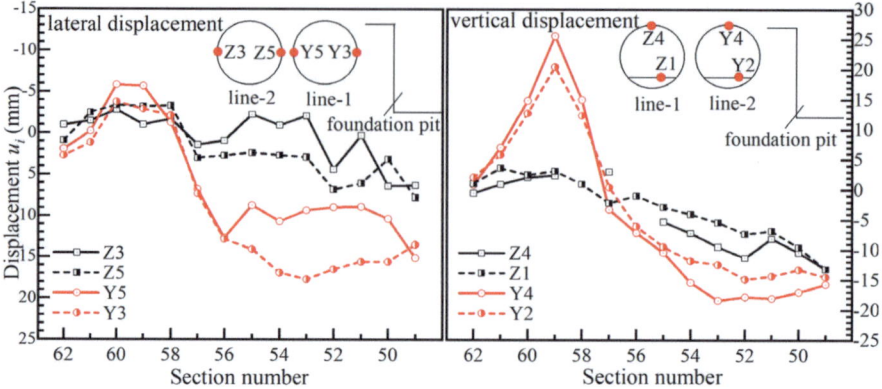

Fig. 5 Monitoring displacement of characteristic points of existing line tunnels

the pit side is significantly greater than that of the left tunnel far away from the foundation pit. Both the lateral and vertical displacement directions of the tunnel changed significantly near section 58. The lateral displacement of the tunnel changed from deformation away from the foundation pit to deformation in the direction of foundation pit excavation, and the vertical displacement changed from uplift to settlement. Combined with the monitoring points in Fig. 1, it can be seen from the plan that the cross section width of the foundation pit here suddenly changes, and the width of the "interval soil" between the foundation pit and the tunnel increases, which affects the overall tunnel deformation. Excavation of the foundation pit, combined with the hollowness of the existing tunnel, creates a phenomenon of "empty space" on both sides of the "interval soil mass." The interval soil mass deforms downwards and squeezes the tunnel, resulting in tunnels within the 58–62 section range, deformation toward the side away from the foundation pit.

4 Control Method for Deformation of Adjacent Existing Tunnels Caused by Foundation Pit Excavation

It can be seen from the above actual monitoring data that foundation pit excavation has a relatively sensitive impact on the deformation of adjacent existing tunnels. For this type of working conditions, in order to further study the tunnel deformation control effect under different measures, a standard section was selected (Fig. 1, 2–2′), establishing a two-dimensional finite element model for analysis.

4.1 Numerical Analysis Model

The total size of the model is 150.0 m × 70.0 m. The dimensions of the foundation pit tunnel are shown in Fig. 6. The inner diameter of the existing subway lining adjacent to the foundation pit is 5.7 m, outer diameter is 6.3 m, the lining wall thickness is 0.3 m, using C50 concrete, the center distance between the two tunnels is 13 m, tunnel depth 6.3 m, the clear distance between the tunnel and the foundation pit is 8.35 m; the excavation depth of the foundation pit is 30.8 m, excavation width 24.65 m, adopt diaphragm wall and internal support enclosure.

The constitutive structure of the model rock and soil adopts the HSS small strain hardened soil model, and its parameters are determined as shown in Table 1. Among them, E_{50}^{ref} is the reference secant modulus of the triaxial consolidation drainage shear test; $E_{\text{oed}}^{\text{ref}}$ is the reference tangent modulus of the consolidation test; $E_{\text{ur}}^{\text{ref}}$ is the reference loading and unloading of the triaxial consolidation drainage unloading and reloading test. Modulus K_0 is the static side pressure coefficient under normal consolidation conditions. The values of other parameters are Poisson's ratio $\mu = 0.3$, failure rate $R_f = 0.9$, stiffness stress level correlation index $m = 0.5$, and reference pressure $p^{\text{ref}} = 100$ kPa. See Table 2 for other relevant material parameters such as diaphragm walls, isolation piles, linings, and internal supports.

Based on the construction conditions of the project site, the finite element analysis steps are set as initial ground stress balance, tunnel excavation and lining construction, and diaphragm wall construction. Then, in accordance with the principle of "support first and then dig," the steps are excavated step by step to each internal support elevation to the bottom of the foundation pit.

The horizontal displacement cloud diagram of the diaphragm wall and tunnel calculated by the model is shown in Fig. 7. Due to unloading during foundation pit excavation, the soil on both sides deforms in the direction of foundation pit excavation, and the maximum deformation of the diaphragm wall is located in the middle. Affected by the excavation of the foundation pit, the deformation of tunnel 1 closer to the foundation pit is more obvious, and the overall horizontal displacement and settlement toward the foundation pit direction.

In order to verify the rationality of the numerical model, the horizontal displacement monitoring data of the diaphragm wall closest to the section and close to the

Fig. 6 Schematic diagram of model dimensions

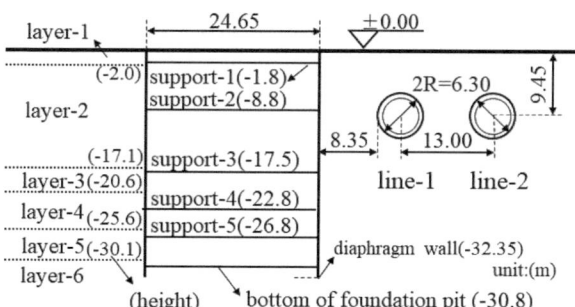

Table 1 Physical and mechanical parameters of materials in the model

Property	Fill soil	Mixed granite residual soil	Completely weathered mixed granite	Strongly weathered mixed granite	Moderately Weathered mixed granite	Lightly weathered mixed granite
Density γ/ kN m^{-3}	18.8	18.1	18.1	23.5	25.5	25.6
Cohesion c/ kPa	26.1	26.0	28.6	52.0	200.0	500.0
Internal friction angle φ/(°)	15.5	19.5	22.6	29.5	42.0	48.0
Dilation angle/(°)	0	0	0	0	12	18
Secant stiffness E_{50}^{ref}/MPa	2.8	2.9	3.0	25.0	84.0	350
Tangential stiffness E_{oed}^{ref}/MPa	3.6	3.8	3.9	32.0	108.0	450.0
Unloading elastic modulus E_{ur}^{ref}/MPa	28	29.4	30	500	840	3500
Lateral soil pressure coefficient K_0	0.733	0.666	0.616	0.508	0.331	0.257

tunnel side were selected for comparison, as shown in Fig. 8. The maximum horizontal displacement of the numerical simulation is slightly larger than the actual measured data, but the overall change trend is consistent with the actual monitoring data, and the compliance is good. The maximum horizontal displacements are all located at an altitude of −15 m, along the depth direction, the horizontal displacement of the diaphragm wall first increases and then decreases. Among them, the numerical simulation displacement of the middle and upper part of the diaphragm wall is significantly different from the monitored displacement. At the bottom of the diaphragm wall and near the bottom of the foundation pit, the numerical simulation and the difference in monitoring displacement results are relatively small.

Based on the above numerical model, the effects of isolation piles, internal supports, unloading, and other measures on tunnel deformation control were studied, respectively, and the effective control rate of tunnel deformation was defined as:

$$P_s = \frac{S_0 - S_1}{S_0} \times 100\% \tag{1}$$

Table 2 Support structures parameters

Structure type	Dimensions (m) @ spacing (m)	Density γ/ (kN m^{-3})	Elastic modulus E/ MPa	Poisson's ratio μ
Diaphragm wall	1	25	20,000	0.3
Isolation pile	Φ0.8@0.8	25	20,000	0.3
Tunnel lining	0.3	25	20,000	0.3
The 1st strut	0.7 × 0.9@9	25	20,000	0.3
The 2nd strut	Φ0.609(wall thickness 0.016)@3	78	200,000	0.3
The 3rd strut	1.3 × 1.3@9	25	20,000	0.3
The 4th strut	Φ0.8(wall thickness 0.020)@3	78	200,000	0.3
The 5th strut	Φ0.8(wall thickness 0.020)@3	78	200,000	0.3

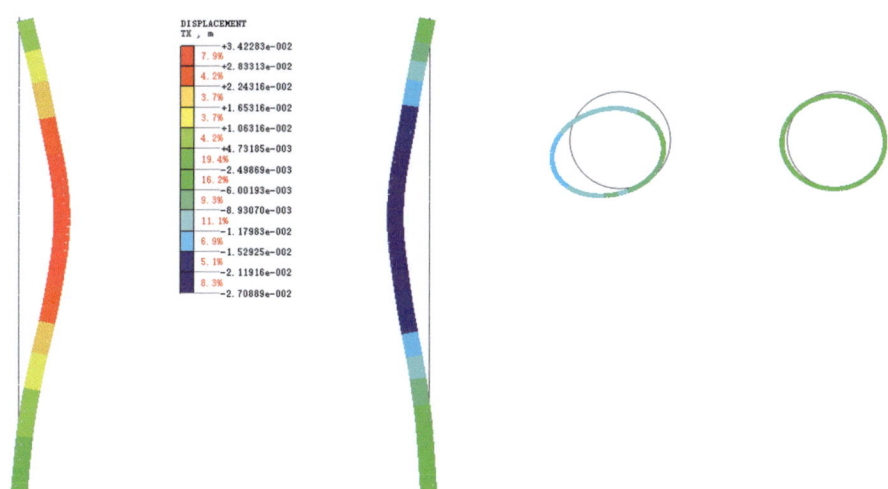

Fig. 7 Numerical model diaphragm wall and tunnel horizontal displacement cloud diagram

In the formula, P_s is the tunnel deformation control rate under foundation pit excavation, S_0 is the displacement of the tunnel lining when no control measures are taken, and S_1 is the displacement of the tunnel lining after taking corresponding measures. When P_s less than 0, it means that the control measures increase the tunnel displacement (the absolute value of the displacement increases) and have a negative effect on tunnel deformation control; when P_s equal to 0, it means that the control

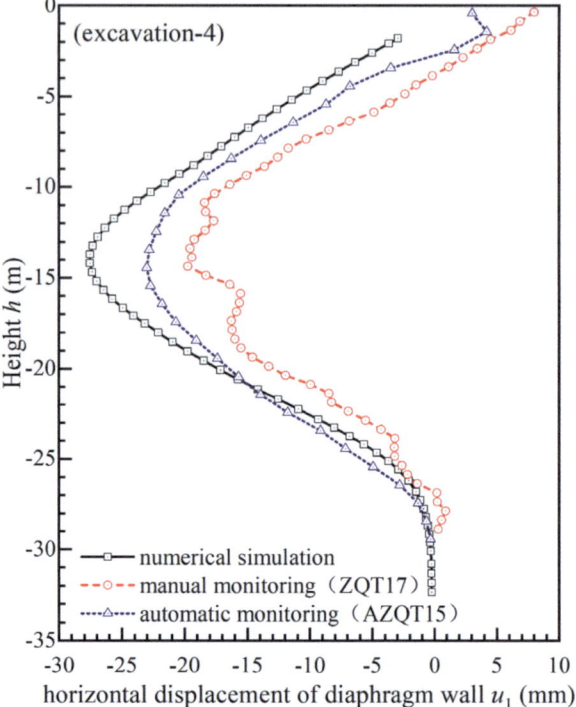

Fig. 8 Comparison of horizontal displacement of diaphragm walls

measures are ineffective; when P_s greater than 0, it means that the control measures It is effective in controlling tunnel deformation (the absolute value of displacement is reduced), and the larger the value, the better the control effect.

4.2 Installation of Isolation Piles

As shown in Fig. 9, the influence of the horizontal and vertical layout of the isolation piles is considered, respectively. When considering the horizontal layout, the horizontal clear distance between the diaphragm wall and the tunnel is H_0, and the horizontal distance between the isolation piles and the diaphragm wall is h. When calculating the ratios h/H_0 are taken to be 0.2, 0.4, 0.6, and 0.8, respectively; when considering the vertical arrangement, the vertical unit length V_0 of the isolation pile is taken to be 0, 1, 2, 4, 6, and 8 m, respectively, where $V_0 = 0$ m represents no isolation pile; let the vertical distance between the center of the vertical unit of the isolation pile and the center of the tunnel be v_0. When the center point of the isolation pile is above the center of the tunnel, v_0 is positive, and vice versa.

Taking the vault, left and right arch halves, and arch bottom of tunnel line-1 and line-2, respectively, as tunnel deformation characteristic points, and taking the tunnel

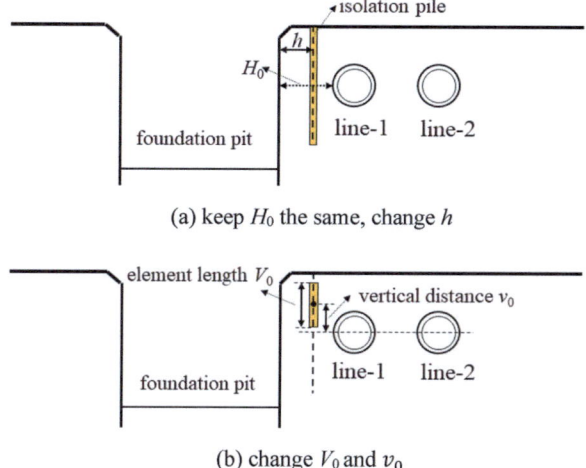

Fig. 9 Schematic diagram of isolation pile layout and tunnel-foundation pit geometry

(a) keep H_0 the same, change h

(b) change V_0 and v_0

deformation when isolation piles are not installed as the benchmark, the tunnel deformation control rate under different horizontal positions of isolation piles is calculated as follows: As shown in Fig. 10, the setting of isolation piles reduces the overall horizontal displacement of the tunnel and increases the vertical displacement of the tunnel. For tunnel line-1, which is closer to the foundation pit side, its deformation is larger, but the deformation control rate is smaller; for tunnel line-2, the absolute value of the deformation is small, and the deformation control rate changes greatly. Therefore, under the condition of ensuring the safe distance of the tunnel, the isolation piles or grouting bodies should be placed as close to the tunnel as possible and controlled outside the scope of influence of foundation pit construction.

Taking the horizontal displacement of the left and right arch halves of tunnel line-1 as the characteristic value, Fig. 11 shows the tunnel deformation control rate

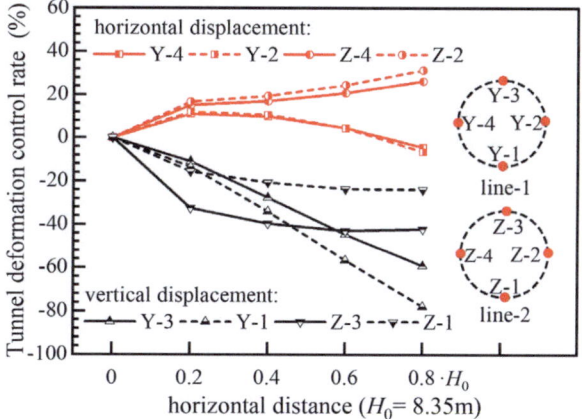

Fig. 10 Tunnel deformation control rate under different horizontal positions of isolation piles

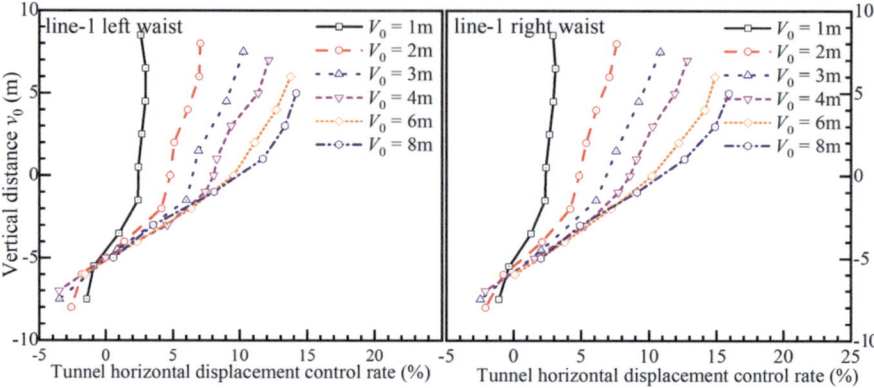

Fig. 11 Tunnel deformation control rate under vertical arrangement of isolation piles of different lengths

considering the vertical arrangement of isolation piles. It can be seen that when the tunnel burial depth is within the excavation depth range of the foundation pit, the displacement control rate decreases with the decrease in vertical distance and increases with the increase in isolation pile length. When the pile length exceeds the tunnel diameter, the deformation control effect of the increase in pile length weakens. Therefore, the center points of isolation piles should be mainly distributed near the center point of the tunnel, with the top and bottom exceeding the tunnel diameter range of 1–3 m, at this time, the deformation of the tunnel under foundation pit excavation can be better controlled.

4.3 Increase of Support Stiffness in the Foundation Pit

In the above model foundation pit support system, the first and third supports are concrete supports, and the second, fourth, and fifth supports are steel supports. In order to study the influence of support position on tunnel deformation, the contribution of supports at different positions to tunnel deformation control was analyzed by reducing the support stiffness. In order to unify the benchmark, the support system is adjusted to a full concrete support. The stiffness reduction coefficient l_0 is 0.4, 0.6, 0.8, 1.0, and 1.2, respectively. Supports 2–5 are reduced, respectively, and the horizontal displacement of the left arch waist of the tunnel is calculated. Corresponding deformation control rate: according to the calculation results (Fig. 12), it can be seen that for tunnel line-1 and tunnel line-2, support 2 and support 3, which are closer to the tunnel, have a greater impact on the tunnel deformation. Support 2 has the greatest impact, and the deformation control rates of tunnel line-1 and tunnel line-2 are for the sensitivity of the reduction factors is roughly the same. Supports 4 and 5 are located in the middle and lower part of the foundation pit, relatively far from

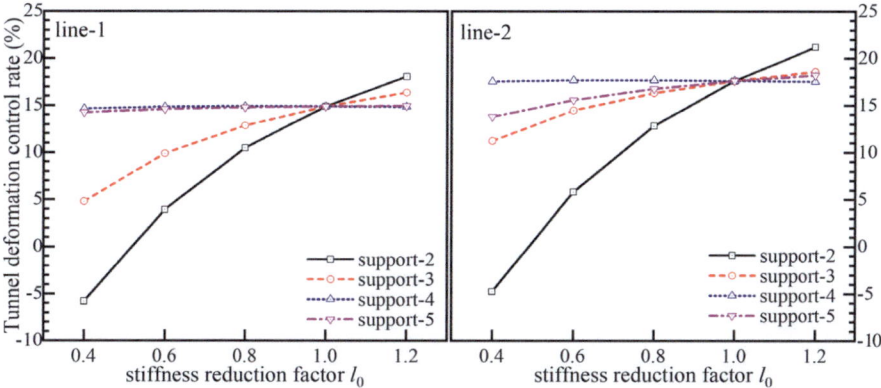

Fig. 12 Tunnel deformation control rate under different support stiffness reduction coefficients

the tunnel, and the overall deformation of the soil is small, resulting in the setting of supports 4 and 5 having little impact on the tunnel deformation. Therefore, adding supports or increasing the stiffness of supports within the tunnel elevation range is beneficial to controlling tunnel deformation, and optimizing supports at deeper locations has less impact on tunnel deformation.

4.4 Unloading of Soil Above the Tunnel

In actual projects, the settlement of the tunnel caused by the excavation of the adjacent foundation pit can be buffered by unloading the soil above the tunnel. Considering that the soil unloading rebound is more sensitive to vertical deformation, the vertical displacement changes of tunnel line-1 and tunnel line-2 are mainly analyzed. As shown in Fig. 13, the unloading block is set to be a rectangle, its size is controlled by the unloading width and unloading depth, and the center of the block is located directly above the center of the tunnel. Considering that the net burial depth of the tunnel is 6.45 m, the horizontal distance between tunnel line-1 and tunnel line-2 is 13 m, take the unloading depth X_h as 1, 2, and 3 m, respectively, take the unloading width X_v 2, 4, 6, and 8 m, respectively.

Calculations are performed for the unloading above tunnel line-1 and tunnel line-2, respectively. In the analysis step, the foundation pit is first excavated to the bottom of the foundation pit, and then the earthwork above the tunnel is unloaded.

Considering the vertical deformation of the tunnel's characteristic points (vault, vault bottom, left arch waist and right arch waist), the deformation control rate is calculated based on the vertical deformation of the tunnel before unloading. Figures 14 and 15 show the tunnel. The relationship between the deformation control rate and the unloading depth and width of the soil above the tunnel can be used as a reference when similar projects meet the unloading conditions. Combining the

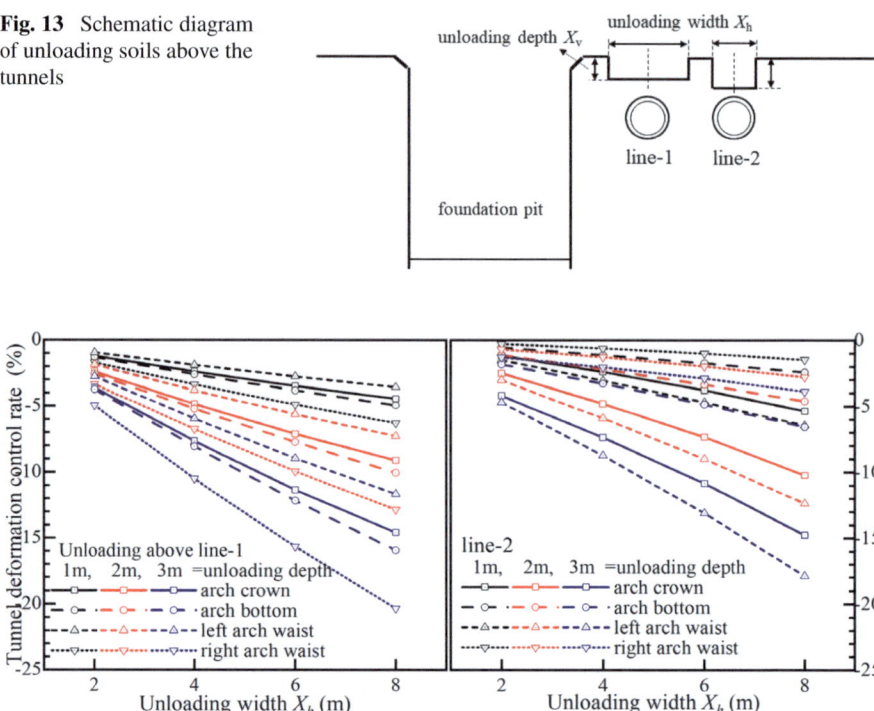

Fig. 13 Schematic diagram of unloading soils above the tunnels

Fig. 14 Tunnel deformation control rate under different unloading widths and depths (unloading of soil above tunnel line-1)

two figures, it can be seen that the unloading deformation control rate increases linearly with the increase in unloading width and unloading depth, and the overall vertical deformation of the tunnel is more sensitive to the influence of unloading. Among them, the right arch waist of tunnel line-1 and the right waist of tunnel line-2 The deformation of the left arch waist is most affected by unloading, but the change amplitudes under different unloading widths and unloading depths are roughly equal. Therefore, when controlling the vertical deformation of the tunnel through earthwork unloading, the amount of unloading should be carefully controlled to avoid excessive bulging of the tunnel caused by excessive unloading.

5 Conclusions

(1) Setting isolation piles between the foundation pit and the adjacent existing tunnel can effectively reduce the deformation of the existing tunnel caused by foundation pit excavation, and the length and relative position of the isolation piles have a greater impact on deformation control. In the horizontal direction,

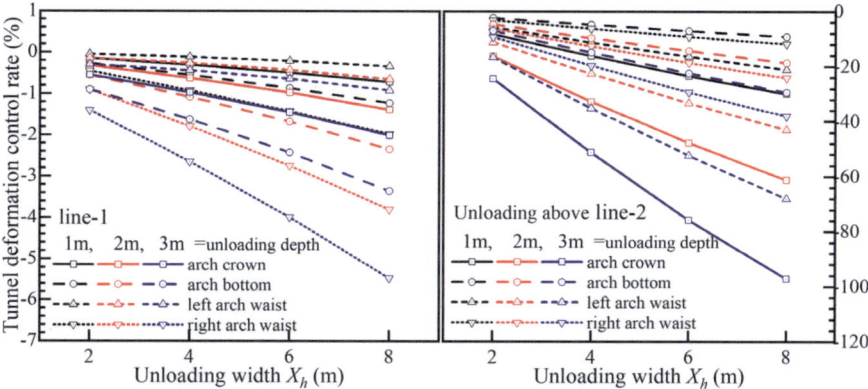

Fig. 15 Tunnel deformation control rate under different unloading widths and depths (unloading of soil above tunnel line-2)

the closer the isolation pile is to the side of the tunnel (but subject to meeting the minimum safety distance requirements of the tunnel), the better the deformation control effect of the tunnel has; in the vertical direction, when the length of the isolation pile covers the bottom of the tunnel or its below, the deformation control effect will be better.

(2) Keep the number of supports in the foundation pit unchanged and increasing the stiffness of the internal supports can effectively control the deformation of the adjacent tunnels, and the overall stiffness of the internal supports within the tunnel section elevation has a greater impact on the deformation of the tunnel, especially when the tunnel is shallowly buried.

(3) The unloading of soil above the tunnel can also effectively control the displacement of the tunnel (in the vertical direction). The depth and width of earth unloading have a significant impact on the vertical displacement of the tunnel. During the implementation of specific projects, it is recommended to give priority to unloading the soil 0.5 times of the diameter of the tunnel directly above the tunnel, and the unloading should be carried out multiple times in stages to avoid unloading a large amount soil at one time.

Acknowledgements This work was supported by the National Natural Science Foundation of China (grants number: 42377150) and the National Key Research and Development Program, Ministry of Science and Technology of China (grants number: 2022YFB2603000).

References

1. Wang P, Ren J, Zhai JQ (2022) Research on displacement control measures of lateral tunnels in foundation pits in soft soil areas in Shanghai. J Undergr Space Eng 18(S1):466–471
2. Lu X, Zheng MF, Liu QB et al (2021) Model test of tracking grouting protection for subway tunnels adjacent to foundation pits. J Undergr Space Eng 17(06):1839–1846
3. Qiao JG, Peng R, Li JW et al (2022) Research on the impact of foundation pit excavation on tunnel safety based on modified Mohr-Coulomb. China Sci Technol Product Saf 18(02):177–183
4. Xu S, Li GX, Wang LC et al (2023) Research on the superposition effect of unloading and blasting vibration in close tunnel excavation. China Product Saf Sci Technol 19(09):96–102
5. Pu JY, Liu B (2021) Control effect of foundation pit soil reinforcement in weak strata on the deformation and excavation affected areas of underlying subway tunnels. Chin J Geotech Eng 43(S2):146–149
6. Liu B, Fan XH, Wang YY et al (2021) Research progress on the impact of foundation pit excavation on adjacent existing subway tunnels. Chin J Geotechn Eng 43(S2):253–258
7. Wang LX, Shi WSY, Xu SS et al (2021) Deformation rules and reinforcement measures of subway tunnels under unloading conditions near foundation pits. Railway Constr 61(08):68–72
8. Li L, Liu QB, Lu X et al (2021) Grouting protection test and numerical simulation of subway tunnels excavated adjacent to foundation pits. J Undergr Space Eng 17(S1):387–396
9. Xu SF, Zhou QH, Zheng WH et al (2021) Full-process measured analysis of the impact of foundation pit construction on the deformation of adjacent operating tunnels. Chin J Geotechn Eng 43(05):804–812
10. Tian S (2020) Protection scheme for running metro tunnels under disturbance of upper foundation pit in complex environment. Tunnel Constr 40(S2):196–203
11. Ding Z, Zhang X, Liang FY et al (2021) Research and prospects on the impact of soft soil foundation pit excavation on adjacent existing tunnels. J Chin Highw 34(03):50–70
12. Zeng XX, Ding WX, Peng L et al (2017) Effect of isolation pile diameter on the displacement of existing subway tunnels caused by foundation pit excavation. Railway Constr 7(06):95–97
13. Wei G, Zhao CL (2016) Calculation method of additional load on adjacent subway tunnels caused by foundation pit excavation. Chin J Rock Mech Eng 35(S1):3408–3417
14. Zheng G, Du YM, Diao Y et al (2016) Research on the influence zone of adjacent existing tunnel deformation caused by foundation pit excavation. Chin J Geotech Eng 38(04):599–612
15. Liang FY, Chu F, Song Z et al (2012) Centrifugal model test study on deformation characteristics of deep foundation pits adjacent to subway hubs. Rock Soil Mech 33(03):657–664

Open Access This chapter is licensed under the terms of the Creative Commons Attribution 4.0 International License (http://creativecommons.org/licenses/by/4.0/), which permits use, sharing, adaptation, distribution and reproduction in any medium or format, as long as you give appropriate credit to the original author(s) and the source, provide a link to the Creative Commons license and indicate if changes were made.

The images or other third party material in this chapter are included in the chapter's Creative Commons license, unless indicated otherwise in a credit line to the material. If material is not included in the chapter's Creative Commons license and your intended use is not permitted by statutory regulation or exceeds the permitted use, you will need to obtain permission directly from the copyright holder.

Analysis of Stability of Surrounding Rock in the Blasting Excavation of Pumped Storage Power Station Powerhouse

Xiji Li, Shuangquan Xu, Chuncheng Ma, and Fan Zhang

Abstract Cave excavation in the pumped storage power plant building is usually carried out by the technique of drilling and blasting. In this study, a part of the pumped storage power plant building was analyzed by the blasting simulation with the Ls-Dyna. Three different explosive arrangements were considered, and comparative analyses of the damage and cumulative effects on the surrounding rock were carried out, including blasting effects, damage analysis, and stress response. The results show that the blasting effects of three arrangements are basically similar and meet the requirements of excavation blasting for the project. However, compared with Cases 1 and 3, the damage level of Case 2 (cruciform arrangement) on the perimeter rock was reduced by 22.3 and 29.5%, respectively. The pressure waves generated by the blast were less superimposed and caused less damage to the upstream and downstream protective layers. Therefore, Case 2 represents the best explosive arrangement for blasting machinery room, which can minimize damage and ensure safety and stability during excavation and blasting.

Keywords Excavation by blasting · Drilling and blasting method · Explosive layout · Ls-Dyna · Stability

1 Introduction

Blasting is a traditional and commonly used method for rock fragmentation, widely employed in the construction of underground powerhouses for pumped storage power stations due to its strong adaptability and relatively low cost. The energy released by explosive during blasting efficiently fractures the rock. It also disturbs the surrounding rock that needs to be preserved. During the excavation of underground

---

X. Li · S. Xu · C. Ma
State Grid Xinyuan Shandong Weifang Pumped Storage Co., Ltd, Weifang, China

F. Zhang (✉)
School of Electrical and Power Engineering, Hohai University, Nanjing, China
e-mail: 1933477567@qq.com

powerhouses using blasting techniques, the blasting vibration can lead to degradation of the mechanical properties of surrounding rock, concrete cracking, loss of bonding strength, and other safety hazards [1, 2]. Hence, study on the effect during the explosive process and the load-pressure changes during column charge borehole basting has been conducted [3–5]. Mambetov and Mosinets [6], and Olsson et al. [7] consider that explosive stress wave is the primary cause leading to rock tension cracking and fragmentation. However, Rossmanith et al. [8] suggest that explosive stress wave only plays a triggering role, and the explosive gas is the main factor driving crack propagation, with its effect far outweighing that of the explosive stress wave. Additionally, researchers such as He [9], Jayasinghe et al. [10], Qiang et al. [11], and Zhiyong and Hui [12] believe that the effect of the stress wave and explosive gas varies depending on the properties of the rock mass. The destruction of high-impedance rock mass primarily depends on the stress wave, while the destruction of medium-impedance rock mass relies mainly on the combined effects of the stress wave and explosive gas. Low-impedance rock mass is predominantly damaged by the explosive gas.

Overall, extensive research on the effects of blasting vibrations induced by drilling and blasting has been conducted, as well as the damage to surrounding rock mass. However, there is still a lack of comprehensive analysis on the damage mechanism of surrounding rock under different explosive arrangements for the same charging method.

This paper conducted blasting simulations on a certain section of a pumped storage power station using the dynamic finite element software Ls-Dyna [13]. Three different borehole arrangement schemes were considered, and simulation results were compared with engineering practice data in terms of the blasting effect, damage analysis, and stress response. The most suitable explosive arrangement scheme for engineering blasting excavation was selected, providing theoretical basis and technical guidance for similar engineering blasting excavations.

2 Engineering Background

The blasting excavation of the powerhouse of a certain pumped storage power station is the engineering background. The structure is complex and requires high safety requirements for the power station. Therefore, it is necessary to ensure that the excavation quality of the II level of the powerhouse meets the design and technical requirements. During the construction process of the powerhouse, it is required to have multiple working faces and multiple working procedures operating concurrently. On the other hand, it is necessary to prevent damage to the rock platform and ensure the stability and safety of the surrounding rock. However, existing theories and experimental results generally cannot accurately describe the blasting damage characteristics of surrounding rock under specific geological conditions. Therefore, it is necessary to assess the key technical issues and damage situation of the blasting excavation of the powerhouse.

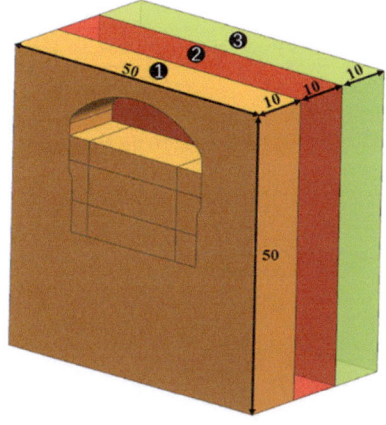 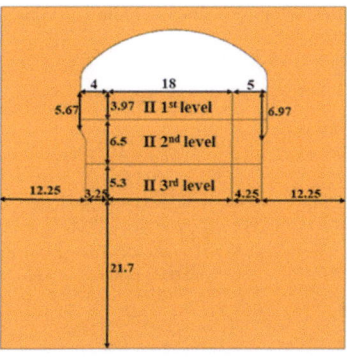

Fig. 1 Numerical model of the underground plants with dimensions (unit: m)

3 Model and Calculation Scheme

3.1 Model Introduction

The section from 0 + 20.00 to 0 + 50.00 on the right side of the underground powerhouse, with a depth of 30 m, width of 50 m, and height of 50 m, is selected as the numerical simulation research object to investigate the blasting excavation of rock. During the numerical simulation, the II level of the powerhouse is divided into three parts. The numerical model and corresponding dimensions are shown in Fig. 1.

3.2 Numerical Simulation Parameter Setting

Material Parameters

Considering engineering experience and the on-site geological conditions, an isotropic homogeneous rock damage model, HJC constitutive model, is adopted. 21 parameters in the HJC constitutive model of rock are detailed in Table 1.

Null material model is employed to describe the air with the keywords of *MAT_NULL [14]. The density of air is 1.2 kg/m^3. TNT emulsion explosive is used with a density of 1150 kg/m^3, detonation velocity of 3500 m/s, detonation pressure of 9.7 GPa, with the keywords of *MAT_HIGH_EXPLOSIVE_BURN. For artillery mud, soil material is chosen with a density of 1800 kg/m^3, using *MAT_SOIL_AND_FOAM.

Table 1 The values used for rock material simulation

Parameters	Values	Units	Parameters	Values	Units	Parameters	Values	Units
ro	2600	kg/m^3	pc	49.6 × 10^6	Pa	k1	13.01 × 10^9	Pa
g	5.33 × 10^9	Pa	uc	4.27 × 10^{-3}	/	k2	9.30 × 10^9	Pa
a	0.27	/	pl	1.2 × 10^9	Pa	k3	75.34 × 10^9	Pa
b	0.144	/	ul	0.1215	/	epsilon	1 × 10^{-6}	/
c	0.0201	/	d1	0.04	/	efmin	0.01	/
n	1.779	/	d2	1.0	/	sfmax	15.0	/
fc	148.9 × 10^6	Pa	t	12.2 × 10^6	Pa	fs	0.0	/

Equation of State

Based on the Ls-Dyna platform, a user subroutine is developed to couple the rock damage model with the HJC model. To eliminate the influence of the reflected wave at the artificial boundary of the rock structure, nonreflecting boundary conditions is set on the six faces of the model. The rock is defined as solid using the Lagrange algorithm, while the air, stemming material, and explosive are defined as fluid using the multi-material ALE algorithm.

Jones–Wilkens–Lee (JWL) state equation is employed to calculate the pressure generated by the detonation of emulsion explosives. The pressure, denoted as P, is expressed as,

$$P = A\left(1 - \frac{w}{R_1 V}\right) e^{-R_1 V} + B\left(1 - \frac{w}{R_2 V}\right) e^{-R_2 V} + \frac{W E_0}{V} \quad (1)$$

where P represents the detonation pressure, A, B, R_1, R_2, and W are explosive detonation parameters, V denotes the relative volume of detonation products, and E_0 is the initial specific internal energy.

EOS_LINEAR_POLYNOMIAL state equation is used to describe the relationship between pressure, density, and specific internal energy in the air model. Its linear polynomial state equation is given by:

$$P = C_0 + C_1 \mu + C_2 \mu^2 + C_3 \mu^3 + (C_4 + C_5 \mu + C_6 \mu^2) E \quad (2)$$

where P is the detonation pressure. E is the specific internal energy. $\mu = \rho/\rho_0 - 1$, and ρ/ρ_0 is the ratio of current fluid density to initial fluid density. C_0 to C_6 are parameters related to air properties.

CONSTRAINED_LAGRANGE_IN_SOLID is used to control the coupling between fluid elements and solid elements. The complete restart technique in Ls-Dyna is utilized to cumulatively calculate various physical quantities during the excavation process. Set the interval between the three excavation steps as 0.015 s, and re-drill holes and charge at each stage to achieve the simulation analysis of rock blasting step-by-step excavation by modifying the finite element model.

3.3 Blasting Design Case

To study how to crush the rock at the slot while minimizing the damage to the upstream and downstream protective layers with the least number of boreholes and the least amount of explosives, three different borehole arrangements are considered during the blasting excavation simulation. The explosive charge structure for the three cases is the same, consisting of artillery mud—explosive—artillery mud cylindrical charge structure. In Case 1, a row of 7 blast holes is arranged in 5 rows, with a hole spacing of 2.5 m and a row spacing of 2 m, totaling 35 blast holes. In Case 2, a cruciform arrangement is adopted, with hole spacing and row spacing the same as Case 1. In the first, third, and fifth rows, seven blast holes are placed per row, while in the second and fourth rows, six blast holes are placed per row, totaling 33 blast holes. In Case 3, a cruciform arrangement is also adopted, with seven columns arranged, each column with a hole spacing of 2.5 m and a column spacing of 2.5 m. In the first, third, fifth, and seventh columns, five blast holes are placed per column, while in the second, fourth, and sixth columns, four blast holes are placed per column, totaling 32 blast holes. The three different borehole arrangements are shown in Fig. 2.

Hexahedral element type (SOLID164) is used for the three-dimensional excavation model of machinery room of the factory building. The mesh of the overall section and the mesh at the palm surface (taking Case 1 as an example) are shown in Fig. 3.

4 Results and Analyses

4.1 Damage Analysis

When excavating underground powerplant in layers, the blasting at the midsection grooves in each stage will have a certain degree of impact on the upstream and downstream protective layers as well as the surrounding rock mass. To investigate which blasting scheme yields better results, the damaged state of rock blocks after blasting at the midsection grooves of layers II first, II second, and II third of the underground powerplant (at 0.015, 0.03, and 0.045 s after detonation stability, respectively) and the rock damage state after completion of excavation in each layer were further analyzed. The rock damage cloud maps during blasting and excavation are shown in Fig. 4.

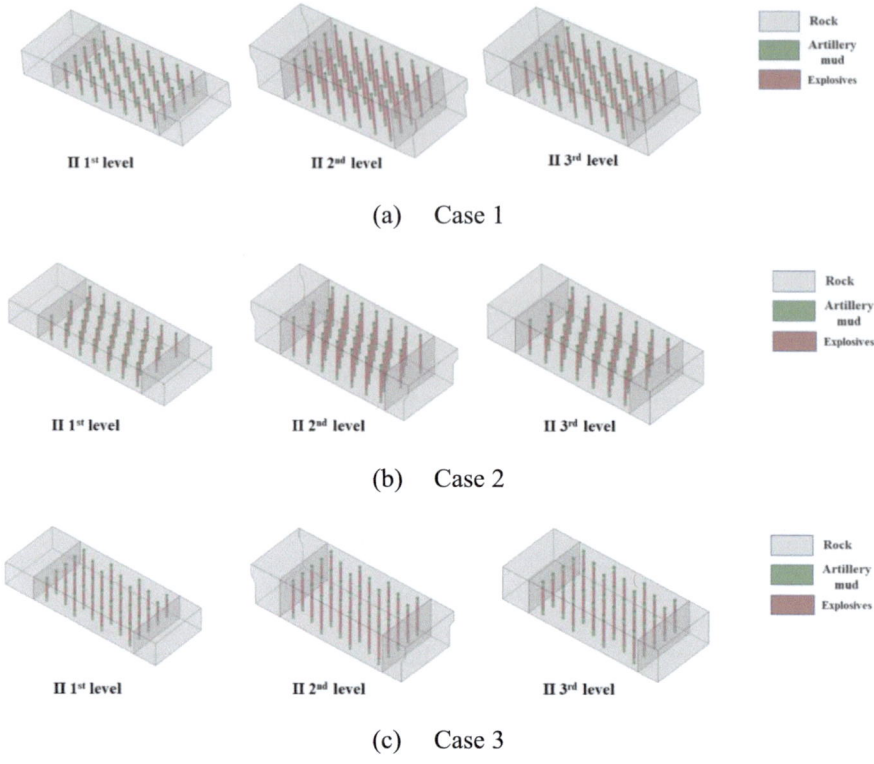

Fig. 2 Schematic diagram of three explosive layouts and charge structures

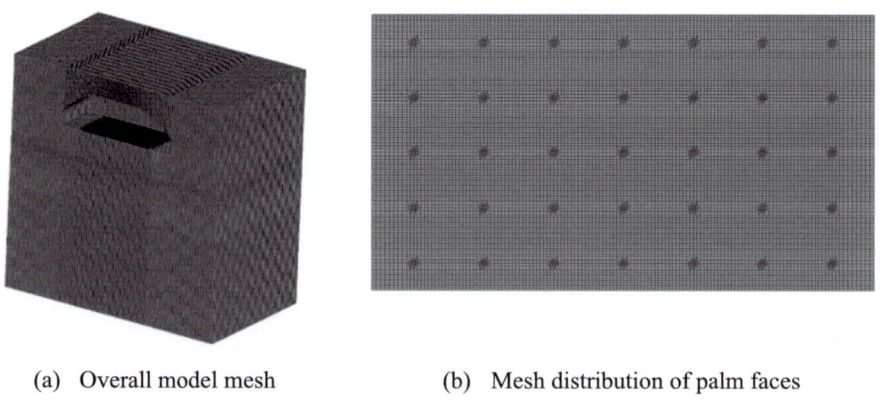

(a) Overall model mesh   (b) Mesh distribution of palm faces

Fig. 3 The schematic diagram of the blasting grid model (case 1)

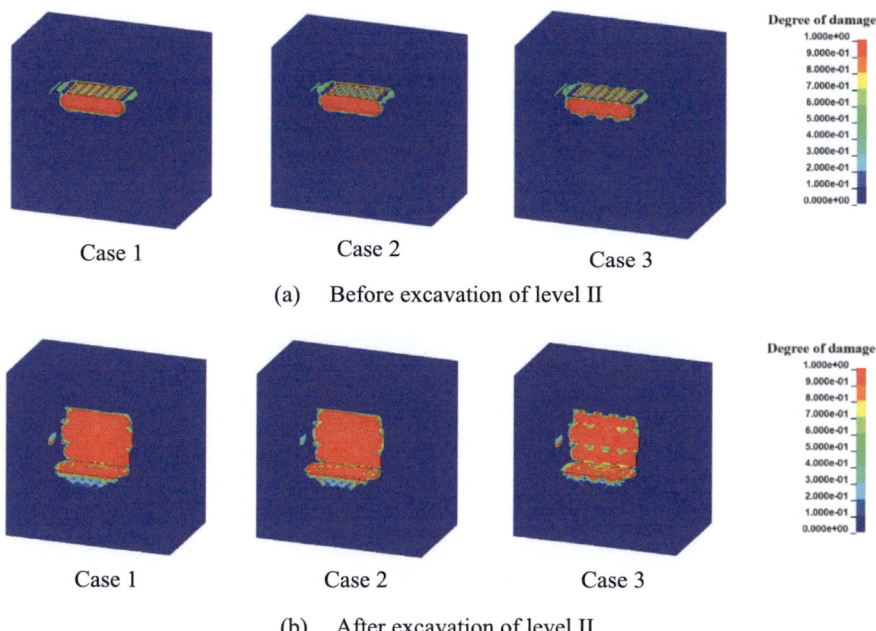

Fig. 4 Damage cloud maps before and after excavation of the underground powerplant

From the blasting effect on level II first of the underground powerplant, it's evident that in all three cases, the rock around the blast hole is completely damaged, indicated by $D = 1$ (red area in Fig. 4). However, the degree of damage between blast holes varies for each case. Case 1 has the best blasting effect, with minimal damage between blast holes at $D = 0.7$ (green area in Fig. 4). In cases 2 and 3, there are still parts of the rock between blast holes that remain undamaged, indicated by $D = 0$ (blue area in Fig. 4). After blasting level II first of the underground powerplant, the impact of the three cases on the protective layers and level II second of the underground powerplant is similar and meets the engineering standards. Secondary blasting causes fragmentation damage to the rock of level II second of the underground powerplant. All three cases cause some degree of damage to the protective layers, with case 2 having the least damage to the protective layers, especially at the left side of the powerplant, and the magnitude of dynamic disturbances and minor cracks for all three cases is within an acceptable range. The third blasting round causes fragmentation damage to the level II third rock of the underground powerplant. Among the three cases, each causes complete damage to the rock below along the y-axis and x-axis at distances of 0.7, 17.5; 0.7, 16; and 1.5, 17 m, respectively. Therefore, Case 2 causes the least damage to the rock of the lower level.

The blasting of level II will indeed cause some damage to the rock behind. The extent of this damage is illustrated in Fig. 5.

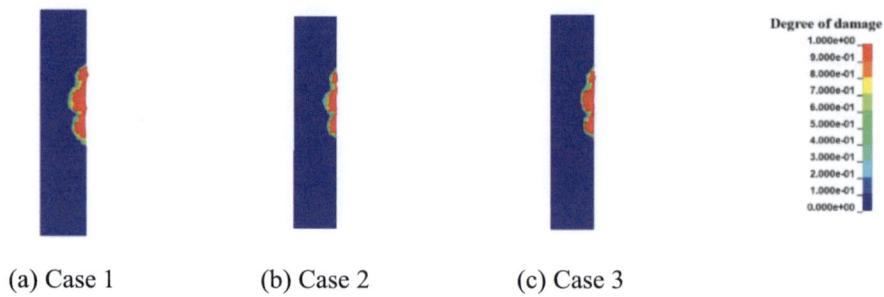

(a) Case 1   (b) Case 2   (c) Case 3

Fig. 5 Damage cross-sectional diagram caused by the blasting of level II

Table 2 Damage zone caused by the blasting of level II (unit: m)

	y-axis (m)	z-axis at level II 1st	z-axis at level II 2nd	z-axis at level II 3rd
Case 1	15	2.0	3.0	2.5
Case 2	15	1.0	3.0	1.5
Case 3	15	1.5	2.4	2.0

From Fig. 5, it can be observed that cases 1, 2, and 3 cause fragmentation damage to the rear rock along the y-axis and along the z-axis at levels II first, second, and third, as detailed in Table 2.

Therefore, after three rounds of blasting, Case 2 has the least impact on the rock behind, while Case 1 has the greatest.

4.2 Stress Response

Stress wave undergoes nonlinear superposition with an increase of the number of times the blast load is applied. The pressure perspective diagrams at the time of 0.005, 0.02, and 0.045 s during the pressure wave propagation process are shown in Figs. 6, 7, and 8.

From Fig. 6 and 7, it can be observed that at $t = 0.005$ and $t = 0.02$ s, the propagation range of pressure waves and the degree of stress wave superposition of the three cases are almost identical. From Fig. 8, at $t = 0.045$ s, Case 2 exhibits a lower degree of pressure wave superposition compared to the other two cases, causing lesser pressure damage to the surrounding rock, and is safer.

To compare the effects of different schemes on the blasting outcomes at different locations, two points, P0 and P1, at the II first level are selected (refer to Fig. 9). The pressure and vibration velocity response at P0 and P1 under different explosive arrangement schemes are depicted in Figs. 10 and 11, respectively.

From Figs. 10a and 11a, it is evident that during the blasting simulation, both Case 1 and Case 2 exhibit simultaneous pressure and vibration velocity peaks at P0,

Analysis of Stability of Surrounding Rock in the Blasting Excavation … 105

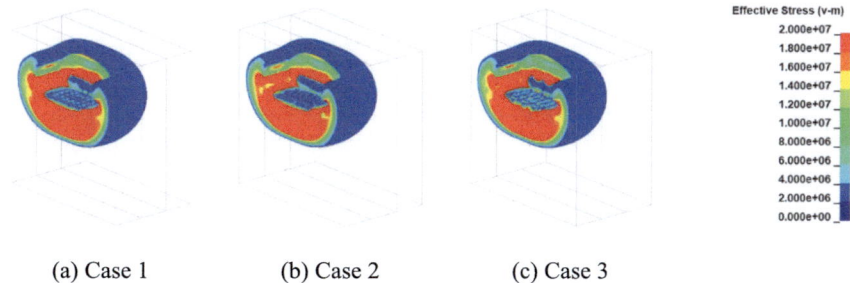

(a) Case 1   (b) Case 2   (c) Case 3

Fig. 6 The equivalent stress cloud maps at $t = 0.005$ s

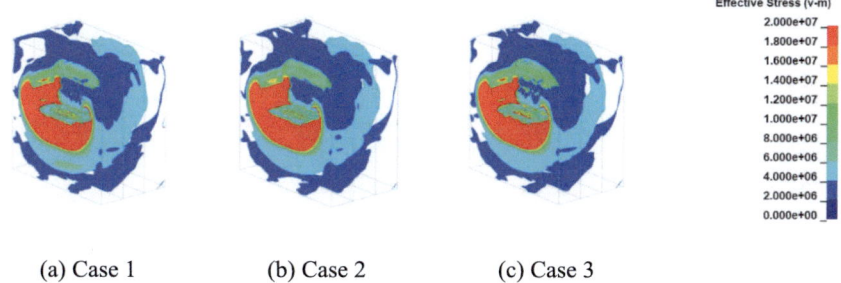

(a) Case 1   (b) Case 2   (c) Case 3

Fig. 7 The equivalent stress cloud maps at $t = 0.02$ s

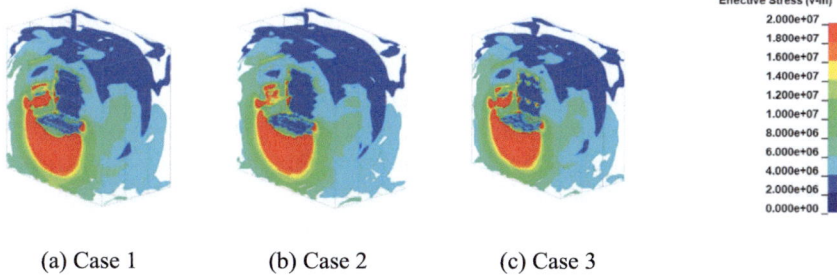

(a) Case 1   (b) Case 2   (c) Case 3

Fig. 8 The equivalent stress cloud maps at $t = 0.045$ s

Fig. 9 Diagram of monitoring points on layer II first (example using case 1)

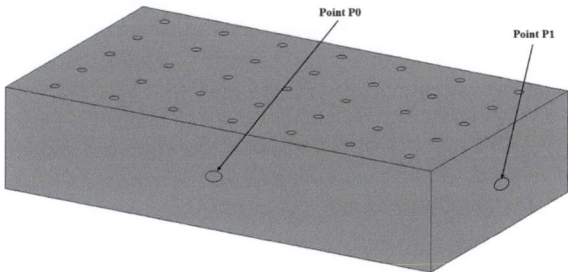

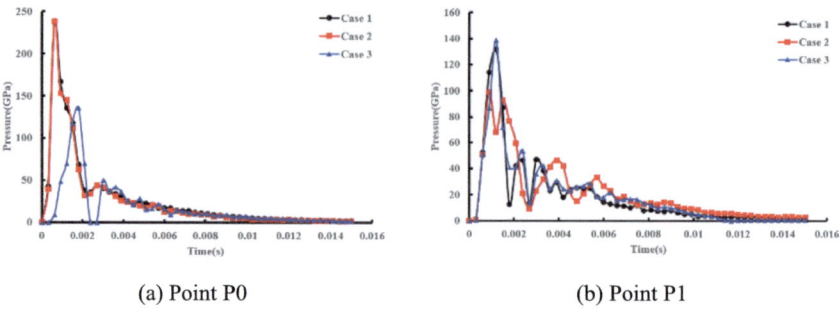

Fig. 10 Pressure–time curves under different explosive arrangement schemes

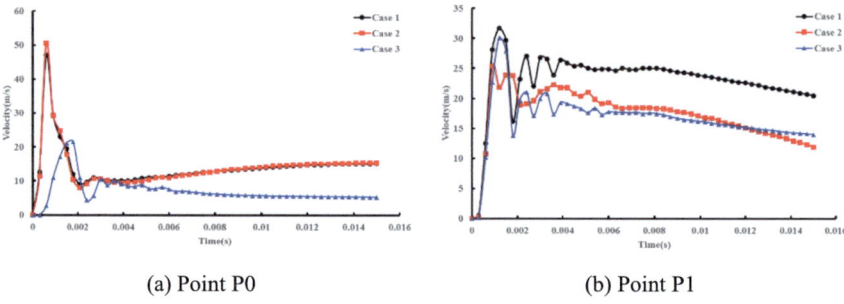

Fig. 11 Velocity–time curves under different explosive arrangement schemes

with identical pressure peaks of 237 GPa. However, Case 2 shows a slightly higher vibration velocity peak of 50.6 m/s compared to 47 m/s in Case 1. In Case 3, the pressure and vibration velocity peaks lag behind Cases 1 and 2, with peak values of 136 GPa and 21.5 m/s, respectively. This indicates that the rock mass experiences a greater impact load at P0 in Case 2, resulting in the most effective blasting outcome. From Fig. 10b and 11b, Case 2 exhibits pressure and vibration velocity peaks first, at 98.9 GPa and 25.6 m/s, respectively, followed by another set of peaks at 92.6 GPa and 23.9 m/s at P1. The time for the pressure and vibration velocity peaks in Cases 1 and 3 lie between the two peaks of Case 2. The pressure and vibration velocity peaks for Cases 1 and 3 are 132 GPa and 31.7 m/s, and 139 GPa and 30.1 m/s, respectively. Therefore, although the pressure and vibration peaks at P1 in Case 2 are lower than in Cases 1 and 3, the occurrence of two peaks in Case 2 leads to an equally effective blasting outcome as in Cases 1 and 3.

5 Conclusion

Based on the dynamic finite element software platform Ls-Dyna, blasting simulations were conducted on the main section between host units of a pumped storage power station in East China, leading to the following conclusions:

(1) After three rounds of blasting, the rock around blast holes in all three cases exhibited complete damage, though the extent of damage between blast holes varied. Case 1 demonstrated the most favorable blasting effect, yet all three explosive arrangements met the engineering standards.

(2) After experiencing three rounds of step blasting, Cases 1, 2, and 3 caused complete damage to the rock below along the y-axis by ~0.7 m, and along the x-axis by about 17.5 m; 0.7 m along the y-axis, and 16 m along the x-axis; and ~1.5 m along the y-axis, and 17 m along the x-axis, respectively. Compared to Cases 1 and 3, Case 2 reduced the degree of damage to the rock surrounding the host units by 22.3 and 29.5%, respectively, suggesting higher safety and stability.

(3) At 0.045 s after blasting completion, Case 2 exhibited lower pressure wave superposition compared to the other two cases, resulting in lesser pressure-induced damage to the surrounding rock, thus enhancing safety. Additionally, analysis of the pressure and vibration time history curves at P0 and P1 indicated similar blasting effects across all three cases. Therefore, considering both blasting effectiveness and overall engineering safety, Case 2 emerged as the optimal explosive arrangement for machinery room, meeting the requirements for rock fragmentation while ensuring safety. This blasting design can provide theoretical support and technical guidance for similar engineering projects in the future.

Acknowledgements This research was funded by State Grid Xinyuan Group Co, Ltd Technology Project (No. 462152-9003001-F078).

References

1. Ke W, Wang X, Yan C (2022) Numerical study of rock damage mechanism induced by blasting excavation using finite discrete element method. Appl Sci 12(15):7517–7517
2. Wan S, Huang J, Luo Y et al (2021) Research on cumulative damage characteristics of rock anchor beam concrete supporting structure by blasting vibration of underground powerhouse. Shock Vibr 2021:5421
3. Zuo J, Yang R, Ma X et al (2020) Explosion wave and explosion fracture characteristics of cylindrical charges. Int J Rock Mech Min Sci 12:135104501
4. Chen H, Yin J, Li X et al (2022) A numerical study on the blast wave distribution and propagation characteristics of cylindrical explosive in motion. Math Probl Eng 2022:59
5. Gao C, Kong X, Fang Q (2023) Experimental and numerical investigation on the attenuation of blast waves in concrete induced by cylindrical charge explosion. Int J Impact Eng 15:174

6. Mambetov SA, Mosinets VN (1965) Kinetics of the propagation of fractures in the process of explosive breaking of rocks. Soviet Min 1(3):232–239
7. Olsson M, Bavik SO, Ouchterlony F (2000) Perimeter blasting in a 130 m road cut in gneiss with holes with radial bottom slots. In: Explosives and blasting technique, pp 225–234
8. Rossmanith HP, Knasmillner RE, Daehnke A et al (1996) Wave propagation, damage evolution, and dynamic fracture extension: Part II—blasting. Mater Sci 32(4):403–410
9. He R (2024) Research on cumulative damage induced by cyclic blasting of tuff. J Phys Confer Ser 2694(1):17
10. Jayasinghe B, Zhao Z, Chee TGA et al (2019) Attenuation of rock blasting induced ground vibration in rock-soil interface. J Rock Mech Geotech Eng 11(4):770–778
11. Pan Q, Zhang J, Wang J et al (2023) Deep hole smooth blasting technology for excavation of large-span arch bridge foundation pits. Blasting 40(3):85–150
12. Deng Z, Liu H (2000) Discussion on new blasting technology for gas tunnel excavation. Eng Blast 6(3):62–64
13. Su D, Zheng D, Zhao L (2019) Experimental study and numerical simulation of dynamic stress–strain of directional blasting with water jet assistance. Shock Vibr 2019:1–15
14. Yu L, Wang Z, Wang S et al (2023) Numerical analysis of surrounding rock damage and disturbance induced by deep tunnel blasting excavation. Hydrogeol Eng Geol 50(5):117–123

Open Access This chapter is licensed under the terms of the Creative Commons Attribution 4.0 International License (http://creativecommons.org/licenses/by/4.0/), which permits use, sharing, adaptation, distribution and reproduction in any medium or format, as long as you give appropriate credit to the original author(s) and the source, provide a link to the Creative Commons license and indicate if changes were made.

The images or other third party material in this chapter are included in the chapter's Creative Commons license, unless indicated otherwise in a credit line to the material. If material is not included in the chapter's Creative Commons license and your intended use is not permitted by statutory regulation or exceeds the permitted use, you will need to obtain permission directly from the copyright holder.

Discussion of the Collapsibility of Loess after Pre-Immersion Water Treatment

Bai He, Bin Zhi, and Tiantian Wei

Abstract To understand the effect of eliminating collapsibility and the change of soil structure after soaking treatment in the process of treating collapsible loess foundation with pre-soaking method, the indoor collapsibility test and scanning electron microscope (SEM) scanning test were carried out after washing and losing water treatment of soil at different depths, and the collapsibility and SEM images of undisturbed loess and soaking losing water loess at various depths were obtained. The results show that the shallow and deep strata can be divided according to whether the overlying strata's saturated self-weight pressure can reach the strata's initial collapsible pressure. The place where the saturated self-weight pressure of the overlying strata can go is deep strata, and vice versa. However, with the increase of burial depth, the decrease of porosity continues to increase, and the increase of initial collapsibility pressure tends to a fixed value and does not increase.

Keywords Undisturbed loess · Microstructure · Soggy—dehydrated loess · Initial subsidence pressure · Porosity

1 Introduction

The pore structure characteristics of wet subsidence loess determine that it will rapidly reduce the consolidation capacity after the moisture is immersed, and the strength will be significantly reduced; that is to say, it produces wet subsidence deformation, which will seriously affect the regular use of the building, so it is necessary to eliminate the treatment of damp subsidence of loess foundation before the construction of the building. Piles are difficult to penetrate because of the enormous thickness, the low water content of the wet subsidence stratum, the use of extruded piles, and other treatments. Hence, it is necessary to use the pre-soaking method of its treatment.

Scholars have done research on the effect of pre-soaking on wetted loess foundations and loess's wetting and pore structure characteristics [1–21]. For using the pre-soak method in engineering, scholars mainly research through the field pre-soak test. It was found that the treatment effect of the pre-submergence method is closely related to the range of submergence [1–9], the permeability characteristics of the soil, the water transport law in loess [10–13], the pore ratio [14–16, 21, 22], and the depth of the wet loess stratum [20], and other factors. Shao et al. [22] studied the variation of wetting characteristics of remodeled loess in the saturated state as well as remodeled loess in the unsaturated state; an indoor test was carried out, and X-ray scanning was performed to study the microstructure of the soils that were not affected by water immersion and the soils after wetting treatment in the chamber. In summary, it can be seen that the change of wetting subsidence characteristics of the loess after leaching treatment is mostly unfolded by field test or wetting treatment of the soil, and the comparative study of the in situ loess and the leaching-loss treatment of the loess is lacking.

However, these studies mainly focus on the structural change characteristics of the soil after submerged treatment or on the diffusion pattern of water during submerged treatment and less on the comparative analysis of in situ loess and submerged-water-loss loess. Whether the wetted loess foundation treated by the pre-soaking method is still wetted after water dissipation is a major problem in later engineering construction.

This study compares the in situ loess of a project and its corresponding loess taken from the same stratum depth with the same water content, analyzes the pore characteristics and wetting of loess before and after the pre-soaking treatment of wetted foundations, explores the treatment effect of the pre-soaking method at different stratum depths, and gives a prediction model of the wetting coefficient of loess after the treatment of the pre-soaking method to provide a reference for the use of the pre-soaking method in the project. It gives a model for predicting the wetting coefficient of loess after treatment by the pre-soaking method and provides a reference for using it in engineering.

Table 1 Physical properties of loess

Number	Depth (m)	Moisture content (%)	Density (g/cm³)	Dry density (g/cm³)	d_s	e	Collapsibility coefficient
Y3	3	10.15	1.46	1.31	2.71	1.21	0.140
Y5	5	10.80	1.49	1.32	2.71	1.13	0.080
Y10	10	13.70	1.52	1.31	2.71	1.07	0.037
Y15	15	18.61	1.65	1.33	2.71	0.96	0.039
Y20	20	18.89	1.75	1.42	2.71	0.84	0.034

2 Experimental Method

2.1 Physical Parameters of Loess

In this study, loess samples were taken from the foundation treatment site of a project using a DPP-100 vehicle-mounted drilling rig and Φ127 mm auger rotary drilling tool, and Φ120 mm loess thin-walled soil extractor was used for static compression sampling. The physico-mechanical properties of the soil samples are given in Table 1.

2.2 Methods of Preparation of In Situ and Immersion-Water Loss Specimens

This study deals with two structural states of loess, in situ and submerged-without-water. When making the in situ soil samples, the tin bucket containing the in situ loess was placed horizontally on the cutting table. The label on the top of the bucket prevailed in place, and the labeled end was kept facing upwards. The sealing tape was opened to take out the specimen, remove the floating soil on the surface, and observe the specimen's disturbed condition and whether there was any fissure due to the packing or transportation. If there was any severe disturbance, it was not allowed to be used as the soil for the test. The geotechnical test determined the physical and mechanical properties of the in situ specimen, as given in Table 2.

The immersion-loss specimen can be obtained using the in situ specimen after immersion and water loss treatment. Immersion treatment refers to simulating the pre-infiltration treatment foundation, each stratum of the soil body immersion wetting process, that is, respectively, in its corresponding stratum of the overburden saturated

Table 2 Physical properties of soil samples

Depth (m)	3	5	10	15	20
Moisture content %	10.53	10.80	13.70	18.61	18.89
Dry density (g/cm³)	1.31	1.36	1.31	1.33	1.42

Table 3 Flooding pressure

Depth (m)	3	5	10	15	20
P_Z (kPa)	53.47	89.11	180.52	274.25	370.82
$P_{Z'}$ (kPa)	50	87.5	175	275	375

soil under the pressure of self-weight immersion, so that the specimen wetting occurs. Combined with the ground investigation report, it can be seen that the saturated self-weight pressure of the overburden borne by the soil body at 3, 5, 10, 15, and 20 m is given in Table 3. The specifications of the weights used in the laboratory are 12.5, 25, 100, and 200 kPa. In combination with the experimental conditions, the saturated self-weight pressure of the overburden in each stratum is taken as an approximation of the saturated self-weight pressure of the overburden at the stratum P_Z, which is the immersing pressure $P_{Z'}$. for the soakage pressure. Under the action of submergence pressure $P_{Z'}$, use W-G single-lever consolidator to simulate the submergence, and after the wetting, deformation is stabilized, use a syringe to discharge the excess water in the consolidator, and take out the ring knife specimen. The mass of the specimen m_1 and m_2 2 before and after the water immersion treatment was recorded, respectively.

After the pre-soak treatment, the water loss treatment simulates the process of water dissipation and water content recovery of the loess foundation. Put the ring knife specimen after water immersion treatment into the electric heat-blast constant temperature drying oven, drying at 105 °C, every 0.5 h to take out the weighing, and the specimen mass m_1 before water immersion treatment for comparison, to be its mass deviation of about 0.2 g, put into the humidification cylinder static for 24 h, so that the specimen moisture is uniform. Record the specimen mass m_3, compare m_1 and m_3, the quality deviation is not more than 0.2 g, before continuing the next test.

2.3 Test Scheme

This test was carried out in 2 groups, and the purpose of each group and the information of the specimens included in the test are given in Table 4. The first group of these tests aimed to investigate the wet subsidence of the in situ soil. This group of tests consisted of five specimens (Y3, Y5, Y10, Y15, Y20), of which there were five samples each of 3 and 5 m soil samples, and seven samples each of 10, 15, and 20 m. To carry out this group of tests, the in situ soil samples prepared in Sect. 2.2 were directly placed into the consolidation vessel with a retaining ring, porous stones, and a pressurized cover, and the consolidation-wet subsidence tests were carried out at the pressure settings in Table 5. The second group of tests investigated the wetting subsidence of soil samples with infiltration-loss of water, which consisted of five specimens (J3, J5, J10, J15, and J20), and the prepared infiltration-loss of water specimens in Sect. 2.2 was taken, with five each of the soil samples at 3 and 5 m.

Discussion of the Collapsibility of Loess after Pre-Immersion Water ...

Table 4 Test plan and purpose

Number	Purpose of the test	Specimen number
1	Collapsibility of undisturbed soil	Y3, Y5, Y10, Y15, Y20
2	Collapsibility of immersion-dehydration soil	J3, J5, J10, J15, J20

Table 5 Undisturbed sample pressure setting

Depth (m)	Pressure (kPa)							
3	12.5	3	12.5	3	12.5		
5	25	5	25	5	25		
10	25	10	25	10	25	10	25
15	100	15	100	15	100	15	100
20	100	20	100	20	100	20	100

Table 6 Water immersion—loss sample pressure setting

Depth (m)	Pressure (kPa)						
3	50	75	100	125	150		
5	75	100	125	150	175		
10	125	150	200	225	250	10	125
15	200	300	325	350	375	15	200
20	200	300	400	425	450	20	200

Seven soil samples at 10, 15, and 20 m were subjected to the consolidation wetting subsidence test under the pressure settings in Table 6. The soils at 3 and 20 m were also taken, and an electron microscope scanned their in situ specimens and specimens after immersion and loss of water to analyze their pores and structural changes.

3 Results and Analyses

3.1 Analysis of the Change of Self-Weight Collapsibility Coefficient

Figure 1 compares collapsible coefficient changes of undisturbed and soaking-dehydration loess under consolidation collapsible test. The initial collapsible pressure of samples in each stratum can determine the collapsible coefficient changes of undisturbed and soaking-dehydration loess in each stratum.

According to Fig. 1a and b, the growth of the self-weight collapsible coefficient of the undisturbed loess and the loess after immersion-dehydration in each stratum

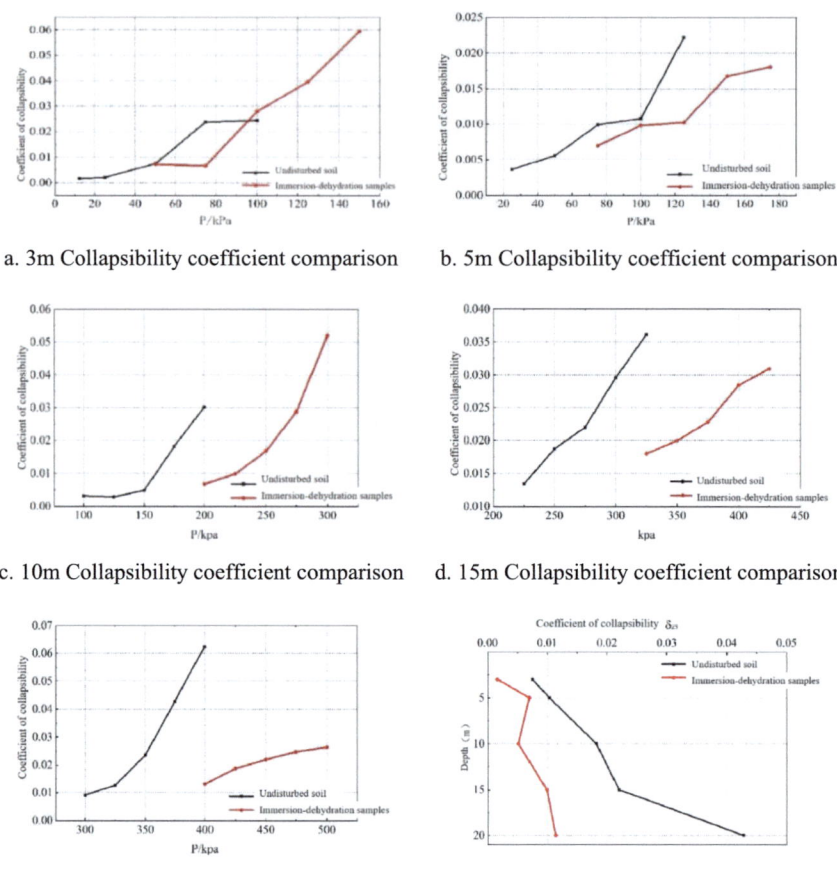

Fig. 1 Comparison of the collapsibility coefficients of undisturbed and submerged-dehydrated loess

is given in Table 7. With the increase of burial depth, the self-weight collapsible coefficient of the undisturbed loess gradually increases. The loess at 3 and 5 m did not show a collapsible phenomenon under the saturated self-weight pressure of the corresponding overlying soil. It can be seen from the corresponding $\delta_s - P$ curve that the initial collapsible pressure of the soil in these two strata is greater than the saturated self-weight pressure of the overlying soil. The loss has specific structural strength and iswetted by water under the upper load. Because the shear stress generated by particle contact is less than its structural strength, It is the same as the general cohesive soil and only produces compressive deformation. Therefore, when the pre-immersion method is used to treat the collapsible loess foundation, achieving the purpose of existing collapsible deformation and eliminating collapsibility is impossible at 15, and 20 m, it can be seen from the corresponding $\delta_s - P$ curves that the

Table 7 Change of collapsibility coefficient of dead weight

Depth (m)	3	5	10	15	20
Undisturbed soil	0.0075	0.0104	0.0182	0.0220	0.0428
Immersion-dehydration soil	0.0016	0.0070	0.0051	0.0099	0.0114

Table 8 Variation of initial pressure in collapsibility

Depth (m)	3	5	10	15	20
Undisturbed soil	61.75	109.41	169.20	231.87	331.64
Immersion-dehydration soil	84.21	143.38	243.30	314.19	409.50
Value added	22.47	33.97	74.10	82.32	77.86

initial collapsible pressure of these three soils is less than the self-weight pressure of the overlying soil. Under this pressure, the structure of the loess is destroyed, and collapsible deformation occurs with large deformation and fast speed. Under the pre-immersion method, the collapsibility of the stratum can be eliminated in advance.

At the same time, the changes in the onset pressure of wet depressions of the loess at each stratum in its original state and after immersion-loss of water can be obtained, as given in Table 8.

Figure 1f compares the collapsible coefficient of undisturbed loess and loess after soaking-dehydration with the stratum change. For undisturbed soil, with the increase of buried depth, the void ratio decreases, the water content increases, and the collapsibility gradually increases. The soil also shows the same characteristics after immersion-dehydration. At the same buried depth, the self-weight collapsibility coefficient of loess after immersion-dehydration is significantly reduced compared with undisturbed soil. The reduction degree of 3 and 5 m buried depth is considerably less than that of 10, 15, and 20 m.

3.2 Analysis of the Change of Initial Collapsible Pressure

Figure 2 shows the change of the initial collapse pressure of undisturbed loess and loess after soaking-dehydration. At the same buried depth, the initial collapsible pressure of loess after water immersion-dehydration is significantly higher than that of undisturbed loess. The undisturbed loess is soaked by water under the saturated self-weight pressure of its overlying soil, and the structure is rapidly destroyed. The overhead pores collapse, and the large and medium pores are filled with soil particles, forming many tiny pores. The number of pores decreases, the soluble salt content decreases, the cementing material between soil particles decreases, and the soil is closer. Therefore, the loess after water immersion-dehydration at the same depth shows smaller collapsibility, decreases collapsibility coefficient, and increases the initial collapsible pressure significantly. According to the relationship between

Fig. 2 Initial sag pressure change curve

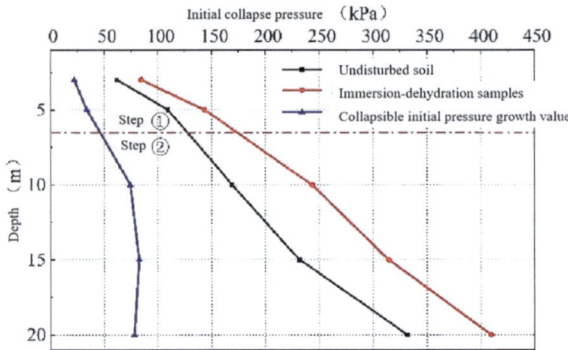

the saturated self-weight pressure of overlying soil and its corresponding initial collapsible pressure, the stratum is divided into stages 1 and 2.

Stage 1 is the accelerated growth stage of the initial collapsible pressure. In this stage, the initial collapsible pressure of loess is greater than the saturated self-weight of the overlying soil (such as 3 and 5 m). When it is immersed in water under this pressure, part of the overhead pores collapse, the collapsibility of the soil is partially eliminated, and the initial collapsible pressure is increased slightly. As the buried depth increases, the saturated self-weight pressure of the overlying soil gradually approaches the initial collapsible pressure of the corresponding soil layer. When the water is immersed, the collapse degree of the pores gradually increases, the large and medium pores are significantly reduced, the compactness of the soil gradually increases, and the variation of the void ratio increases, showing a gradual increase in the initial collapsible pressure. Stage 2 is the uniform growth stage of the initial collapsible pressure. In this stage, the initial collapsible pressure is less than the saturated self-weight of the overlying soil (such as 10, 15, 20 m). When the soil is soaked under this pressure, the collapsibility of the soil is eliminated to a certain extent, and the increase in the initial collapsible pressure tends to be constant.

3.3 Analysis of Pore Change

To study the change of soil structure after immersion treatment in shallow and deep strata, undisturbed loess and immersion-dehydration loess samples at 3 and 20 m were selected for scanning electron microscopy, and the SEM images were obtained, as shown in Fig. 3.

The SEM images of undisturbed and soaked-dehydration loess at 3 and 20 m are analyzed. It can be seen that when the buried depth is 3 m, the skeleton particles of undisturbed loess with a dry density of 1.31 g/cm^3 are grain collection, with more cement blocks, more pores, relatively more large and medium pore contents, and oapparentedges and corners of soil particles. With the increase of buried depth, the

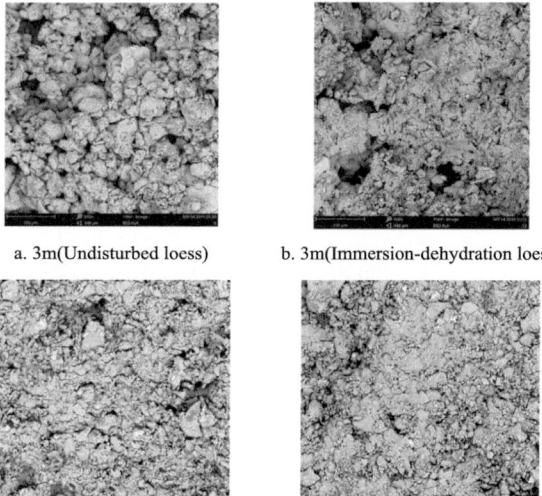

Fig. 3 SEM images of undisturbed and waterlogged-lost loess

a. 3m(Undisturbed loess)   b. 3m(Immersion-dehydration loess)

c. 20m(Undisturbed loess)   d. 20m(Immersion-dehydration loess)

large and medium pore contents of undisturbed soil with dry density of 1.42 g/cm^3 are significantly reduced at 20 m, with a large number of tiny pores, and the skeleton particles are relatively small. This is because, with the increase of buried depth, the pressure of overlying soil gradually increases. The compaction degree of undisturbed loess at 20 m is relatively high.

Compared with the undisturbed soil at the corresponding buried depth, the pores in the soil gradually change from large and medium pores to medium and small pores. With immersion and water loss, the edges and corners of the soil particles gradually decrease. The soil at a buried depth of 3 m does not reach the initial pressure of collapsibility due to its slightly saturated self-weight pressure of the corresponding overlying soil. Therefore, the structure of the loess is destroyed by soaking treatment, and the loess's large and medium pore content is still high after soaking-dehydration. The loess at the corresponding buried depth of 20 m belongs to the deep stratum, where the saturated self-weight pressure is greater than the initial collapsible pressure, and the soaking treatment effect is better. After soaking-dehydration, the loess's large and medium pore content is small, and the compactness is high.

In summary, the pre-immersion method is suitable for eliminating the collapsibility of the foundation at the deeper layer of the large-thickness self-weight collapsible loess. For the shallower layer, the collapsibility cannot be eliminated. Therefore, the Building Standards for Collapsible Loess Areas (GB_50025-2018) states that after the foundation immersion, the upper collapsible loess layer that has not been eliminated should be treated with the application cushion or other treatment methods. For the deep soil with a pre-soaking method for collapsibility elimination, the degree of collapsibility elimination will not increase with the increase of soil compaction degree. The specific performance is that the initial pressure of collapsibility does not increase with the increase of void ratio; that is, there is a certain 'bottleneck elimination' when the pre-soaking method is used to eliminate collapsibility.

After soaking treatment, the structure of the soil was destroyed, the large and medium pores collapsed, and the void ratio of the loess in each stratum changed to varying degrees. Figure 4 shows the change in the void ratio of undisturbed loess and soil after immersion-dehydration at different depths. With the increase of buried depth, the void ratio of undisturbed loess and loess after immersion-dehydration decreases to a gradually increasing trend. This is because the soil at the deeper stratum bears a considerable consolidation pressure and has a relatively small number of large and medium pores. This is consistent with the fact that the self-weight collapsibility coefficient of soil decreases with the increase of buried depth.

At the same stratum, the void ratio of loess after soaking-dehydration is significantly reduced compared with that of undisturbed loess. The void ratio of undisturbed loess is e. After soaking-dehydration treatment, consolidation and collapsible deformation are produced. The pore volume deformation is $\Delta V_V = \Delta S \cdot A$, where ΔS is the average settlement of the sample during the soaking-dehydration process, and A is the horizontal cross-sectional area of the ring knife sample, 30 cm^2. The void ratio of loess after soaking-dehydration is $e\prime = e - \frac{\Delta V_v}{V_s}$, and the decrease value of void ratio is $\Delta e = \frac{\Delta V_v}{V_s}$, where V_s is the volume of soil particles. The calculation results of the loess void ratio and void ratio reduction after immersion-dehydration are given in Table 9.

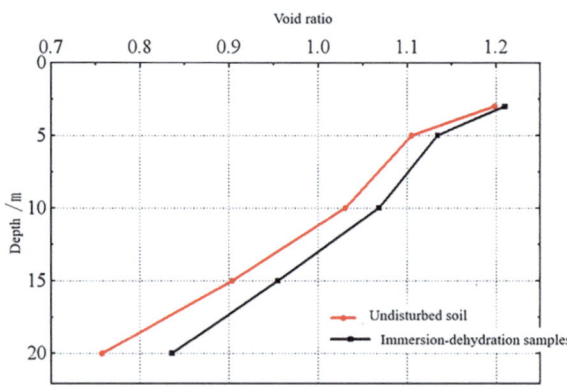

Fig. 4 Porosity comparison curve

Table 9 Porosity variation comparison

Depth (m)	3	5	10	15	20
e	1.21	1.13	1.07	0.96	0.84
$e\prime$	1.15	1.07	0.93	0.84	0.63
Δe	0.06	0.03	0.04	0.05	0.08

Fig. 5 Change curve of porosity decrease value

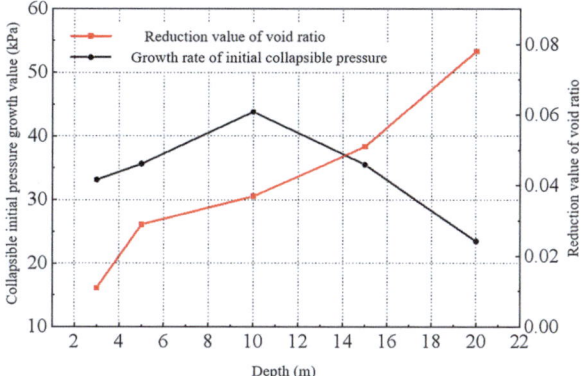

As shown in Fig. 5, the void ratio decreases gradually with the increase in formation depth. The porosity ratio of the soil decreases, which makes the structure of the loess more compact, the compactness of the soil increases, and the collapsibility decreases. When the saturated self-weight pressure of the overlying soil is less than the initial collapsible pressure of the corresponding soil layer, the increase of the initial collapsible pressure increases with the decrease of the void ratio. This is because, in this stage, the loess is immersed in water and can only eliminate part of the collapsibility of the stratum. As the overlying pressure gradually approaches the initial collapsible pressure of the corresponding stratum, the degree of collapsibility elimination gradually increases. When the overlying pressure is equal to the initial collapsible pressure, the collapsibility of the soil is eliminated. At this time, with the decrease of the void ratio, the increase of the initial collapsible pressure tends to a fixed value, which is about 78 kPa.

4 Prediction of Self-Weight Collapsibility Coefficient

Under the combined action of water immersion and saturated self-weight load of overlying soil, the pores of loess after soaking treatment are recomposed, and the loess structure is reconstructed. The void ratio of the soil after soaking treatment is e'. When the collapsibility test is carried out, it is assumed that the void ratio of the sample is $e_a\prime$ when the consolidation is completed. The void ratio of the sample is $e_b\prime$ when the collapsibility is completed. According to the definition of the self-weight collapsibility coefficient, the relationship between the self-weight collapsibility coefficient and the void ratio of the soil after soaking treatment can be expressed as

$$\delta_{zs} = 1 - \frac{e_b\prime}{e_a\prime} \qquad (1)$$

According to the collapsible test of the immersion-dehydration sample above, it can be seen that the void ratio $e_a{\prime}$, $e_b{\prime}$ of the samples at different depths after consolidation and collapsibility is given in Table 10.

The calculation model of the void ratio of soil after immersion treatment is established when the consolidation and collapsibility are completed, such as Eqs. (2) and (3).

$$e_a{\prime} = a * \exp(b * P_Z) \tag{2}$$

$$e_b{\prime} = c * \exp(d * P_Z) \tag{3}$$

where P_Z is the saturated self-weight of the overlying soil, and a, b, c, and d are the test parameters. In this paper, $a = 1.252$, $b = -1.693 \times 10^{-3}$, $c = 1.249$, $d = -1.720 \times 10^{-3}$.

The combined Eqs. (1)–(3) are used to calculate the collapsible self-weight collapsible coefficient of the loess after soaking treatment. The results are given in Table 11.

The calculation results are compared with the experimental results, as shown in Fig. 6. It can be seen that the calculation model can reasonably predict the self-weight collapsibility coefficient of loess after soaking treatment and provide a basis for pre-judgment of the treatment effect of the pre-soaking method.

Table 10 Porosity ratio of consolidation and collapse complete

Depth (m)	3	5	10	15	20
$e_a{\prime}$	1.154	1.066	0.925	0.836	0.628
$e_b{\prime}$	1.152	1.059	0.921	0.828	0.621

Table 11 Collapsibility coefficient of dead weight

Depth (m)	3	5	10	15	20
Experimental values	0.0016	0.0070	0.0051	0.0099	0.0114
Values	0.0037	0.0048	0.0071	0.0098	0.0125

Fig. 6 Calculation result of the self-weight wetting

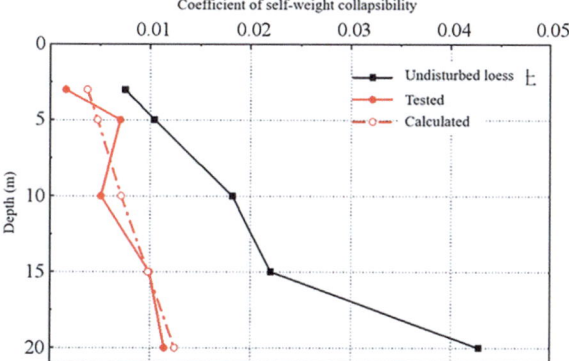

5 Conclusion

In this study, the following conclusions are obtained:

(1) Shallow and deep strata should be divided according to whether the saturated self-weight stress of the overlying soil can reach its corresponding collapsible initial pressure. The saturated self-weight pressure of the overlying soil cannot reach the collapsible initial pressure of the corresponding stratum, which is a shallow stratum; otherwise, it is a deep stratum.

(2) In the shallow strata, the collapsibility is not eliminated when the pre-immersion method treats the collapsibility. With the increase of buried depth, the overlying pressure gradually approaches the initial pressure of collapsibility, the decreased value of void ratio gradually increases, the increased value of the initial pressure of collapsibility also increases, and the degree of collapsibility elimination gradually increases. For this stratum, collapsibility should be eliminated by applying a cushion or other methods after pre-soaking treatment.

(3) In the deep strata, when the pre-immersion method is used to eliminate the collapsibility of the loess foundation, when the collapsibility is eliminated to a certain extent, the initial pressure of collapsibility cannot continue to increase. With the increase of buried depth, the decreased value of the void ratio continues to grow, the increased value of collapsible initial pressure tends to be a fixed value, and the growth rate of collapsible initial pressure gradually decreases. The treatment effect of the pre-soaking method is discussed.

(4) Combined with the pore change characteristics of loess after soaking treatment, the calculation model of the self-weight collapsibility coefficient is given, which can effectively predict the self-weight collapsibility coefficient of loess after soaking treatment and better the treatment effect of the pre-soaking method.

References

1. Song X, Zhang J, Sun J et al (2023) Discussion on the treatment method of collapsible loess foundation in municipal engineering. Sichuan Cement 2023(01):268–270
2. Xu X, Fan J, Yuan K et al (2023) Field immersion test on self-weight collapsibleloess site with large thickness. J Xi'an Univ Architect Technol 55(06):849–857
3. Zhang Z, Yang X (2023) Measurement method of collapse coefficient and initial pressure of collapsible loess in Helinger New Area. J Inner Mongolia Univ Technol 42(06):555–560
4. Liu Y, Luo X, Wang Y (2022) Analysis on field immersion test of collapsible loess in Guanzhong area. Site Invest Sci Technol 2022(02):22–26
5. Wang Q, Fan H, Liu Y et al (2022) Study on field immersion test of large thickness self-weight collapsible loess sites. Geotechn Eng Techn 36(05):409–416
6. Huang X, Zhang G, Yao Z et al (2011) Research on deformation, permeability regularity and foundation treatment method of dead-weight collapse loess with heavy section. Rock Soil Mech 32(S2):100–108
7. Wang F, Liu D, Cheng F et al (2023) Immersion test and collapsibility evaluation of deep loess site. Hydro-Sci Eng 2023(05):131–138
8. Yang J, Yan G, Hu X et al (2022) Experimental study on permeability characteristics of collapsible loedd in Guanzhong Plain. Ground Water 44(05):160–169
9. Wang Q, Li K, Gu H et al (2019) Application of field test pit immersion test in subway engineering in collapsible loess regions. Ground Water 41(1):115–120
10. Li X, Hong B, Li L et al (2017) Experimental research on permeability coefficient under influence of loess collapsibility. China J Highw Transp 30(06):198–222
11. Fan W, Wei Y, Yu B et al (2022) Research progress and prospect of loess collapsible mechanism in micro-level. Hydrogeol Eng Geol 49(05):144–156
12. Chen C, Zhang D, Zhang J et al (2017) Compression and wetting deformation behavior of intact loess under isotropic stresses. Chin J Rock Mech Eng 36(07):1736–1747
13. Li X, Liu J, Zhang K et al (2018) Study on relationship between pore structure parameters and permeability of Malan loess. J Eng Geol 26(06):1415–1423
14. Feng L (2023) Influencing factors and evaluation methods of collapsibility of large thickness loess. Develop Guide Build Mater 21(08):53–55
15. Zhu F, Nan J, Wei Y (2019) Mathematical statistical analysis on factors affecting collapsible coefficient of loess. Chin J Geol Hazard Control 30(02):128–133
16. Mu Q, Dang Y, Dong Q et al (2019) Water-retention characteristics and collapsibity behaviors: comparison between intact and compacted loesses. Chin J Geotechn Eng 41(08):1496–1504
17. Xiao W, Li J, Wu K et al (2024) Micromechanical characteristics of collapsible loess and its subgrade settlement law. J Shandong Univ 2024(02):163–173
18. Li Z, Gao X (2018) Study on collapsibility of remolded loess by model test and evaluation of loess collapsibility. Sci Technol Eng 18(03):319–327
19. Zheng W, Dou H, Liang W et al (2023) Study on collapsibility mechanism based on microstructure evolution of loess. Highway 68(08):322–325
20. Xie X, Hou X, Zhao Q et al (2016) Analysis of the relationship between the pore distributions and collapsibility loess. J Eng Geol 24(s1):362–368
21. Zhang Y (2023) Study on collapsibility characteristics and evaluation method of large thickness loess field. Lanzhou Jiaotong University
22. Shao X, Zhang H, Tan Y (2018) Collapse behavior and microstructural alteration of remolded loess under graded wetting tests. Eng Geol 233:11–22

Open Access This chapter is licensed under the terms of the Creative Commons Attribution 4.0 International License (http://creativecommons.org/licenses/by/4.0/), which permits use, sharing, adaptation, distribution and reproduction in any medium or format, as long as you give appropriate credit to the original author(s) and the source, provide a link to the Creative Commons license and indicate if changes were made.

The images or other third party material in this chapter are included in the chapter's Creative Commons license, unless indicated otherwise in a credit line to the material. If material is not included in the chapter's Creative Commons license and your intended use is not permitted by statutory regulation or exceeds the permitted use, you will need to obtain permission directly from the copyright holder.

Application of Building Information Modeling + Algorithm Model in the "Four Preparations" for Seepage Safety of Concrete Face Rockfill Dam

Yi Hou, Bin Mei, Jian Zhang, Wenyang Lin, and Hong Yu

Abstract Not only does China possess more than half of the total number of concrete face rockfill dams (CFRD) worldwide, but it has also experienced painful lessons regarding seepage stability control. Particularly, the continuous advancement of digital twin technology within the water conservancy industry offers diverse possibilities for emerging technologies to address traditional engineering difficulties. Against this backdrop, in combination with the engineering practice of Kaihua Reservoir, this paper creatively proposes a "four-preparation" method for the seepage safety of CFRDs based on the building information modeling+ (BIM+) algorithm model. Specifically, to begin with, this research constructs the BIM model applicable to the dam body and foundation in the design stage. Meanwhile, by integrating various information such as material parameters, this research further extracts the calculated cross-section from the model for the finite element seepage stability calculation and slice-method-based dam slope stability analysis, thereby improving the efficiency of design safety calculation. Furthermore, during the operation and management stage, by virtue of diverse technologies such as BIM, GIS, and IoT, this research implements the integration of security monitoring data with digital scenarios. In this foundation, this research establishes a confidence interval statistical model and corresponding graded warning indices through reorganization and analysis, thus forming a technical framework with algorithms for prediction, indices for pre-warning, scenarios for preview, and templates for pre-arranged planning, which furnishes support for engineering safety management.

Keywords Concrete-faced rockfill dam · Data backplane · Algorithm model · Seepage safety · "Four-pre"

Y. Hou · B. Mei · J. Zhang · H. Yu
Zhejang Desicn Instute of Water Conservancy and Hydro-Electric Power CO. Ltd., Hangzhou, Zhejiang, China

W. Lin (✉)
Zhejiang Tongji Vocational College of Science and Technology, Hangzhou, Zhejiang, China
e-mail: 601081728@qq.com

© The Author(s) 2026
G-F. Zhao (ed.), *Geotechnical Modeling and Intelligent Systems*,
https://doi.org/10.1007/978-981-96-6925-7_10

1 Introduction

With a host of advantages such as making full use of local materials, simple construction, and low-cost, concrete face rockfill dam (CFRD) is regarded as one of the most common dam patterns in hydropower construction in China [1]. Presently, the number of CFRDs in China exceeds half of the global total, with its technical difficulty, engineering scale, and dam height being among the highest worldwide [2].

Water load, as the most basic load of CFRDs, is one of the most important factors affecting the overall safety of the dam [3]. The critical performance of various diseases and failures of CFRDs can be summarized as dam leakage, which serves as a significant difference between it and other dam patterns [4]. Moreover, seepage safety control is a key technology of CFRDs. In this regard, China's CFRD has experienced a painful lesson in terms of seepage stability control. For instance, the dam break caused by out-of-control seepage of the Gouhou Reservoir, coupled with the destruction of the face slab of Zhushuqiao Reservoir due to the loss of cushion bedding materials, reveals the necessity of seepage stability control of CFRDs [5].

The seepage control of CFRDs aims to ensure the water storage function of the dam and avoid its serious leakage. It is necessary to continuously monitor the surface and interior of the dam to collect various related effect variables such as environmental quantity, load quantity, and deformation, seepage, stress, strain, and temperature of the dam body and foundation under load. All these efforts are beneficial to grasp its performance changes and determine abnormal situations in time, thereby taking effective countermeasures promptly to ensure engineering safety [6]. Concurrently, these efforts can provide a solid scientific basis for enhancing the level of design, construction, and management.

On these grounds, in combination with the engineering practice of Kaihua Reservoir, this research creatively proposes a "four-preparation" method for seepage safety of CFRDs based on the BIM+ algorithm model, thus forming a technical framework with algorithms for prediction, indices for pre-warning, scenarios for preview, and templates for pre-arranged planning. Briefly, this research is of reference significance to the digital twin construction of similar engineering applications.

2 Engineering Overview

Kaihua Reservoir, the largest reservoir engineering built in Zhejiang Province at present, aims at flood control, water supply, and improving the ecological environment of the basin. On the same note, it serves as a large reservoir with comprehensive utilization of irrigation and power generation. The barrage adopts reinforced CFRD with a maximum dam height of 85.50 m.

This research creatively proposes a "four-preparation" method for seepage safety of CFRDs based on the BIM+ algorithm model, the research content framework is shown in Fig. 1, with finite element seepage algorithm, slice-method-based slope

Application of Building Information Modeling + Algorithm Model ... 127

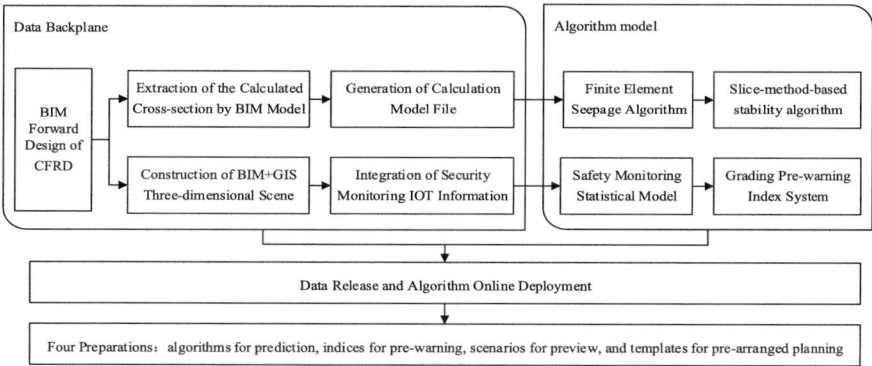

Fig. 1 "Four preparations" logic diagram for safety of CFRDs based on BIM+ algorithm model

stability algorithm, and safety monitoring statistical model as the fundamental algorithm models. More exactly, first and foremost, in the design stage, this research employs BIM forward design to build a BIM model applicable to the dam body and foundation. Simultaneously, by extracting the calculated cross-section encompassing material parameters from the model, this research further implements the analysis of seepage stability and dam slope stability, which improves the design of CFRD and calculates the separation between them, thereby enhancing the efficiency of design optimization. Moreover, in the operation and management stage, this research utilizes BIM, GIS, and IOT technology to build the engineering safety data backplane. Based on the reorganization and analysis of safety monitoring data, statistical models and corresponding grading pre-warning indices are further established. With the online deployment of the foregoing data backplane, mechanism algorithm, and statistical model, this research ultimately plays the role of online safety judgment of the proposed model.

3 Construction of Data Backplane Based on BIM

The BIM model stands as a basic element of digital twin engineering. In this connection, this research initially constructs the CFRD structure and geological model of Kaihua Reservoir through parametric design. Subsequently, this research leverages the model component as the carrier to hook up the safety-related material information [7], thus extracting the calculated cross-section. Lastly, taking the digital scene of BIM + GIS as the visual window, this research forms a seepage safety data backplane with complete information and accurate mapping.

3.1 Parametric Design of BIM

The anti-seepage system of CFRD in Kaihua Reservoir is composed of the concrete face slab, toe slab, waterproof and anti-seepage curtain at its joints, and other components. Taking toe slab as an example, this research illustrates the parametric design process of BIM [8]. The X-shaped line of the toe slab is not only the intersection line between the bottom surface of the face slab and the foundation surface of the toe slab but also the reference line for the design and construction of the toe slab. First of all, this research draws up the X-shaped line of the toe slab as per the position of the dam axis, the slope ratio of the face slab, and the topographic and geological conditions of the dam foundation. As the X-shaped line of the toe slab is a three-dimensional multi-segment line, this research further extracts the spatial coordinates of each inflection point to form a three-dimensional point set indicated as $\{X_i\}$. Based on this, this research automatically creates the relevant parameters of the toe-slab model through visual programming, as shown in Fig. 2. Notably, the specific program logic steps are presented as follows:

Step 1: Designing the cross-section according to the toe slab and calculating the corner coordinates of each turning cross-section of the toe slab according to the geometric relationship to form six three-dimensional point sets denoted as $\{A_i\}$, $\{B_i\}$, $\{C_i\}$, $\{D_i\}$, $\{M_i\}$, and $\{T_i\}$;

Step 2: Creating a toe-slab adaptive cross-section family, wherein six adaptive points correspond to six three-dimensional point sets respectively;

Step 3: Arranging the six corner coordinates of the cross-section of the i-th inflection point of the X-shaped line of the toe slab in the order of $1(A_i)$, $2(B_i)$, $3(T_i)$, $4(M_i)$, $5(D_i)$, and $6(C_i)$;

Step 4: Arranging the six corner coordinates of the cross-section at the $i + 1$-th inflection point of the X-shaped line of the toe slab in the order of $7(A_{i+1})$, $8(B_{i+1})$, $9(T_{i+1})$, $10(M_{i+1})$, $11(D_{i+1})$, and $12(C_{i+1})$;

Step 5: Inputting the coordinates of the foregoing 12 points into the adaptive cross-section family to automatically create a fine toe-slab model.

The above-mentioned process is iterated to create the toe-slab model of each segment in turn.

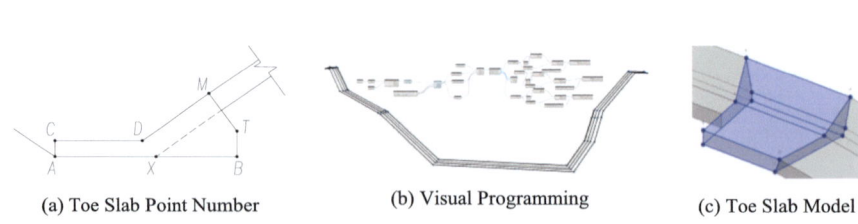

(a) Toe Slab Point Number   (b) Visual Programming   (c) Toe Slab Model

Fig. 2 BIM parametric design of toe slab

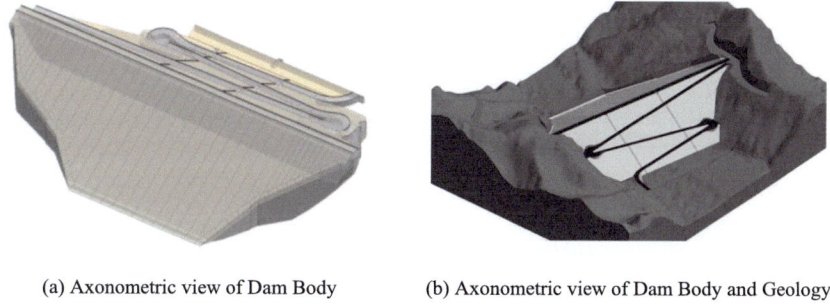

(a) Axonometric view of Dam Body   (b) Axonometric view of Dam Body and Geology

Fig. 3 BIM model of dam body and geology. *Note* The image model was created by the author's team

The toe slab of each segment should be checked whether it is arranged on the bank slope with suitable geological conditions, small engineering quantity, and convenient construction. In cases where these conditions fail to be satisfied, the new BIM model can be generated by adjusting the X-shaped line of the toe slab and re-running the program.

Based on the parametric modeling method, this research establishes the models of face slab, rockfill body, and downstream slope protection in turn, ultimately forming the BIM model of the rockfill dam body as shown in Fig. 3a. Moreover, as per the geological drilling data within the dam site area, this research further constructs the corresponding three-dimensional geological model, which is assembled with the dam model to complete various operations such as step-slope excavation, as illustrated in Fig. 3b.

3.2 Extraction of the Calculated Cross-Section by BIM Model

With the BIM model as the carrier of information integration [9], a series of engineering data, encompassing design reports and geotechnical test results, are linked to corresponding structures and soil layers to form a three-dimensional visual seepage safety basic database, which is convenient for automatic extraction of the calculated cross-section. Furthermore, the "fixed" calculated cross-section as shown in Fig. 4, including physical and mechanical indices related to structural line material division, geological stratification, and soil layers, is extracted from the BIM model. Based on this, program development is employed to convert the calculated cross-section into a recognizable text file for further modification and reuse. Meanwhile, this research reserves interfaces for dynamic calculation settings of boundary conditions, calculation methods, and calculation accuracy to support online settings. Eventually, a complete calculation model is formed by combining the fixed calculated cross-section with the dynamic calculation settings.

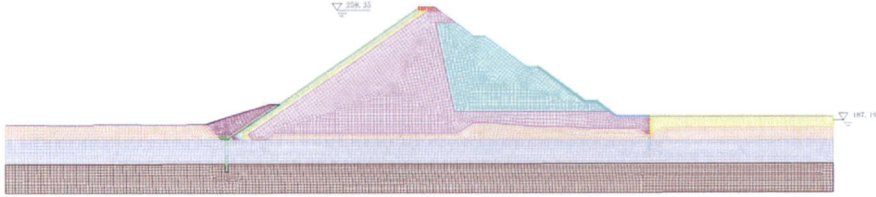

Fig. 4 Cross-section calculated by finite element seepage algorithm

3.3 Construction of BIM + GIS Digital Scene

Through the registration and fusion of multi-source data including BIM model, tilt photography, and images within the hub area, this research implements a high-fidelity characterization of entities and information of Kaihua Reservoir in digital space. The specific technical route of BIM and GIS in this research is described as follows. First, a utility tool is developed to facilitate the model perspective of BIM data, thereby checking whether there is any problem between the texture mapping and topology. Subsequently, the plane rectangular coordinate system of the project base points of the BIM model is transformed into a geographical coordinate system to realize the accurate registration of BIM and GIS data, intending to ensure the correct position of the model on the earth. Lastly, the scene rendering of the proposed model is published as a service to realize its transition from offline to online and from visible to usable.

As illustrated in Fig. 5, the BIM + GIS digital scene serves as an intuitive expression of the digital twin engineering [10], which not only constitutes the space base for IOT data integration but also provides a solid basic environment for diverse simulations.

(a) BIM Model of the Hub Area   (b) BIM+GIS model within the Hub Area

Fig. 5 Digital scene within the hub area. *Note* The image model and scene were created by the author's team

4 Research on Seepage Safety Algorithm Model

The algorithm model, being the core of the digital twin, stands as the key to realizing intelligent simulation based on data backplanes. During the design stage, the common strategy is to establish a mathematical model based on the mechanism of seepage generation, development, and action, thus supporting online calculation through algorithm program development and online deployment. Taking into account the changes in various factors such as material rheology, the parameters in the mechanism algorithm need to be recalibrated over time, indicating higher professional requirements. Hence, it is imperative to construct an additional safety monitoring statistical model during the operation stage, which is capable of predicting the future development trend as per the correlation between environmental quantities and effect quantities.

4.1 Mechanism Algorithm During the Design Stage

The steady seepage conforms to the heterogenous anisotropic two-dimensional seepage field in Darcy's law. In this regard, the hydraulic head potential function satisfies the following differential equation:

$$\frac{\partial}{\partial x}\left(k_x \frac{\partial \phi}{\partial x}\right) + \frac{\partial}{\partial y}\left(k_y \frac{\partial \phi}{\partial y}\right) + Q = 0 \tag{1}$$

where $\varphi = \varphi(x, y)$ represents the hydraulic head potential function to be solved; x and y represent plane coordinates; and, K_x and K_y represent the seepage coefficients in the x-axis and y-axis directions.

Based on the discretization of the seepage field by the finite element method, this research assumes that the hydraulic head potential function φ of the unit seepage field is polynomial. Furthermore, differential equations and boundary conditions can be employed to determine the variational form of the problem, from which the following linear equation can be derived:

$$[H]\{\varphi\} = \{F\} \tag{2}$$

where $[H]$ denotes the seepage matrix; $\{\varphi\}$ denotes the hydraulic head of seepage field; and, $\{F\}$ denotes the seepage flow of the node.

Through solving the foregoing equations, the hydraulic head of the node can be obtained, thus determining the physical quantities such as hydraulic gradient and velocity of the unit element. Additionally, the node flow balance method can be utilized to automatically determine the position of the saturation line and seepage flow through iterative calculation.

The primary mechanism of dam landslide can be summarized as the emergence of massive seepage caused by abnormal deformation of the dam, such as joint opening and panel fracture, which leads to the significant increase of pore pressure, seepage scouring force, and sliding force within the dam, ultimately making the dam tend to be unstable [11]. In accordance with the requirements of relevant design specifications [12, 13], the simplified Bishop method is adopted, which is given by:

$$K = \frac{\sum\{[(W \pm V)\sec a - ub \sec a]\tan \varphi' + c'b \sec a\}/(1 + \tan a \tan \varphi'/K)}{\sum[(W \pm V)\sin a + M_C/R]} \quad (3)$$

According to the cross-section calculated in Sect. 3.2, the seepage stability calculation is initially implemented. Specifically, the seepage stability calculation is based on the working condition that the design flood level is 235.85 ($P = 0.5\%$) on the waterfront side and the corresponding low water level on the land side, provided that the seepage stability of the dike body and foundation must satisfy $J \leq [J]$. Under the design working conditions of the normal operation of the anti-seepage system, the seepage calculation results are illustrated in Fig. 6a. It can be observed that the concrete face, toe slab, and upstream curtain grouting undertake ~50% of the hydraulic head, which is in line with the engineering experience.

This research further checks the safety of the design cross-section under extreme working conditions that the rockfill body does not present obvious seepage failure when the concrete face slab fails to prevent seepage. To this end, a crack is set on the face slab at ~2/3 of the dam height to simulate the seepage situation when the face slab cracks here. The calculation result is outlined in Fig. 6b. Regarding the specific parameter settings, the maximum seepage gradient within the cushion and transition-layer material is less than the critical seepage gradient, complying with the requirements of seepage stability.

Based on the seepage calculation, the stability calculation of dam slope is conducted in this research, with the natural and saturated volume-weight being employed respectively above and below the saturation line. Notably, the calculation results are depicted in Fig. 7. The minimum safety coefficients of upstream and

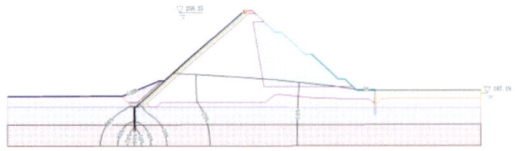

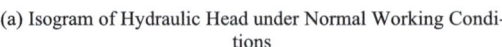

(a) Isogram of Hydraulic Head under Normal Working Conditions

(b) Concentrated Area of Hydraulic Gradient under the Working Slab Crack Condition

Fig. 6 Results of Kaihua dam seepage calculation

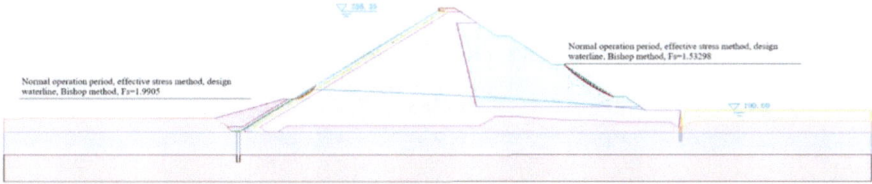

Fig. 7 Calculation results of dam slope stability under the design condition

downstream dam slopes are greater than the minimum safety coefficient in relevant specifications, satisfying the requirements of stability security.

4.2 Safety Monitoring Statistical Model During the Operation Stage

The anti-seepage monitoring cross-section of the CFRD riverbed section of Kaihua Reservoir is depicted in Fig. 8.

In combination with the actual situation of Kaihua Reservoir as well as the seepage characteristics of CFRD, this research further analyzes diverse factors such as the seepage flow of the dam foundation, reservoir water level, rainfall, and time, thereby establishing the following statistical model:

$$\begin{aligned} Q &= Q_H + Q_P + Q_\theta \\ &= a_0 + a_1 H + a_2 H^2 \\ &\quad + a_3 P + a_4 \theta + a_5 \ln \theta \end{aligned} \quad (4)$$

where Q, Q_H, Q_P, and Q_θ
respectively represent the total seepage, reservoir water level component, rainfall component, and time-dependent component; H represents the reservoir water level, with m as its unit; P represents rainfall, with mm as its unit; θ represents the result of

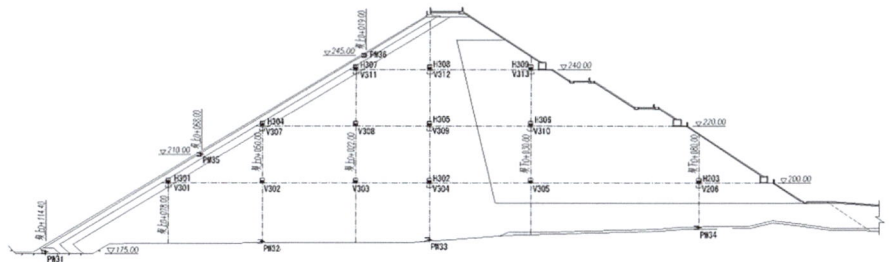

Fig. 8 Safety monitoring cross-section of CFRD riverbed section of Kaihua reservoir

Table 1 Measured value, fitting value, and proportions of various components of seepage flow

Items	Measured value	Fitted value	Reservoir water level component Q_H	Rainfall component Q_P	Time-dependent component Q_θ
Seepage flow (L/s)	17.63	17.51	51.00	0.65	−34.15
Proportion of relative measured values	–	99.32%	289.31%	3.71%	−193.70%

dividing the accumulated days from the initial measurement date by 100; a_0 stands for the constant term; a_1 and a_2 represent reservoir water level components, respectively; a_3 stands for rainfall regression coefficient; and, a_4 and a_5 respectively represent the time-dependent regression coefficient.

Building upon this, stepwise regression calculation is leveraged to derive the regression statistical model as outlined in Eqs. (3–5). The negative correlation coefficient R of the regression model is 0.874, whereas its residual standard deviation S is 6.58 L/s, demonstrating a higher accuracy of the regression model.

$$Q = 116.6 - 6.57H + 0.16H^2 + 0.01P \\ - 0.97\theta - 2.8\ln\theta \tag{5}$$

This research takes the average seepage flow of 17.63 L/s with an excellent fitting degree as an example for further analysis. The actual measured value, fitting value, and the proportion of each component are shown in Table 1. As can be seen from Table 1, the reservoir water level exerts a significant influence on the seepage flow, with the reservoir water level component QH of 51.0 L/s, accounting for 289.31% of the measured value. Second, rainfall generates a relatively limited influence on the monthly average seepage, with the rainfall component QP of 0.65 L/s, solely accounting for 3.71% of the measured value. Last but not least, the time-dependent component exerts a remarkable influence on the seepage flow, and the annual average seepage flow exhibits a downward trend year by year, with its value lower than the initial stage of operation.

5 "Four Preparations" Research on the Seepage Safety

Given the normal and extreme operation scenarios of the dam, with the aim of predicting and forecasting the seepage safety issues according to the changing trend of water levels, this research establishes a "four-preparation" system for the CFRD seepage safety based on the BIM+ algorithm model. Utilizing the "four-preparation" technical system, this research further transforms the passive risk investigation into active safety judgment, thereby realizing the predictive engineering safety management strategy featuring prediction with professional analysis, pre-warning with real-time dynamics, preview with intelligent auxiliary, and pre-arranged planning with the digital process.

5.1 Seepage Safety Prediction

The aforementioned mechanism algorithm and statistical model are deployed in the server and receive the predicted values of environmental quantities from data sources such as weather forecasting systems and flood forecasting systems, thus furnishing a solid data and algorithm foundation for the safety-related "four preparations". With the field safety monitoring data and finite element calculation results as the sample database, the proposed model implements nonlinear learning involving extensive relevant factors [14], supporting the prediction of seepage flow change trend, thereby discovering hidden danger in time.

5.2 Seepage Safety Pre-Warning

In combination with the characteristics of engineering structure, potential safety hazards, and weak links, this research constructs a dam seepage safety grading pre-warning system covering point location, cross-sections, and the dam body. Specifically, the seepage flow indices calculated by mathematical statistical model and confidence interval method are set as the tertiary monitoring indices, with the safety state defined as a slight abnormality. Furthermore, the seepage flow indices calculated by structural simulation under the seepage failure and extreme working conditions are set as the primary monitoring indices, with the safety state defined as a severe abnormality. Moreover, the secondary monitoring indices are eclectically determined between the primary and tertiary monitoring indices, with the safety status defined as a moderate abnormality.

5.3 Seepage Safety Preview

Within the BIM + GIS digital scene, this research performs a preview of the whole process involving meteorological forecast, flood forecast, seepage forecast, pre-warning information release, and emergency response through simulation, visually presenting the process of seepage generation and development in various ways such as thermodynamic diagram. In this way, this research realizes the interaction between physical engineering and digital twin engineering.

5.4 Seepage Safety Pre-arranged Planning

Based on the preview results as well as the emergency response conditions and procedures in the engineering safety emergency plan, a host of response measures, such as patrol inspection and emergency water release, are elaborated in this research.

6 Conclusions and Outlook

Regarding the "four preparations" of CFRD seepage safety in Kaihua Reservoir, this paper implements various investigations encompassing BIM parametric design, calculation model extraction, mechanism algorithm development, statistical model and pre-warning index construction, digital preview, etc., offering strong data and model algorithm support for seepage simulation calculation and safety management of CFRD engineering. Overall, this research primarily draws several conclusions as follows.

To begin with, BIM is essentially a database built according to the intuitive physical form of buildings. Based on "BIM+" technology, this research constructs a digital twin data backplane, which marks the concrete practice of BIM's whole life cycle. Hence, this research is of practical reference significance.

Second, the original intention of digital twins is to build a digital, high-fidelity test system in virtual space. Based on this system, floods, structural damage and other extreme situations can be simulated for low-cost. Therefore, the algorithm model serves as the core of digital twinning.

Third, given the inherent advantages and disadvantages of the mechanism algorithm and mathematical statistics model, it is imperative to verify their effects by comparison, thereby providing more scientific and reliable algorithm support for the "four preparations" of engineering safety.

The current "four preparations" construction of drainage basins is superior to that of engineering. This can be attributed to the early promotion and relatively solid foundation of the practice related to the integration of hydrological and hydrodynamic models concerning drainage basins with diverse information technologies. It follows that, the combination of professional models within various fields such as engineering safety and the new generation of information technology represented by BIM possesses the value of in-depth research.

Acknowledgements This paper is a phased achievement of the Major Science and Technology Projects of the Ministry of Water Resources (KS-2022152) and the Zhejiang Province Water Conservancy Technology Project (RB-2418).

References

1. Hou Y, Zeng Y, Yu H et al (2023) Research and practice of dynamic design method for panel dams based on BIM. Hydroelectr Energy Sci 41(11):60–64
2. Yang Z, Zhou J, Wang F et al (2017) 30 years of development of concrete faced rockfill dams in China. Hydroelectr Pump Storage 3(01):1–12
3. Xu Z (2021) Key technologies and research progress of concrete faced rockfill dams collection of 2019 papers on earth rock dam technology state key laboratory of basin water cycle simulation and regulation. China Academy of Water Resources and Hydropower Sciences, Beijing, p 18
4. Niu X, Tan J, Tian J (2016) Characteristics and reinforcement of concrete face rockfill dam diseases. People's Yangtze River 47(13):1–5
5. Suo L, Liu N (2014) Hydraulic design manual. China Water Resources and Hydropower Press, Beijing
6. Jiang G, Fu Z, Feng J (1996) Concrete panel dam engineering. Hubei Science and Technology Press, Wuhan
7. Xu Z, Fang Z, Li D et al (2024) Research progress of BIM technology in the field of building carbon emission calculation. Build Struct 54(05):138–148
8. Guo W, Zheng X, Jie S et al (2023) Research on building intelligence and greenization based on BIM technology. Build Struct 53(S1):2367–2370
9. Li Y, Zhou J (2022) The application of parameterization technology in structural design. Build Struct 52(10):142–147
10. Li H, Li H, Zhang L et al (2020) Research on key technologies for information conversion between BIM models and structural analysis platforms. Build Struct 50(S1):649–653
11. Li J (1994) Analysis of the causes of dam failure in the Gouhou reservoir in Qinghai Province. J Geotech Eng 14(06):1–14
12. SL228-2013 design specification for concrete faced rockfill dams
13. SL274-2020 design specification for rolled earth rock dams
14. Wen P, Liu B, Li Z et al (2023) Exploration and practice of digital twin dateng gorge China. Water Resour 15(21):80–84

Open Access This chapter is licensed under the terms of the Creative Commons Attribution 4.0 International License (http://creativecommons.org/licenses/by/4.0/), which permits use, sharing, adaptation, distribution and reproduction in any medium or format, as long as you give appropriate credit to the original author(s) and the source, provide a link to the Creative Commons license and indicate if changes were made.

The images or other third party material in this chapter are included in the chapter's Creative Commons license, unless indicated otherwise in a credit line to the material. If material is not included in the chapter's Creative Commons license and your intended use is not permitted by statutory regulation or exceeds the permitted use, you will need to obtain permission directly from the copyright holder.

Study on the Secondary Lining Force of Soft Rock Tunnel with Large Deformation

Guize Liu, Changyi Yu, Shigang Liu, Yonghua Cao, and Binbin Xu

Abstract The secondary lining is not considered as a load-bearing structure in tunnel design. However, the current field measurements of tunnels show that the secondary lining is subjected to some load. Therefore, it is necessary to study the load sharing ratio of the secondary lining in detail. This paper analyzes the load sharing ratio of the secondary lining based on the thick-walled cylinder theory. Based on this theory, an observational design method and process for soft rock high-deformation tunnels are proposed, which integrate field measurements and consider the construction duration, facilitating design and construction. Additionally, the contact pressures between the surrounding rock, the initial support, and the secondary lining, were statistically analyzed for more than 60 tunnels. The statistical analysis shows that the worse the surrounding rock, the higher the contact pressure between the surrounding rock, the initial support, and the secondary lining. Finally, combining the field measurements of the Jingzhai tunnel and the thick-walled cylinder theory, the contact pressure of the secondary lining was determined to be 6110 kPa in the observational design, which is very close to the subsequent measured value of 6690 kPa, validating the effectiveness and accuracy of the proposed observational design method. The results of this paper provide reliable theoretical references and empirical guidance for the design and construction of tunnels.

G. Liu (✉) · C. Yu · Y. Cao
China Communications Construction Company Tianjin Port Engineering Institute, Co., Ltd., Tianjin, China
e-mail: 1143097881@qq.com

G. Liu · C. Yu · S. Liu · Y. Cao · B. Xu
China Communications Construction Company First Harbor Engineering Company, Co., Ltd., Tianjin, China

G. Liu · C. Yu · Y. Cao
Key Laboratory of Geotechnical Engineering, Ministry of Communications, Tianjin, China

Key Laboratory of Geotechnical Engineering of Tianjin, Tianjin, China

Key Laboratory of Geotechnical Engineering, CCCC, Tianjin, China

S. Liu · B. Xu
No.3 Engineering Company Ltd. Of CCCC First Harbor Engineering Company, Dalian City, Liaoning, China

© The Author(s) 2026
G-F. Zhao (ed.), *Geotechnical Modeling and Intelligent Systems*,
https://doi.org/10.1007/978-981-96-6925-7_11

Keywords Load sharing ratio of the secondary lining · Thick-wall cylinder · Soft rock large deformation · Observational design

1 Introduction

The engineering geological characteristics of soft surrounding rock are that the rock mass is broken and loose, the bonding force is poor, the water softening is fast, the bearing capacity is weak, and the rheological effect is strong [1, 2]. During the construction period, the deformation and deformation rate of the surrounding rock are large, and the cave collapse and lining cracking are easy to occur [3]. Therefore, the lining construction should be completed as soon as possible to improve the bearing capacity of the surrounding rock within the bearing ring. In addition to the form of support, the timing of secondary lining application in soft rock tunnel is very important for the overall stability of the tunnel, and it is not beneficial to the safety of the tunnel if it is too early or too late [4]. Since the introduction of the new Austrian tunneling method into China, its concept is clear, but from the measured data of tunnel construction from the 1980s to now [5], the viewpoint that the secondary lining is only used as a safety reserve has not been fully implemented in the actual tunnel. However, some domestic specifications have not fully considered the secondary lining as part of the load-bearing structure, leading to the cracking of the secondary lining in some operation periods, and especially in the bad geological tunnel, the cracking of the secondary lining is particularly serious [6, 7]. At present, the research on load sharing ratio of the secondary lining mainly focuses on two aspects [8]. One is to analyze the thick-walled tube theory, but the theoretical conditions are relatively ideal, far from the measured data, which is not suitable for the sharing ratio research of grade III and IV surrounding rock, and the V and VI surrounding rock needs to be "discounted" before guiding the actual. The other is based on the measured data [9], focusing on the secondary lining sharing ratio of different parts of the tunnel, such as the arch waist and the arch top, but the sharing ratio is very discrete, which is difficult to explain the problem. In essence, the primary support and the secondary lining of the tunnel as an organic whole play a role together, but the force distribution mechanism is not completely clear. This paper analyzes the contact pressure characteristics between surrounding rock and primary support, and between primary support and second lining from the theoretical analysis and measured data statistics. Through theoretical analysis, the load sharing ratio of the secondary lining is not only related to the tunnel structure parameters, but also related to the site construction process. The reasonable range of the load sharing ratio of the secondary lining is given from the field measured data statistics.

2 Theory of Load Transfer Between Primary Support and Secondary Lining

The study of the load sharing ratio of the secondary lining of the tunnel has been partially completed [10, 11]. The theory of thick-walled tube has the following assumptions: (1) the displacement between each layer of lining changes continuously; (2) each layer of support is only subject to uniform normal force. As shown in Fig. 1, the tunnel has two layers of lining. The outer diameter and inner diameter of the primary support are r_1 and r_2, respectively. The normal forces of the external (contact pressure of the surrounding rock) and the internal (contact pressure of the primary support and the second lining) are P_1 and P_2, respectively. The outer diameter and inner diameter of the second lining are r_2 and r_3, respectively.

According to the theory of thick-walled tube, under the action of contact pressure P_1 and P_2, the elastic theory solution of displacement U_{21} is [12, 13]:

$$U_{21} = \frac{1+\mu_1}{E_1}\left[\frac{(1-2\mu_1)r_1^2 r_2 + r_1^2 r_2}{(r_1^2 - r_2^2)}P_1 - \frac{(1-2\mu_1)r_2^3 + r_1^2 r_2}{(r_1^2 - r_2^2)}P_2\right] \quad (1)$$

where μ_1 and E_1 are the Poisson's ratio and elastic modulus of the primary support.

The displacement U_{23} of the outer wall of the secondary lining is

$$U_{23} = \frac{1+\mu_2}{E_2}\frac{(1-2\mu_2)r_2^3 + r_3^2 r_2}{(r_2^2 - r_3^2)}P_2 \quad (2)$$

Fig. 1 Theory model of thick-walled tube

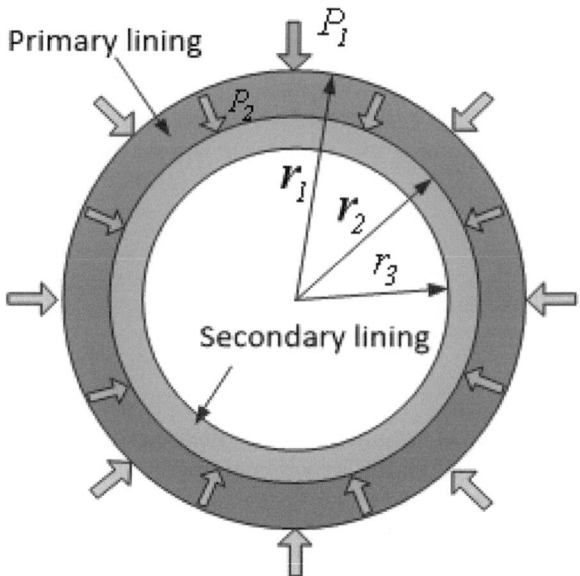

where μ_2 and E_2 are the Poisson's ratio and elastic modulus of the second liner.

Based on the aforementioned assumptions, the load sharing ratio G_0 of the second liner can be obtained as follows:

$$G_0 = \frac{P_2}{P_1} = \frac{2r_2 r_1^2}{\frac{E_1}{1-\mu_1^2}(r_1^2 - r_2^2)} \bigg/ \left[\frac{(1-2\mu_1)r_2^3 + r_2 r_1^2}{\frac{E_1}{1+\mu_1}(r_1^2 - r_2^2)} + \frac{(1-2\mu_2)r_3^3 + r_2 r_3^2}{\frac{E_2}{1+\mu_2}(r_2^2 - r_3^2)} \right] \quad (3)$$

When the surrounding rock is poor, the first support will be applied during construction; when the first support is basically stable, the second lining will be applied; that is, the deformation of the primary support and the second lining is not coordinated, and then, the accuracy of formula (3) will decline, resulting in a large error between the results and the field measurement.

Now we are going to start over by dividing both sides of formula (1) by P_1, and we get

$$\frac{U_{21}}{P_1} = \frac{1+\mu_1}{E_1} \left[\frac{(1-2\mu_1)r_1^2 r_2 + r_1^2 r_2}{(r_1^2 - r_2^2)} - \frac{(1-2\mu_1)r_2^3 + r_1^2 r_2}{(r_1^2 - r_2^2)} \frac{P_2}{P_1} \right] \quad (4)$$

When the pressure P_1 is constant, the larger the displacement U_{21} released by the initial support, according to formula (4), the smaller the pressure P_2. During tunnel construction, it is required that after the primary support is basically convergent, the second lining is applied, and the above theory also explains this requirement.

However, soft rock large deformation tunnels are limited by the pressure of the construction period, which is difficult to converge in a short time, and may begin to perform secondary lining when the initial support is not completely stable. At present, the observational design of soft rock tunnel with large deformation is required. The basis of observational design is not very clear, but the empirical evaluation is made according to the effect of on-site support, and then, the feedback is given. The following will be discussed according to formula (4), the observational design calculation of soft rock tunnel with large deformation. The surrounding rock pressure P_1 of the tunnel can be determined according to the ground stress test or the earth pressure theory. First, the preliminary dimension parameters (r_1 and r_2) and the physical parameters (μ_1 and E_1) of the primary support are given, and the displacement U_t of the specified section is measured continuously for t days. Using exponential function fitting to determine the final deformation of the section U_∞, then

$$U_{21} = U_\infty - U_t \quad (5)$$

According to formula (4), the pressure P_2 of the second liner can be determined after t days when the second liner is made, and then, it can be used for the structural design. The above process includes the basic logic process of observational design and the timing of the implementation of the second lining. It can be seen that observational design should not only consider the structural design of the tunnel, but also consider the construction period. The above process explains this view theoretically.

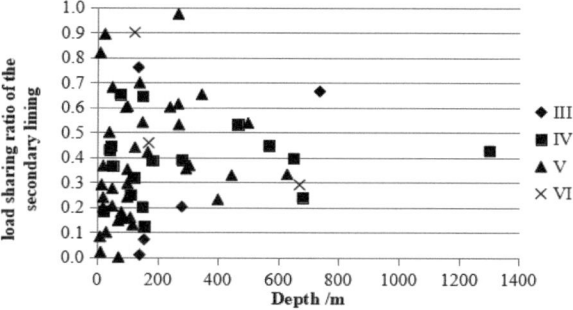

Fig. 2 Statistics of depth and load sharing ratio of the secondary lining

3 Contact Pressure of Primary and Secondary Support

Generally, the residual deformation U_{21} of the primary support is difficult to be accurately measured and calculated, so the load sharing ratio G_0 of the primary branch and the secondary liner should be further studied according to engineering practice. The most direct way to calculate G_0 is to determine the contact pressure P_1 of surrounding rock and the contact pressure P_2 of primary support and secondary liner. Among them, the contact pressure P_1 is mainly based on the relaxation pressure of the surrounding rock in lake experiment, such as Xie Jiajie formula, Terzaghi formula, and Platts theory, its quasi-determination cannot be well combined with the field, and the contact pressure P_2 has no reliable theoretical calculation support. Therefore, it is important to analyze the field measured data. Figure 2 shows the buried depth and the share ratio of tunnels with different perimeter levels. Among them, there are 5 tunnels in grade III surrounding rock, 19 tunnels in grade IV surrounding rock, 45 tunnels in grade V surrounding rock, and 3 tunnels in grade VI surrounding rock.

According to Fig. 2, the relationship between buried depth and share ratio of tunnels is unclear. The average pressure P_1 of the III level surrounding rock is 68.89 kPa, and the average pressure P_2 is 18.7 kPa. Most of the load sharing ratio G_0 of the IV level surrounding rock is between 0.1 and 0.5 (74%), the average pressure P_1 is 183.29 kPa, and the average pressure P_2 is 93.89 kPa. The load sharing ratio G_0 of V grade surrounding rock is 84.4% between 0.1 and 0.7, the average pressure P_1 is 265.72 kPa, and the average pressure P_2 is 106.29 kPa. The load sharing ratio G_0 of grade VI is 100% between 0.3 and 0.9, the average pressure P_1 is 372.37 kPa, and the average pressure P_2 is 272.1 kPa. It can be seen that the poorer the surrounding rock is, the larger the load sharing ratio is.

4 Example Analysis

Jingzhai tunnel is located in the middle line of the Trans-Asian Railway from Kunming to Bangkok, and its width and height are 10.05 m and 8.16 m, respectively. Section DK409 + 037 is a large deformation section of soft rock with high

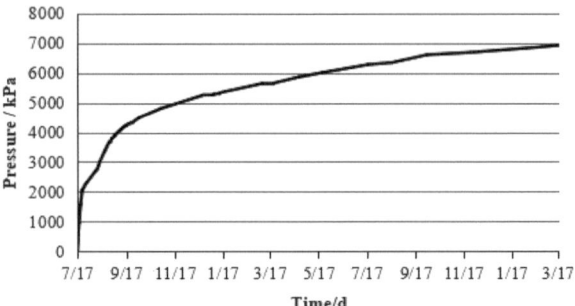

Fig. 3 Section secondary lining pressure of DK409 + 037

ground stress level III. The measured ground stress is about 15 MPa, the rock strength is about 9.8 MPa, and the surrounding rock strength is 0.98 MPa. In order to ensure smooth traffic and fast construction progress of the tunnel, through monitoring the convergence of the primary support, the secondary lining is started when the residual deformation of the primary support is about 7 mm. It can be obtained that the theoretical contact pressure between the primary support and the secondary lining can reach a maximum of 6110 kPa according to formula (1). As shown in Fig. 3, the maximum contact pressure between section DK409 + 037 secondary liner and primary support is 6690 kPa, which is very close to the theoretical calculated value.

5 Conclusions

Based on the theory of thick-walled tube, this paper analyzes that the ratio of secondary lining is not only related to the size and physical parameters of the tunnel, but also to the residual deformation of the primary support of the tunnel in the field construction. At the same time, according to the theory, the observational design method and process of soft rock large deformation tunnel are proposed. The method is closely combined with the field measurement data and can consider the construction period, which is convenient for design and construction.

Secondly, the contact pressure between the surrounding rock, the primary branch, and the secondary lining of multiple tunnels is calculated. Under the existing design and construction conditions, the load sharing ratio of the secondary lining of the surrounding rock of the surrounding rock class III is recommended to be less than 0.2, IV less than 0.5, V less than 0.7, and VI less than 0.9.

Finally, combined with the measured data of Jingzhai tunnel and the thick-walled tube theory, the contact pressure of the secondary lining is determined to be 6110 kPa in observational design, which is very close to the subsequent measured 6690 kPa, which verifies that the observational design method of large deformation of soft rock proposed in this paper is effective and accurate.

References

1. Sun X, Zhao C, Tao Z et al (2021) Failure mechanism and control technology of large deformation for Muzhailing Tunnel in stratified rock masses. Bull Eng Geol Env 80:4731–4750
2. Tang B, Ren Y, Lin Y (2021) Study on seismic response and damping measures of surrounding rock and secondary lining of deep tunnel. Shock Vib 2021:1–9
3. Zhang X, He M, Wang F et al (2020) Study on the large deformation characteristics and disaster mechanism of a thin-layer soft-rock tunnel. Adv Civ Eng 2020:1–15
4. Wu F, Miao J, Bao H et al (2015) Large deformation of tunnel in slate-schistose rock. 17–23
5. Xu G, He C, Wang J et al (2020) Study on the mechanical behavior of a secondary tunnel lining with a yielding layer in transversely isotropic rock stratum. Rock Mech Rock Eng 53:2957–2979
6. JI Y (2013) Study on control of concrete backfilling amount of secondary lining of large-deformation soft-rock tunnels. Tunn Constr 33:664–667
7. Xu G, He C, Yang Q et al (2019) Progressive failure process of secondary lining of a tunnel under creep effect of surrounding rock. Tunn Undergr Space Technol 90:76–98
8. Rao J, Tao Y, Xiong P et al (2020) Research on the large deformation prediction model and supporting measures of soft rock tunnel. Adv Civ Eng 2020:1–13
9. Xue Y, Ma X, Qiu D et al (2021) Analysis of the factors influencing the nonuniform deformation and a deformation prediction model of soft rock tunnels by data mining. Tunn Undergr Space Technol 109:103769
10. Zhang J, Wang X, Wang S et al (2016) Study on Construction Time of Secondary Lining for Large Section Shallow Buried Tunnel in Soft Rock 2016 International Conference on Smart City and Systems Engineering (ICSCSE)
11. Chen W, Wan W, Lian S et al (2020) Mechanical properties and failure modes of thick-walled cylinder granites with different apertures under triaxial compression. Adv Civ Eng 2020:1–18
12. Zhang C, Lv W, Liu L et al (2018) Study on the reserved deformation of large section soft rock highway tunnel based on monitoring measurement. IOP Conf Ser: Mater Sci Eng 452:022099
13. Wang R, Liu Y, Deng X et al (2020) Analysis on loose circle of surrounding rock of large deformation soft-rock tunnel. Adv Civ Eng 2020:1–11

Open Access This chapter is licensed under the terms of the Creative Commons Attribution 4.0 International License (http://creativecommons.org/licenses/by/4.0/), which permits use, sharing, adaptation, distribution and reproduction in any medium or format, as long as you give appropriate credit to the original author(s) and the source, provide a link to the Creative Commons license and indicate if changes were made.

The images or other third party material in this chapter are included in the chapter's Creative Commons license, unless indicated otherwise in a credit line to the material. If material is not included in the chapter's Creative Commons license and your intended use is not permitted by statutory regulation or exceeds the permitted use, you will need to obtain permission directly from the copyright holder.

Construction Risk Assessment of Biogas Layer under Shield of Subway Tunnel

Dongyin Qi and Fei Yu

Abstract The risk factors in the construction of subway shield tunnels have complex and variable characteristics. To address the impact of these characteristics in the evaluation process, a risk assessment method for subway tunnel shield tunneling through the biogas layer construction using WBS-RBS, AHP combined with cloud models is proposed. The WBS-RBS method is used to analyze the construction risk factors from the construction stage and 4M1E, and 24 risk indicators are identified. Use AHP to calculate the weight of risk indicators. Construction risk assessment model is constructed by cloud model theory to evaluate construction risk. Based on the construction project of Hangzhou–Haining Intercity Railway under biogas layer, this paper verifies the construction risk assessment method of subway tunnel under biogas layer with shield. The case study shows that the construction risk of biogas layer under the shield of Hangzhou–Haining Intercity Railway tunnel is medium risk. This paper provides reference value for the analysis and control of the construction risk factors of biogas layer under the shield of subway tunnel.

Keywords Shield tunnel construction · Construction through biogas layer · WBS-RBS · Cloud model · Risk assessment

1 Introduction

In recent years, rail transit construction has become a key means to promote the development of urban agglomerations, and subway construction projects are gradually increasing. Shield tunneling is often used in subway tunnel construction. This is because the shield tunnel construction has the advantages of small impact on climate and high construction accuracy, but at the same time it has the requirements of large impact on geological environment and strong uncertainty. This makes there are more potential risks in the process of construction [1, 2]. One of the more difficult geological conditions in tunnel shield construction is biogas geology. Once the

---

D. Qi · F. Yu (✉)
Zhejiang Haining Rail Transit Operation Management Co., Ltd, Haining, China
e-mail: sattuo@163.com

biogas is released, it may lead to combustion and explosion, causing damage to buildings and casualties [3–6]. Therefore, conducting a risk assessment of tunnel shield construction on the biogas layer is of great significance for the construction of tunnel shields.

Regarding the research on the construction risks of shield tunnels, scholars mainly study the construction process and construction risk factors of shield tunnel. Zhang, YQ et al. proposed the method of using the operating rules of interval number combined with membership function to calculate the risk level of shield tunnel construction and verified the feasibility of the method by applying Shenzhen Metro Line 12 [7]. Xu, N et al. established a Bayesian dynamic model to dynamically evolve the safety risks caused by different construction activities during shield tunnel construction [8]. Fu, HL et al. established a subway tunnel shield risk assessment model based on cloud theory model and applied it to Shiyang shield tunnel to verify its effectiveness [9]. Cao, L et al. conducted three-dimensional numerical modeling under representative conditions, established risk analysis standards for tunnel shield construction, and proposed countermeasures for important risks [10]. When only the construction content of shield method is studied, the particularity of different construction environments is not considered, resulting in the lack of accuracy and comprehensiveness of risk assessment results.

Tunnel shield construction is greatly affected by geological environment, so it is necessary to carry out risk assessment under different special environment to ensure the safety of construction. Some scholars have assessed the risk of tunnel shield construction in karst area and put forward countermeasures against the key risks [11–13]. Most scholars have studied the situation of shield tunnel construction under existing buildings, analyzed the construction risks, and put forward prevention and control measures [14–16]. Although scholars have made some progress in the study of some special construction geology, the geological environment of tunnel shield tunneling through the biogas layer was not studied, which makes the current shield construction risk assessment results limited.

In view of the current research status of shield tunneling through biogas layer construction, the risk assessment of shield tunneling through biogas layer construction is carried out by using WBS-RBS combined with cloud model, in order to propose a scientific risk assessment method for shield tunneling through biogas layer construction risk assessment.

2 Theoretical Model

2.1 Establish Assessment Indicator System

The shield construction of subway tunnel is complicated, coupled with the biogas layer under the construction, resulting in more risks in construction. In order to accurately identify the construction risk factors, according to the Design Standards of Shield Tunnel Engineering and the Code for the Construction and Acceptance of Shield Tunnel and other relevant standards and specifications, WBS-RBS was used to decompose the shield construction project from two aspects: construction process and 4M1E (man, machine, material, and method and environment), and Delphi method was used to screen the main risk factors [17, 18].

Spss software is used to verify the reliability and reliability of index factor screening [19]. A total of 120 questionnaires were issued, and 109 were recovered, of which 89 were valid. The results show that the Alpha coefficients of the questionnaire are all greater than 0.8, the measured values of KMO samples are all greater than 0.7, and the significance of Bartlett sphericity test is less than 0.001, which indicates that the data obtained from the risk factors questionnaire statistical table at each stage of shield tunnel construction are reliable and effective.

According to the main construction contents of the shield method, the risk factors are comprehensively identified for the six construction stages of the working well construction stage, the preparation stage before starting, the shield starting stage, the shield driving stage, the shield arrival stage, and the shield crossing the special terrain, as well as the construction risks of the five aspects of 4M1E. The shield tunnel construction risk assessment system is established, which includes 3 primary indicators and 24 secondary indicators, and the details are in Table 1.

2.2 AHP Calculates the Indicator Weight

AHP is practical and effective in dealing with the security problems of complex systems and is widely used in the analysis, assessment, and decision of security system elements. The main steps are as follows:

Step 1: Establish a risk assessment hierarchy. According to the objectives and principles of risk assessment, the risk factors are classified hierarchically, including the target layer A, primary decision level B, and the secondary decision level C.

Step 2: Construct the judgment matrix. According to the scale method of judgment matrix 1–9, the judgment matrix is constructed.

Step 3: Perform consistency test on the judgment matrix $CR = \frac{CI}{RI}$, when $CR < 0.1$, the calculation results are valid.

$$CI = \frac{\lambda_{max} - n}{n - 1} \quad (1)$$

wherein RI is the standard value of average random consistency.

Step 4: Get the weight value of each indicator in the matrix $W = (W_1, W_2, ..., W_n)$.

Through discussion with ten experts, the opinions of experts on the ranking of the importance of influencing factors were sorted out, the scores of each indicator were allocated according to the results, and the judgment matrix was constructed:

Table 1 Metro tunnel shield construction risk assessment system

Construction Risk A	Work well construction B1	Foundation pit dewatering C1
		Well-confined water treatment C2
		Construction of safety facilities C3
	Preparation before departure B2	Tunnel door soil reinforcement C4
		Hoisting the shield machine down the well C5
		Personnel operation C6
		The shield machine is installed C7
	Shield initiation B3	Grouting pressure C8
		Chisel out the cave door C9
		Assembly of negative ring tube C10
		Tunnel door waterproof facilities C11
		Stability of scaffold C12
	Shield tunnelling B4	Pipe grouting pressure control C13
		Velocity of shield propulsion C14
		Shield tunneling parameters C15
		Earthwork excavation and transportation C16
	Shield arrive B5	Chisel out the cave door C17
		Install the tunnel door seal C18
		End reinforcement C19
		Receiver bracket positioning accuracy C20
		Shield disassembly and lifting C21
	Crossing special terrain B6	Soil reinforcement C22
		Earth pressure control C23
		Device status C24

$$A = \begin{bmatrix} 1 & 3 & 1/3 & 1/8 & 1/6 & 1/5 \\ 1/3 & 1 & 1/5 & 1/7 & 1/4 & 1/6 \\ 3 & 5 & 1 & 1/5 & 3 & 1/3 \\ 8 & 7 & 5 & 1 & 5 & 2 \\ 6 & 4 & 1/3 & 1/5 & 1 & 1/4 \\ 5 & 6 & 3 & 1/2 & 4 & 1 \end{bmatrix}$$

$$B_1 = \begin{bmatrix} 1 & 1/3 & 3 \\ 3 & 1 & 5 \\ 1/3 & 1/5 & 1 \end{bmatrix} \quad B_2 = \begin{bmatrix} 1 & 3 & 7 & 5 \\ 1/3 & 1 & 6 & 3 \\ 1/7 & 1/6 & 1 & 1/3 \\ 1/5 & 1/3 & 3 & 1 \end{bmatrix}$$

Table 2 Results of consistency calculation of judgment matrix

Judgment matrix	A	B_1	B_2	B_3	B_4	B_5	B_6
CR	0.099	0.037	0.047	0.086	0.087	0.077	0.051
Test result	Yes	Yes	Yes	Yes	Yes	Yes	Yes

Table 3 Weight calculation results

Judgment matrix	ω_1	ω_2	ω_3	ω_4	ω_5	ω_6
B_1	0.261	0.633	0.106			
B_2	0.553	0.274	0.054	0.119		
B_3	0.094	0.134	0.479	0.257	0.036	
B_4	0.285	0.041	0.548	0.126		
B_5	0.239	0.508	0.135	0.077	0.041	
B_6	0.639	0.274	0.087			
A	0.052	0.033	0.137	0.410	0.112	0.255

$$B_3 = \begin{bmatrix} 1 & 1/3 & 1/5 & 1/4 & 5 \\ 3 & 1 & 1/5 & 1/3 & 4 \\ 5 & 5 & 1 & 3 & 8 \\ 4 & 3 & 1/3 & 1 & 7 \\ 1/5 & 1/4 & 1/8 & 1/7 & 1 \end{bmatrix} \quad B_4 = \begin{bmatrix} 1 & 7 & 1/3 & 4 \\ 1/7 & 1 & 1/9 & 1/5 \\ 3 & 9 & 1 & 5 \\ 1/4 & 5 & 1/5 & 1 \end{bmatrix}$$

$$B_5 = \begin{bmatrix} 1 & 1/4 & 3 & 4 & 6 \\ 4 & 1 & 3 & 6 & 7 \\ 1/3 & 1/3 & 1 & 3 & 4 \\ 1/4 & 1/6 & 1/3 & 1 & 3 \\ 1/6 & 1/7 & 1/4 & 1/3 & 1 \end{bmatrix} \quad B_6 = \begin{bmatrix} 1 & 3 & 6 \\ 1/3 & 1 & 4 \\ 1/6 & 1/4 & 1 \end{bmatrix}$$

Consistency test is carried out on the judgment matrix, and the test results are obtained as Table 2.

According to Table 2, CR values are all less than 0.1, it can be seen that the judgment matrix has passed the consistency test, and the weight of each risk factor has been obtained as Table 3, which provides the basis for construction risk assessment.

2.3 Build a Risk Assessment Cloud Model

The risk matrix method is adopted to classify the probability of risk occurrence and the severity of potential consequences caused by risks, and the risk levels are divided into four levels in Table 4 [20, 21].

Table 4 Standard cloud model

Risk level	Minimum score	Maximum score	(Ex, En, He)
Low risk	0	2.5	(1.25, 0.420.1)
Lower risk	2.5	5	(3.75, 0.420.1)
Medium risk	5	7.5	(6.25, 0.420.1)
High risk	7.5	10	(8.75, 0.420.1)

Assessment Criteria Cloud

Formulas (2)–(4) are used to calculate the parameters of the standard cloud model, and the results are shown in Table 4.

$$Ex_i = \frac{x_i^{\min} + x_i^{\max}}{2} \tag{2}$$

$$En_i = \frac{x_i^{\max} - x_i^{\min}}{6} \tag{3}$$

$$He_i = 0.1 \tag{4}$$

wherein $x_i^{\min}$ is the minimum score obtained from the risk score, $x_i^{\max}$ is the maximum score obtained from the risk score, Ex_i stands for expected value, and En_i stands for entropy of standard cloud model parameters.

Standard cloud maps are drawn according to standard cloud model parameters as Fig. 1.

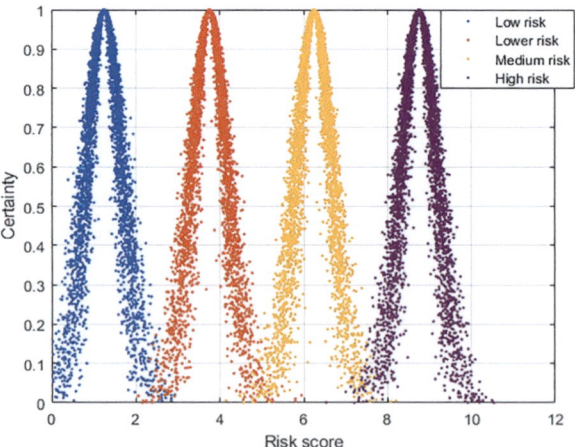

Fig. 1 Standard cloud chart

Assessment Indicator Cloud

According to the risk scoring results of the secondary indicator, the parameters of the cloud model of the secondary indicator are calculated using Eqs. (5)–(7) [22–24].

$$Ex = \bar{x} = \frac{1}{n}\sum_{i=1}^{N} x_i \tag{5}$$

$$En = \sqrt{\frac{\pi}{2}}\frac{1}{n}\sum_{i=1}^{n}|x_i - \bar{x}| \tag{6}$$

$$He = \sqrt{(S^2 - En^2)} \tag{7}$$

According to the weight of the secondary indicator, the cloud model parameters of the primary indicator are calculated using Eqs. (8)–(10).

$$Ex = \sum_{i=1}^{n}(Ex_i w_i) \tag{8}$$

$$En = \sqrt{\sum_{i=1}^{n}(Ex_i w_i)^2} \tag{9}$$

$$He = \sqrt{\sum_{i=1}^{n}(He_i w_i)^2} \tag{10}$$

After obtaining the parameters of the cloud model, MATLAB software is used to draw the cloud map. The closest standard cloud in the assessment cloud map is determined, the grade of this standard cloud is determined as the assessment cloud grade, and the assessment result of the construction project of subway tunnel through the biogas layer under the shield is obtained.

3 Case Analysis

3.1 Basic Situation of Hangzhou–Haining Intercity Railway Shield Section

The section between Linpingnan High-speed Railway Station and Xucun Station of Hangzhou–Haining Intercity Railway was constructed by shield method. The left line length of the shield section is 3128.22 m, the right line length is 3128.555 m, and an intermediate air shaft is set in the section. There are many pipelines distributed above the new tunnel within the shield section. The buried depth of the tunnel is 8.6–25 m, and the line spacing is 10.8–16 m [24, 25].

3.2 Assessment Result Analysis

By means of questionnaire, the construction risk of the tunnel under the biogas layer in Hangzhou–Haining Intercity Railway shield construction was evaluated. Organize the scores of ten experts, the average value, entropy, and overentropy of the score value were calculated according to formula (5)–(7), and the cloud model parameters of each indicator are obtained as Table 5.

Table 5 Cloud model parameters of primary indicator

Primary indicator	(Ex, En, He)	Secondary indicator	(Ex, En, He)
B_1	(5.023,0.378,0.163)	C_1	(3.8,0.551,0.146)
		C_2	(6.0,0.251,0.249)
		C_3	(2.2,0.551,0.146)
B_2	(5.763,0.347,0.098)	C_4	(6.8,0.551,0.146)
		C_5	(4.8,0.551,0.146)
		C_6	(3.7,0.802,0.238)
		C_7	(4.1,0.451,0.311)
B_3	(5.259,0.249,0.140)	C_8	(4.1,0.451,0.311)
		C_9	(4.6,0.652,0.185)
		C_{10}	(6.0,0.251,0.249)
		C_{11}	(5.0,0.752,0.244)
		C_{12}	(3.7,0.802,0.238)
B_4	(6.257,0.345,0.098)	C_{13}	(4.8,0.551,0.146)
		C_{14}	(2.4,0.401,0.119)
		C_{15}	(7.8,0.551,0.146)
		C_{16}	(4.1,0.451,0.311)
B_5	(6.001,0.326,0.094)	C_{17}	(4.6,0.652,0.185)
		C_{18}	(7.7,0.551,0.146)
		C_{19}	(4.5,0.251,0.249)
		C_{20}	(3.8,0.551,0.146)
		C_{21}	(2.2,0.551,0.146)
B_6	(4.418,0.530,0.169)	C_{22}	(5.0,0.752,0.244)
		C_{23}	(3.7,0.802,0.238)
		C_{24}	(2.4,0.401,0.119)

Fig. 2 Cloud chart of comprehensive assessment

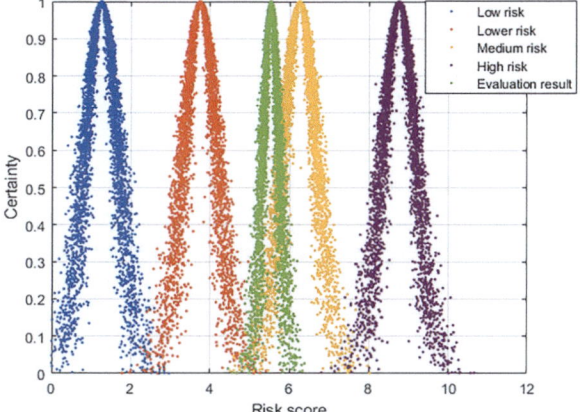

According to the primary indicator cloud model parameters in the above table, combined with the indicator weight in Table 3, the comprehensive cloud model parameters of tunnel construction risk in shield section of Hangzhou–Haining Intercity Railway are calculated using formulas (8)–(10) as (5.536,0.203,0.064), and the construction risk comprehensive assessment cloud map is generated by MATLAB, as illustrated in Fig. 2.

It can be seen intuitively from Fig. 2 that the cloud thickness is relative thick, and the cloud droplet dispersion and span are small, which indicates that the evaluation results are reliable and effective. The construction project risk assessment cloud map is between the medium risk standard cloud map and the lower risk standard cloud map and is closer to the medium risk standard cloud map than the lower risk standard cloud map. Comprehensive analysis can be obtained, the risk level in the section of biogas layer under Hangzhou–Haining Intercity Railway shield construction is determined to be medium risk.

Combined with the results in Table 5, it can be seen that the risk scores of shield tunneling parameters and tunnel door sealing device installation are both bigger than 7.5, which are high risk indicators and important reasons affecting construction safety. It is necessary to control the setting of tunneling parameters and strengthen the sealing effect of tunnel door, so as to reduce the occurrence of methane danger.

4 Results

Considering the complex and uncertain characteristics of risk factors in the construction process of tunnel shield tunneling through biogas layer, a risk factor identification method based on WBS-RBS is proposed, and construct an assessment model by using cloud model to carry out construction risk assessment. The method is also applied to the shield construction of Hangzhou–Haining Intercity Railway under the biogas layer to verify the method. The main results of this paper are as follows:

(1) WBS-RBS is used to identify risk indicators during shield tunnel construction. According to 24 risk indicators of six construction processes, risk assessment indicator system of subway shield tunnel construction is established according to SMART principle.
(2) Due to the complex risk indicators and strong uncertainty in the construction process of biogas layer under the subway tunnel shield, the cloud model method is adopted to establish a risk assessment model.
(3) Combined with the actual situation of shield construction under biogas layer of Hangzhou–Haining Intercity Railway, the applicability of the established subway tunnel shield construction risk assessment system is verified by an example. The expected risk of the construction project in this case is 5.536, which is a medium risk project.

This paper studies the construction risk of shield tunneling through biogas layer in subway tunnel and provides reference value for shield tunneling through biogas layer in the future. However, the risk indicator assessment of shield tunnel construction is not accurate enough, so it can be further refined on the basis of four risk levels.

References

1. Wu X, Wang L, Chen B et al (2022) Multi-objective optimization of shield construction parameters based on random forests and NSGA-II. Adv Eng Inform 1(54):101751
2. Zhu C, Wang S, Peng S et al (2022) Surface settlement in saturated loess stratum during shield construction: numerical modeling and sensitivity analysis. Tunn Undergr Space Technol 1(119):104205
3. He J, Zhu H, Wei X et al (2023) Numerical and experimental analyses of methane leakage in shield tunnel. Front Struct Civ Eng 17(7):1011–1120
4. Li Y (2013) Effect of methane gas on shield tunnel and construction control measures. Urban Rail Transit Res 16(02):95–98+101. https://doi.org/10.16037/J.1007-869x.2013.02.032
5. Yan Q, Yang K, Wu W et al (2020) Prevention and control of gas hazards in a tunnel under construction: a case study. Environ Earth Sci 79:1–7
6. Zhao C (2019) Safety comprehensive prevention and control technology of biogas rich stratum shield tunnel construction. China Stand (12):9–10+12
7. Zhang Y, Zhang J, Guo H et al (2020) A risk assessment method for metro shield tunnel construction based on interval number. Geotech Geol Eng 38:4793–4809
8. Xu N, Guo C, Wang L et al (2024) A three-stage dynamic risk model for metro shield tunnel construction. KSCE J Civ Eng 28(2):503–16.
9. Fu H, Huang Z, Zhang J (2017) Risk comprehensive assessment of shield tunnel construction based on cloud theory. InGeo-Risk 2017, pp 334–345
10. Cao L, Kalinski M (2017) Risk Analysis of Subway Shield Tunneling. InGeo-Risk 2017, pp 309–319
11. Duan GD (2024) Study on the construction risk and control of shield tunnel under the geological condition of high karst development in Shenzhen area. Transp Manager World 07:80–82
12. Yu Q, Yao Q, Li S et al (2024) Karst tunnel construction. J Saf Anal Its Control Strategy and Innov of Sci Technol (4):128–131. https://doi.org/10.15913/j.carolcarrollnkikjycx.2024.04.037

13. Chang Z, Weibin Z, Weiran H et al (2019) Risk analysis and countermeasure study of shield tunnelling in Karst Stratum of China. Geotech Eng (00465828). 50(2)
14. Yang J, Huang S (2024) Construction monitoring analysis of shield tunnel under existing buildings. Transp Sci Technol 01:86–91
15. He K, Zhu J, Wang H et al (2023) Safety risk evaluation of metro shield construction when undercrossing a bridge. Buildings 13(10):2540
16. Xu YJ (2020) Monitoring analysis and construction control of existing buildings above subway shield tunnel and station connection project. Transp Constr Manag 04:120–121
17. Gong W, Huang H, Juang CH et al (2015) Improved shield tunnel design methodology incorporating design robustness. Can Geotech J 52(10):1575–1591
18. Zhu X, Jiang F (2023) Engineering safety supervision 4m1e//China communications construction supervision association 2022 annual academic papers, 2023:5. https://doi.org/10.26914/Arthurc.nkihy.2023.031369
19. Fife-Schaw C (2020) Questionnaire design. Res Methods Psychol, 343–374
20. Yan H, Gao C, Elzarka H et al (2019) Risk assessment for construction of urban rail transit projects. Saf Sci 1(118):583–594
21. Zhang P, Pan HH, Yang ZQ (2023) Research on risk assessment of urban buried gas pipelines based on risk matrix method and Borda ordinal value method. China Prod Saf Sci Technol 19(09):116–122
22. Li X, Ran Y, Zhang G et al (2020) A failure mode and risk assessment method based on cloud model. J Intell Manuf 31:1339–1352
23. Ding T-G, Tan Q-L, Wang X-T et al (2024) Optimization of green low-carbon railway lines in mountain area based on cloud model and cumulative prospect theory. J Railw Sci Eng, 1–10 [2024-05-17]. https://doi.org/10.19713/j.cnki.43-1423/u.T20232056
24. Fan R, Zhao X, Liu X et al (2021) Study on the countermeasures of the Hangzhou-Haining intercity railway shield passing through the biogas formation 42(02):12–16
25. Deng J, Yin S, Rao P (2021) Analysis of BIM application strategy in engineering project based on fuzzy mathematics-SWOT. J Inf Technologyin Civ Eng Archit 13(4):113–119

Open Access This chapter is licensed under the terms of the Creative Commons Attribution 4.0 International License (http://creativecommons.org/licenses/by/4.0/), which permits use, sharing, adaptation, distribution and reproduction in any medium or format, as long as you give appropriate credit to the original author(s) and the source, provide a link to the Creative Commons license and indicate if changes were made.

The images or other third party material in this chapter are included in the chapter's Creative Commons license, unless indicated otherwise in a credit line to the material. If material is not included in the chapter's Creative Commons license and your intended use is not permitted by statutory regulation or exceeds the permitted use, you will need to obtain permission directly from the copyright holder.

Microstructure-Based Dynamic Characterization of Remodelled Loess under Traffic Loading

Hongtao Pei, Kebing Wen, Zezhan Shao, Shenqin Sun, Tianlu Xu, and Junmin Pan

Abstract As a common road subgrade fill in Northwest China, the dynamic properties of remoulded loess are the key to study the structural stability of loess subgrade, which is of great practical significance to the engineering construction and operation safety of roads in the corresponding region. This paper takes the remoulded loess of a roadbed in Shaanxi as the research object, and through the indoor dynamic triaxial test simulating traffic cyclic loading, the dynamic shear stress–shear strain relationship, dynamic shear modulus and damping ratio, and other dynamic characteristics of the soil under different water content, dry density and peripheral pressure are investigated and combined with the SEM test to qualitatively and quantitatively analyse the microstructural characteristics of the soil after vibration, so that the changes in dynamic characteristics of remoulded loess are caused by the microcosmic point of view. The results show that the dynamic shear stress–shear strain curve of the remoulded loess shows a hyperbolic development with an increase first and then a steady development, the dynamic shear modulus shows a sharp decrease at the beginning and a stable trend at the end with the shear strain, the increase of water content will lead to a decreasing trend of both, and the increase of dry density and peripheral pressure will lead to a corresponding increase of both. The test measured that the damping ratio changes in the range of 0.08–0.31 interval, showing a large increase in the initial period, the later interval fluctuation trend; the microstructure of the soil after vibration is studied, comparing the microscopic images and quantitative parameters under different test conditions, explaining the reasons for the change of the dynamics of remoulded loess through the change rule of the pore morphology and distribution, and providing the basis for the actual engineering.

Keywords Remoulded loess · Traffic cyclic loading · Dynamic triaxial test · Dynamic properties · Microstructural characteristics

---

H. Pei (✉) · K. Wen · T. Xu · J. Pan
Xi'an Rail Transit Group Company Limited, Xi'an, Shaanxi, China
e-mail: 59241971@qq.com

Z. Shao · S. Sun
China Railway Xi'an Group Co., Ltd., Xi'an, Shaanxi, China

1 Introduction

China's loess distribution area is very wide, mainly concentrated in the northwestern region, arid and semi-arid areas of the climate conditions gave birth to the formation of the Quaternary sediments, these sediments have become an important witness to the geological history of the region, its structural, water-sensitive, vibration trapping and other characteristics of the outstanding, and the large particles, porous, strong permeability and other characteristics of the significant, in the rainfall, vibration, construction disturbances, etc., are very prone to produce under the action of such as landslides, Subsidence, collapse and other natural disasters, a serious threat to people's lives and property safety. In recent years, with the national western development strategy to vigorously promote the construction of large-scale urban transport in Northwest China, urban road traffic flow is growing, resulting in traffic vehicle load on the roadbed system, and the impact of the roadbed system is intensifying and frequently leads to uneven settlement, cracking and dislocation, roadbed subsidence, and other road disasters [1–3]. Therefore, the in-depth study of the dynamic properties and microstructural characteristic changes of roadbed remodelled loess has become a popular area of concern for related scholars.

For the study of dynamic characteristics of loess, the most important and direct way is to carry out relevant dynamic triaxial tests to explore the changing rules of dynamic characteristics and deformation features of soil body under different soil conditions, static conditions, dynamic conditions, and drainage conditions [4–6], and certain breakthrough results have been achieved. There are also scholars based on dynamic triaxial test to study the dynamic characteristics, strength characteristics, and deformation characteristics of loess under seismic load, train load, and traffic vehicle load [7–9], which closely combines indoor test with actual engineering to provide experimental support and basis. Relevant scholars have explored the dynamic characteristics of loess in different geographical areas, compared and analysed the differences between loess characteristics in several regions by means of dynamic tests, and summarized the regular characteristics of loess dynamic characteristics that change with spatial location [10, 11]. Some scholars have also compared the similarities and differences between original and remoulded loess, saturated and unsaturated loess by means of dynamic triaxial test, so as to describe more comprehensively the changing rules of dynamic constitutive relationship, dynamic strength, dynamic modulus, and other characteristics of this special nature of soil [12–14] and to provide more comprehensive knowledge and experience for the engineering construction.

With the development of soil dynamics, the introduction of microscopic technology has become one of the important means of studying loess dynamics, establishing a close correlation mechanism between macroscopic phenomena and microscopic structures, and carrying out systematic research in a more integrated and comprehensive way [15, 16]. Some scholars analyse the connection between the macro and micro features of soil deformation under the bounded strike number by comparing the strength change law of vibrated loess and investigate the strength increase and decrease law of loess under power cycle [17]; for loess in many regions in Northwest China, some scholars have conducted systematic dynamic triaxial test and scanning electron microscope test, and scholars have carried out detailed comparative analysis of soils in many different areas in the wide range of studies and deeply investigated the soil subsidence characteristics in these areas. In the extensive research, the scholars conducted detailed comparative analyses of soils in many different areas and deeply investigated the seismic subsidence characteristics of soils in these areas and the change rules of their pore structures. Based on these detailed research results, they successfully constructed the formula for calculating the seismic subsidence coefficient of loess. In order to further verify the accuracy and practicability of this formula, the scholars also carried out experimental verification, and the results show that the formula is highly reasonable and applicable in engineering applications [18]. Some scholars have even studied the characteristics of soil microstructure, combining qualitative evaluation and quantitative analysis of structural parameters to explore the influence and connection between the microstructure of loess and the change of its dynamic characteristics [19, 20], and have achieved certain results.

However, the previous research mainly focuses on the dynamic characteristics of loess under the action of earthquake, and most of them are in-situ loess and saturated loess, and the dynamic characteristics of unsaturated remoulded loess under the action of traffic loading are less studied. Based on this, this paper selects a certain roadbed remoulded loess in Guanzhong, Shaanxi Province as the research object, and on the basis of dynamic triaxial test simulating traffic loading, with the help of scanning electron microscope, systematically studies the influence of the dynamic characteristics of remoulded loess under different water content, dry density, and circumferential pressure, and reveals the reasons for the changes from a microscopic point of view.

2 Overview of the Experiment

2.1 Test Soil Samples and Preparation

The specimen studied in this paper was taken from a loess roadbed in Guanzhong, Shaanxi Province, with a sampling depth ranging from 2 to 4 m and belonging to the Q3 loess type. The loess specimen showed a typical brownish-yellow tone, and the colour gradually transitioned from light brown to dark brown as the sampling depth increased. The soil body was relatively homogeneous in texture and contained some notable large particles and large pore structures, but no sand and gravel component was detected. A few calcareous nodules were observed in the specimens. The basic physical properties of the collected loess have been listed in detail in Table 1 for subsequent studies and analyses.

During specimen preparation, different water contents (14, 18, and 22%) and different dry densities (1.25, 1.45, and 1.65 g/cm^3) of the remoulded loess were set

Table 1 Basic physical property indexes of sampled loesses

Density (g/cm^3)	Moisture content (%)	Maximum dry density (g/cm^3)	Liquid limit (%)	Plastic limit (%)	Plasticity index
1.76	18.22	1.68	32.4	19.7	12.7

up according to the natural water content and maximum dry density, the weight of the soil was weighed in each layer, the soil was compacted in 5 layers, and the hair was scraped between the layers to ensure that the contacting surfaces were closely connected. After preparation, the specimens were sealed and stored with cling film for the test.

2.2 Pilot Programme and Steps

As shown in Fig. 1, this test apparatus is a LO7010/5/DYM unsaturated dynamic triaxial instrument built by WILLE, Germany. Its core components include a servo controller, an axial loading control device, a main controller, a body change measuring device, a multi-channel pressure controller, and a pressure chamber, and it is equipped with a complete set of operating system. The instrument is not only capable of performing accurate dynamic and static triaxial tests on conventional soil specimens, but also has the ability to perform a variety of functions such as dynamic cyclic loading, thus meeting the needs of various complex research.

In the course of research, one-dimensional loading forms are widely used when simulating traffic vehicle loads, while for seismic loads, two-dimensional loading forms are more common. In view of this, a one-dimensional sinusoidal waveform was specifically chosen for the selection of the loading pattern for the dynamic triaxial tests in this paper. The purpose of this choice is to accurately simulate and

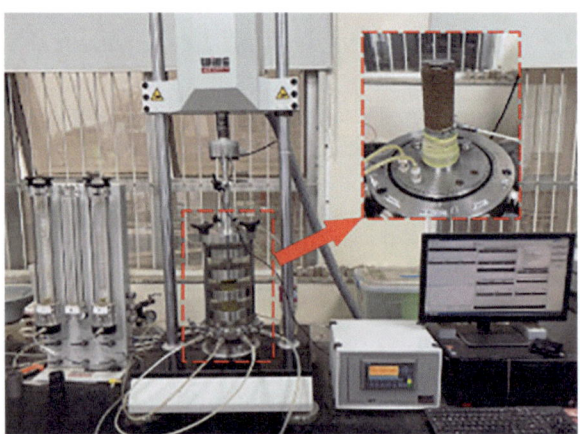

Fig. 1 Dynamic triaxial apparatus (laboratory instrument, obtain a license)

reproduce the vibration characteristics generated by the traffic load to ensure the accuracy and reliability of the test results. Thus, the prevailing conditions of traffic loading are reflected more closely to the actual situation. In conducting the dynamic characteristic test of remoulded loess, 4 Hz was specifically chosen as the loading frequency in this study, aiming to accurately simulate and test its dynamic response. In order to improve the accuracy of the test and reduce the dispersion of the results, we adopted the method of increasing the loading step by step. This loading method can not only effectively reduce the errors that may occur due to direct high load loading, but also shorten the time required for the test appropriately and improve the efficiency of the test while guaranteeing the quality of the test. Therefore, the test was conducted to study the dynamic properties of loess using a multi-stage loading method, with 10 kPa per stage, 20 vibrations per stage, and with the damage criterion of 3% of the axial strain of the specimen as the termination of the test [21].

Taking into account the current experimental research results and the actual soil conditions of the horizontal soil layer at the site, this test was finally decided to be carried out by uniformly adopting the isobaric consolidation method after careful consideration. In particular, the consolidation ratio was set to 1 to ensure the homogeneity and stability of the consolidation process. The standard of consolidation was strictly set as the axial deformation not exceeding 0.01 mm within 30 min, and the change of pore pressure should be kept within the range of less than or equal to 2 kPa, so as to accurately control the physical changes in the consolidation process. The consolidation pressure of loess is mainly based on the actual depth of the soil to reasonably determine, to ensure that the test conditions are consistent with the actual conditions in the field, so as to produce more accurate and reliable test results. Therefore, in this test, we selected three different sets of parameters of 100, 150, and 200 kPa for the test. The specimen is required to open the drain valve during the consolidation process and close the drain valve for the test after the consolidation is completed.

Before performing the SEM test, the selected soil samples first needed to be dried. Then, we took a sample from the centre area of the dried soil sample and cut it through the ring so that it was presented as a small columnar soil sample with a cross-section of about 10 mm × 10 mm and a height of about 20 mm. To ensure that a clear image of the cross-section was obtained when scanning, we also carved a ring of rectangular slots in the longitudinal 1/4 of the soil sample to expose a fresh cross-section of the soil sample for scanning and analysis. The scanned soil samples were carefully wrapped in tinfoil underneath the surface and then placed properly in an airtight plastic bag to ensure they remained dry. Prior to the start of the test, the prepared columns were gently broken into smaller specimens of approximately 5 mm in height along the grooves previously carved, ensuring that the fresh cross-section formed by the break was as flat as possible. Immediately after this, the soil sample was securely attached to the sample base using conductive adhesive. Subsequently, dust particles were carefully removed from the surface of the specimen using suction lugs to ensure a clean surface. Next, gold spraying was performed to enhance the conductivity of the sample. Finally, the treated soil samples were placed into the electron microscope instrument for electron microscope scanning test.

3 Dynamic Characterization

3.1 Dynamic Shear Stress–Shear Strain Relationships

According to the principle of soil dynamics, the relationship between dynamic stress and dynamic strain and dynamic shear stress and dynamic shear strain can be obtained, as shown in Eqs. (1) and (2).

$$\tau_d = \sigma_d/2 \tag{1}$$

$$\gamma_d = \varepsilon_d(1 + \mu) \tag{2}$$

where τ_d is dynamic shear stress; γ_d is dynamic shear strain; μ is Poisson's ratio, and Q3 Poisson's ratio of loess, remoulded loess in this paper takes 0.35 [22].

The dynamic stress–strain relationship curves under different conditions when the axial strain of each specimen reaches 3% are shown in Fig. 2.

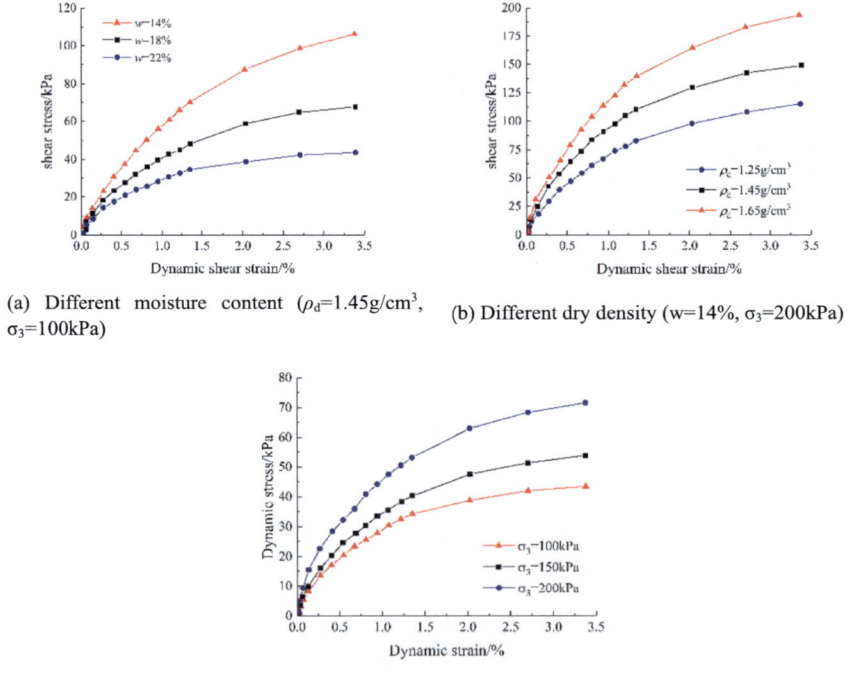

(a) Different moisture content (ρ_d=1.45g/cm³, σ_3=100kPa)

(b) Different dry density (w=14%, σ_3=200kPa)

(c) Different confining pressure (w=22%, ρ_d=1.45g/cm³)

Fig. 2 Curves of dynamic shear stress–shear strain relationship under different influencing factors

Comparative analysis of the dynamic shear stress–shear strain relationship curves of remoulded loess under different water content, dry density, and perimeter pressure conditions shows that the overall dynamic shear stress–shear strain relationship curves in the 2% strain are in line with the H–D model hyperbolic growth pattern, the curve shows a rapid growth in the early stage of the curve, and later development of the development of the trend of gradual development of the development of the shear strain in the interval of 0.5–1% of the inflection point of the weakening of the stress, and then gradually converge to the critical destructive stress value. The shear strain in the interval of 0.5–1% showed a stress weakening inflection point and then gradually converged to the critical destructive stress value, and the growth of shear strain increased with the increase of dynamic shear stress.

Under the same shear strain condition, the dynamic shear stress decreases with the growth of water content, as shown in Fig. 2a, under 3% strain, the water content increases from 14 to 22%, and the corresponding dynamic shear stresses are 104 kPa, 68 kPa, and 43 kPa, respectively, which show a nonlinear decreasing trend, which is in line with the characteristic of the sudden decrease of the structural strength of loess when it meets water. Meanwhile, under the same shear strain, the dynamic shear stress increases with the increase of dry density and peripheral pressure, and the increase of both conditions in a certain range inhibits the deformation of remoulded loess, so that the structural strength of the soil body has been greatly improved.

3.2 Dynamic Shear Modulus

The dynamic shear modulus of soil is one of the most important parameters reflecting its structural properties, and its expression is very similar to the dynamic elastic modulus, with a similar correlation, as shown in Eq. (3).

$$G_d = \frac{\tau_d}{\gamma_d} = \frac{\sigma_d}{2\varepsilon_d(1+\mu)} = \frac{E_d}{2(1+\mu)} \quad (3)$$

where G_d is the dynamic shear modulus, and E_d is the dynamic elastic modulus.

The relationship curves of dynamic shear modulus under different conditions for each specimen with axial strain up to 3% are shown in Fig. 3.

The dynamic shear modulus indicates the degree of difficulty of the soil body subjected to shear deformation, and the structural strength of the soil body is enhanced with the increase of the dynamic shear modulus, which shows the stronger rigidity of the soil body. Overall analysis of the relationship between dynamic shear modulus and shear strain of roadbed loess shown in Fig. 3 shows that the overall curve development tendency is a rapid decrease in the initial period and the later period of gentle development of the law, and the shear strain is at about 0.5% of the obvious shear modulus decay inflection point. This is due to the soil body in the initial loading in the elastic stage, then transition to the elastic–plastic stage, that is, 0.3–1.0% interval, and

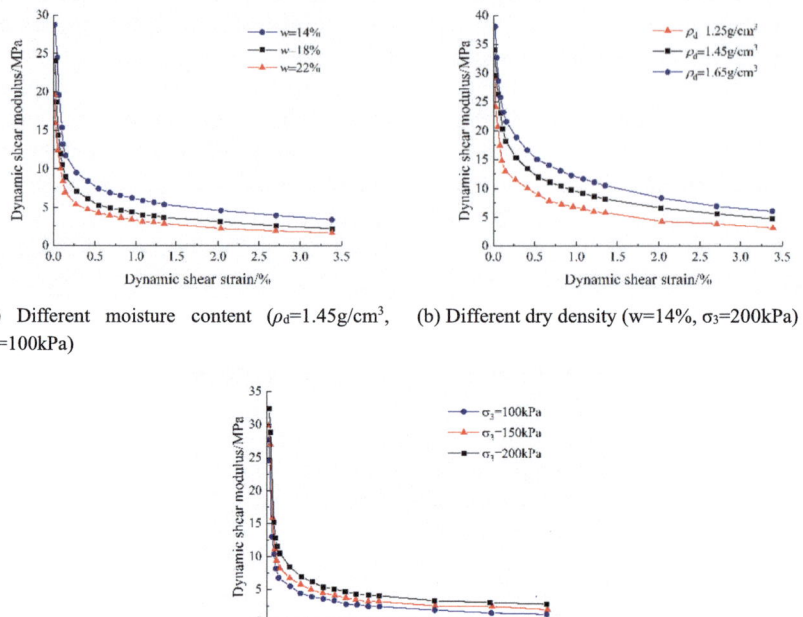

Fig. 3 Curves of relationship between dynamic shear modulus and dynamic shear strain under different influencing factors

finally enter the shear strain is large and stable development of plastic deformation stage, until the soil body by the shear between the complete destruction.

Comparing the dynamic shear modulus under different water content conditions, it can be seen that its value decreases with the increase of water content, the lubrication effect of the water body makes the friction force between particles greatly reduced, and it is easy to produce a large relative slip under the action of loading, which intuitively reflects the weakening of the shear capacity of the soil body. Under the same dynamic shear strain, the dynamic shear modulus increases with the increase of dry density, but the growth rate is nonlinear reduction trend, the growth of dry density makes the proportion of particles in the unit volume of the soil body to improve, the compactness increases, the remoulded loess is subjected to the action of the dynamic loading of the internal friction between the particles to improve the resistance to deformation ability is significantly enhanced. Similarly, the dynamic shear modulus increases with the growth of the perimeter pressure, and it is more difficult for the soil body to produce relative lateral displacement due to the limiting effect of the perimeter pressure. With the increase of the depth of the loess roadbed, the perimeter pressure gradually grows, this lateral limiting effect is more and more obvious, so its structural strength is significantly improved, and it is relatively difficult to deform.

3.3 Damping Characteristics

In practical engineering, the study of the effect of the soil conditions and stress state of remoulded loess on the damping ratio of the soil λ can provide a reliable design basis for the construction of loess roadbeds and embankments. Damping ratio describes the hysteresis of the remoulded loess specimen in the process of reciprocal cyclic periodic vibration, which is mainly manifested in the characteristics of internal energy dissipation due to the deformation of the soil body in the action of vibration cyclic loading, and it is one of the important dynamic characteristics of the soil body. Its calculation formula can be expressed as:

$$\lambda = \frac{1}{4\pi} \frac{A}{A'} \qquad (4)$$

where A is the total area of the hysteresis loop; A' is the area of the right triangle enclosed by the origin, the strain axis, and the fixed point of the hysteresis loop. The change of damping ratio in the formula can also reflect the degree of attenuation of the vibration wave in the transmission process.

The scatter distribution characteristics of the damping ratio of remoulded loess under different conditions are shown in Fig. 4.

Comprehensive analysis of the distribution of the scatter plot of the damping ratio can be seen that the change rule and the dynamic shear stress–shear strain relationship curve of the trend is similar to the initial rapid increase in the development of the late slow mode, but the difference is that the scatter distribution of the damping ratio of the fluctuation of the stronger, more discrete, so Fig. 4 shows that the location of the data points is relatively chaotic, and the phenomenon is also a manifestation of the energy dissipation of the uncontrollable and non-quantifiable. This phenomenon also reflects the uncontrollable and non-quantitative nature of energy dissipation.

From Fig. 4a, it can be seen that the magnitude of damping ratio decreases with the increase of water content, by the influence of water, the cohesion between the remoulded loess particles decreases, the strength between the soil body decreases, so that it produces a larger deformation, so the energy consumed under the same dynamic shear strain decreases, and the damping ratio decreases. In Fig. 4b, the magnitude of damping ratio increases with the increase of dry density, the soil body per unit volume at larger dry density is denser, and the degree of association is enhanced, so the energy dissipated in the deformation of the soil body is increased, and the strength of the structure is relatively elevated under the same strain condition. In Fig. 4c, the damping ratio with the perimeter pressure change law is not obvious, its influence by the perimeter pressure is small, mainly because when the soil body in the consolidation process is compacted to a certain degree, the perimeter pressure continues to increase on the soil structure of the influence of the effect of the shrinkage of the soil body structure, so the low perimeter pressure is higher than the perimeter pressure under the conditions of the remoulded loess of the damping ratio of the change is more significant.

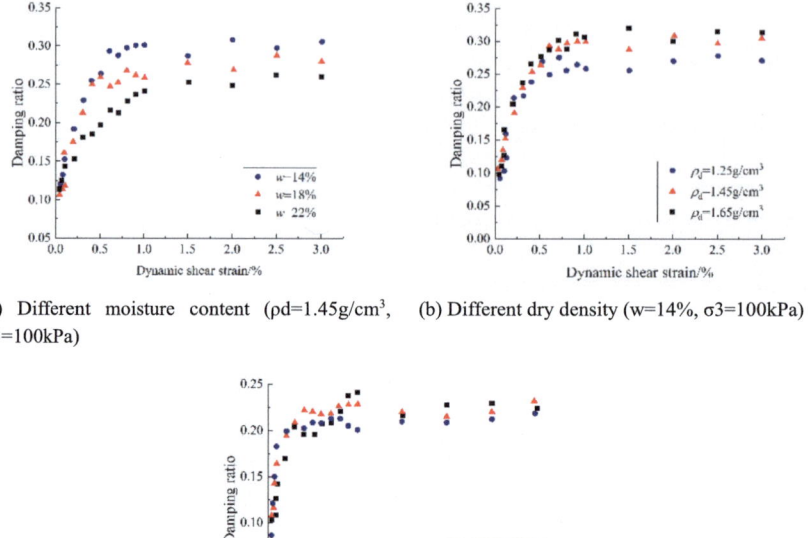

Fig. 4 Distribution of relationship between damping ratio and dynamic shear strain under different influencing factors

Comparative analysis of the relationship between the damping ratio and dynamic shear strain under different conditions shows that the damping ratio of the unsaturated remoulded loess of the specimens varies from 0.08 to 0.31. When the shear strain is about 0.5%, there is an inflection point where the growth of damping ratio changes from sharp to slow, and then it shows the development trend of small fluctuation up and down.

4 Microstructural Characterization

4.1 Qualitative Analysis of Remodelled Loess Microstructure

The study of the microstructural characteristics of remodelled loess is divided into two aspects: qualitative analysis and quantitative analysis, in which the qualitative analysis is mainly carried out with the aid of scanning electron microscope to magnify the observation of the particle units, pore morphology, and association forms of the soil body, comparing the structural characteristics under different conditions and reflecting the trend of the change of the microstructure of the soil body; quantitative analysis needs to be carried out through specific software to process the SEM images, in order to analyse the computational data. The quantitative analysis needs to be processed by specific software to calculate the data to analyse the changing law of soil structure and morphology. In this paper, the SEM images of remoulded loess specimens under 1000× magnification are selected for analysis, and the specific images are shown in Fig. 5.

SEM images of remoulded loess specimens under different vibration stresses were selected for comparative analysis, as shown in Fig. 5. It can be seen that the skeleton of the remoulded loess in this area is coarse, mostly in the form of set grains with clay-grain association, with more clay-grain debris attached to the surface of the granular structure and disorderly distribution. There is a large pore structure between the aggregate particles, which makes the skeleton of the particles clearly visible, and the pore structure is dominated by the intergranular pores, which also contains a certain number of connecting pores and penetrating pores. The contact between the particles is in various forms, mainly in the form of dislocation mosaic, showing the characteristics of point contact and surface contact, which together constitute the complexity of the pore structure, and the linkage mode is mostly lap, bridge state. The overall structure is complex dislocation, strong anisotropy, so the strength of the soil body is closely related to the structural characteristics.

Comparing the microstructure of the soil under different vibration stresses in Fig. 5a–c, it can be seen that along with the growth of vibration stress, the disturbance and compression of the sample loess particles are enhanced, which makes the particles be squeezed and compacted and visually describes the phenomenon that the area of the intergranular pore space is decreasing, the large pore space is deformed by the squeeze and splits into several small pore spaces, and the area of the intergranular pore space is continuously shrinking. The area of the intergranular pores is continuously shrinking, and some originally large pores are deformed and gradually split under external pressure, eventually forming several small pores.

  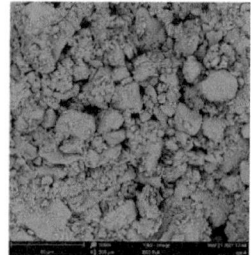

(a) Dynamic stress 20 kPa   (b) Dynamic stress 60 kPa   (c) Dynamic stress 100 kPa

Fig. 5 SEM images of remoulded loess under different vibration stress

At the same time, the agglomerates are affected by vibration compaction, the degree of cementation and adhesion increases, and the contact form of the particles between the skeleton is changed, more from point contact to surface contact, which confirms that the degree of adhesion between the particles is improved and presents more uniform and dense characteristics.

4.2 Quantitative Analysis of Remodelled Loess Microstructure Under Different Conditions

In the process of quantitative analysis of soil specimens, Image-Pro Plus 6.0 is used for binarization, during which the SEM images need to be adjusted by region selection, scale correction, grey scale adjustment, etc. The most important threshold selection is corrected by visual segmentation method by multiple people for many times to get the optimal calculated image and to improve the accuracy of the microstructural geometrical, aligned, and azimuthal features.

In the paper, fractal dimension, average shape coefficient, and oriented probability entropy were chosen as the quantitative research parameters, and the variation curves of pore parameters under different water content, dry density, and peripheral pressure conditions after cyclic vibration are shown in Fig. 6.

From the variation curves of quantitative parameters in Fig. 6, it can be seen that along with the increase of water content of remoulded loess, the structural morphology of soil intergranular pores under the same vibration loading has a large difference. The fractal dimension, average shape coefficient, and oriented probability entropy of the soil pores all show a decreasing trend change, which proves that the increase of water content makes the arrangement form of the remoulded loess pores from relatively disordered to orderly. The adsorption effect of viscous particles under high water content is more and more significant, which makes the soil particles gather and stick to each other, and the degree of grouping is increased, the arrangement of the particles is also tight, the shape of the intergranular pores becomes narrow and long, the form of the large pores is relatively reduced, and the number of the micro- and small pores is increased accordingly.

The effect of different dry densities on the microstructural parameters of remoulded loess has a certain regularity, along with the increase of dry density, the pore structure analysis of the soil body shows that the fractal dimension and the oriented probability entropy show a gradual increase, but this increase is gradually slowing down. When the dry density increases, the number of soil particles per unit volume increases significantly. This increase in the number of particles resulted in the larger pore structure between the original skeleton gradually being filled by an increase in the number of fine particles, thus contributing to the overall densification of the soil body, so that the large pores were divided into multiple small pore forms, the grouping degree of the original pores was reduced, and the distribution of the pores

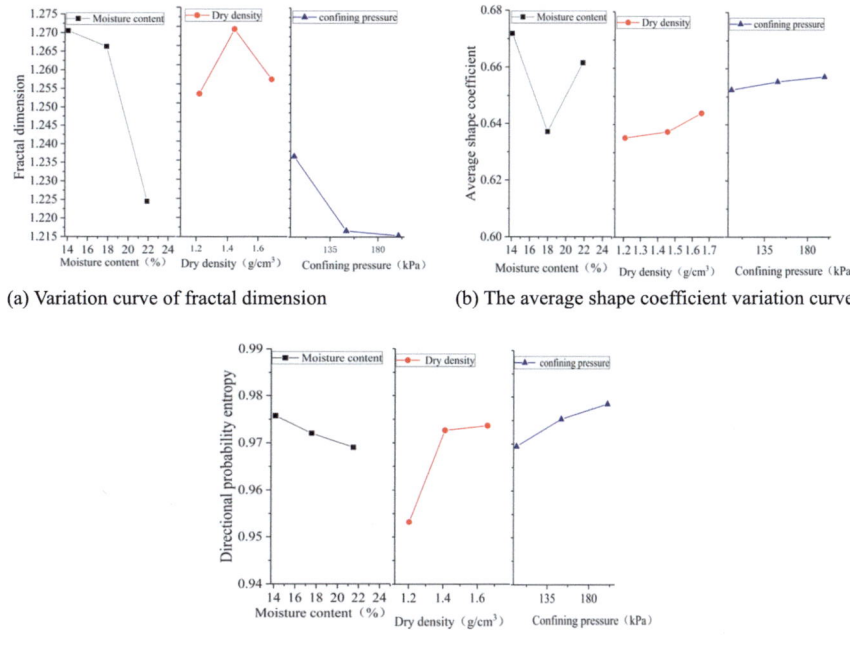

Fig. 6 Curves of quantitative parameters of pore structure of remoulded loess under different conditions

became relatively discrete and chaotic, with the degree of ordering reduced accordingly. At the same time, the average shape coefficient of the pores was improved, indicating that the irregular morphology of the original large pores was changed, and the filling of more particles made the pore structure become smaller and more compact, compared with presenting a more regular and rounded morphology. Therefore, the dry density has a promoting effect on the enhancement of the structural strength of remoulded loess, and the effect is more obvious.

It can also be seen from the variation of the curves shown in Fig. 6 that the effect of the perimeter pressure on the pore structure of the remodelled loess is weaker than that of the water content and dry density. Under the same conditions, along with the increase of dry density, the pore fractal dimension has a tendency to decrease under different perimeter pressures, but there is a phenomenon of recovery under higher perimeter pressures, which is due to the intergranular pores are more obviously constrained by the cyclic constraints during the process of low perimeter pressure increase, but after reaching a certain level of perimeter pressure conditions, the effect of this influence decreases, so if the perimeter pressure continues to grow, the variability of the pore sizes of the soil will be weakened, and the degree of variation will be no longer obvious. The average shape coefficient of the pores increases with the increase of the perimeter pressure, which indicates that the pore structure has no tendency to change narrowly when the soil structure is subjected to axial shear, and the increase of the perimeter pressure has a certain restraining effect on the axial dynamic shear stress, which restricts the transverse deformation of the soil specimen in the macroscopic direction. From Fig. 6c, it can also be seen that the directional probability entropy of the pores increases with the growth of the peripheral pressure, but the increase is small, and the overall value is large, reflecting that the intergranular pores of the remoulded loess before and after the dynamic loading are mostly disordered, and the distribution form is more dispersed.

Through the analysis of the microstructural characteristics of the roadbed remoulding loess, more comprehensively reveals the test dynamic shear stress–shear strain relationship, shear modulus and damping ratio with the soil moisture content, dry density and peripheral pressure change rule, and static and dynamic conditions are in the form of external action to perturb the internal structure of the soil, which in turn has a certain effect on the macroscopic dynamic characteristics of the soil. Therefore, the microstructural characteristics determine the macro-mechanical properties of the soil body, which should be paid attention to in engineering practice.

5 Conclusion

This paper investigates the correlation and influence between the dynamic and microstructural characteristics of the soil by combining macro and microscopic means on the remoulded loess under different water content, dry density, and perimeter pressure conditions by means of dynamic triaxial tests and electron microscope scanning tests. The main conclusions are as follows:

(1) The dynamic shear stress–shear strain relationship curves of the roadbed remoulded loess in Guanzhong region all show a hyperbolic form with rapid growth at the initial stage and slower development at the later stage; the dynamic shear modulus shows a tendency to decrease sharply with shear strain at the initial stage and develop smoothly at the later stage. With regard to the change rule of these two, they show a decreasing trend with the increase of water content; on the contrary, they show an increasing characteristic when the dry density and the perimeter pressure increase.

(2) Under the same conditions of dynamic shear strain, the damping ratio of remoulded loess will gradually increase with the increase of water content, but at the same time, with the increase of dry density and peripheral pressure, the damping ratio will gradually decrease, the changes of the damping ratio are relatively discrete, and the distribution is relatively messy. After the test, the damping ratio of remoulded loess showed a wide range of changes, with specific values between 0.08 and 0.31, showing a large increase in the initial period and an interval fluctuation trend in the later period.

(3) Under the influence of cyclic vibration, the microstructural parameters of remoulded loess pore space under different test conditions are as follows: with the increase of water content, the pore space arrangement tends to be orderly, and the morphology tends to be narrow and long, and its fractal dimensions, average shape coefficient, and directional probability entropy are all decreasing; with the increase of dry density, the structural strength of the soil body is enhanced, the denseness is improved, and the pore space is the opposite of the water content; with the increase of perimeter pressure, the fractal dimension tends to decrease, and the average shape coefficient and directional probability entropy are all slightly increasing. With the increase of perimeter pressure, the fractal dimension shows a decreasing trend, the average shape coefficient and the oriented probability entropy show a small increase trend, and the effect on the intergranular pore space is gradually weakened under high perimeter pressure.

This paper investigates the change of dynamic properties of remodelled loess under traffic dynamic loading, which can effectively provide data support for loess roadbed highway with high traffic flow, reinforce the loess roadbed under long-term vehicle dynamic loading to prolong the service life of the highway, and save the labour and economic costs. However, in this paper, only the role of the vehicle dynamic load under a single condition is studied, and the later can be studied and analysed for loess highway roadbed under multiple loads.

References

1. Shi G, Wang Y, Wu T et al (2020) Experimental study of urban pavement collapse under traffic load. J Undergr Space Eng 16(04):1202–1209
2. Na X, Yu J, Wang B (2019) Influence of additional foundation stress distribution on urban pavement collapse. Highway 64(02):8–12
3. Hu I-H, Bai Y, Xu H (2016) Analysis of the causes of urban road collapse and countermeasures for prevention and control in China in the past 10 years. Highway 61(09):130–135
4. Wang R, Wang L, Hu Z et al (2019) Influence of static bias stress induced by traffic loading on the dynamic properties of compacted loess. J Railway 41(07):110–117
5. Hu Z-P, Wang Q, Zhang Y-H et al (2018) Research on dynamic characteristics of compacted loess under stress and strain control. J Earthq Eng 40(06):1161–1167
6. Ma L, Zhang J, Liu Y (2018) Study on dynamic characteristics of remoulded loess in Shanxi roadbed under vehicle loading. J Earthq Eng 40(01):101–104
7. Liu W, Chen W, Yang F (2021) Influence of long-term effects of earthquakes on mechanical properties of loess. J Earthq Eng 43(04):965–976
8. Wang R, Hu Z, Wang L et al (2022) Dynamic response and long-term settlement of heavy railway embankment in loess area. J Railway Eng 39(01):7–12
9. Wang S, Zhong Z, Liu X et al (2019) Influences of principal stress rotation on the deformation of saturated loess under traffic loading. KSCE J Civ Eng 23(5):2036–2048
10. An L, Dang J, Zheng Z-H et al (2019) Experimental study on resonant columns for dynamic properties of remoulded loess in Guyuan. J Earthq Eng 41(04):949–956
11. Wang Z, Luo Y, Wang R et al (2010) Experimental study on dynamic shear modulus and damping ratio of in-situ loess in different regions. J Geotech Eng 32(09):1464–1469
12. Wen S, Zhang W, Zeng T (2019) Experimental study on dynamic shear modulus and damping ratio of in-situ loess in Haidong area. J Geotech Eng 41(S2):137–140

13. Wei L, Lu Y, Zhou Z et al (2019) Dynamic properties of unsaturated loess and their effects on site ground vibration parameters. J Geotech Eng 41(S2):145–148
14. Wang Q, Ma J, Ma H et al (2019) Experimental study on dynamic shear modulus and damping ratio of saturated loess. J Rock Mech Eng 38(09):1919–1927
15. Li S, Wang J, Li C et al (2021) Experimental study on dynamic properties and microstructural analysis of loess under cyclic loading. Arid Zone Resour Environ 35(12):142–149
16. Chen H, Jiang Y, Niu C et al (2019) Dynamic characteristics of saturated loess under different confining pressures: a microscopic analysis. Bull Eng Geol Env 78(2):931–944
17. Hu R, Li Z, Wang S et al (2000) Strength characteristics and structural change mechanism of loess under dynamic loading. J Geotech Eng 2000(02):174–181
18. Wang L, Dang J, Huang Y (2007) Quantitative microstructural analysis of seismic subsidence of loess. J Rock Mech Eng 2007(S1):3025–3031
19. Chen H, Shan W, Jiang Y (2021) Dynamic characteristics of Xianyang loess based on microscopic analysis: a quantitative evaluation. Bull Eng Geol Env 80(10):8247–8263
20. Wang L, Wang R, Hu Z et al (2018) Influence of large vibration sub-traffic loading on microstructural changes of compacted loess. Highw Traffic Sci Technol 35(06):37–44
21. Zhang R, To Y, Fei W et al (2006) Influence of vibration frequency on dynamic properties of saturated clayey soil. Geotechnics 2006(05):699–704
22. Xue Y, Ma X, Yang W et al (2020) Total deformation prediction of the typical loess tunnels. Bull Eng Geol Env 79(7):3621–3634

Open Access This chapter is licensed under the terms of the Creative Commons Attribution 4.0 International License (http://creativecommons.org/licenses/by/4.0/), which permits use, sharing, adaptation, distribution and reproduction in any medium or format, as long as you give appropriate credit to the original author(s) and the source, provide a link to the Creative Commons license and indicate if changes were made.

The images or other third party material in this chapter are included in the chapter's Creative Commons license, unless indicated otherwise in a credit line to the material. If material is not included in the chapter's Creative Commons license and your intended use is not permitted by statutory regulation or exceeds the permitted use, you will need to obtain permission directly from the copyright holder.

Analysis of Settlement Displacement of Tunnels Traversing Soft Soil Strata Containing Hazardous Gases

Jie He

Abstract The tunnels may traverse soft soil strata rich in hazardous gases such as methane, which poses several adverse effects on construction. These effects primarily include excessive deformation of the tunnel, leakage, or migration of hazardous gases causing soil deformation and settlement, thereby negatively impacting shield tunneling operations; and the ingress of hazardous gases into the tunnel, which could lead to fires or explosions upon encountering an ignition source, as well as endangering the health of construction personnel within the tunnel. Therefore, this paper focuses on the critical technical issues associated with the construction of the B Line tunnel in the presence of hazardous gases. Utilizing numerical simulations, the uneven settlement and reinforcement techniques are analyzed for shield tunneling through hazardous gas-bearing strata. Based on a comprehensive geological investigation report, a robust finite element model of the tunnel has been developed. The implementation of reinforcement measures and strata stabilization techniques has proven effective under identical leakage conditions. The maximum displacement of the tunnel strata was reduced from the initial 2.0–1.5 mm, resulting in a 25% decrease in overall strata displacement. The findings will provide scientific basis and technical support for similar construction projects.

Keywords Tunnel construction · Displacement · Hazardous gases · Numerical simulation · Soft soil strata

---

J. He (✉)
College of Civil Engineering, Tongji University, Shanghai, China
e-mail: hejie@tongji.edu.cn

China Railway 15th Bureau Group Co. Ltd., China construction sci-tech innovation group co., LTD, Shanghai, China

1 Introduction

In the depositional plains along the Yangtze River, southeastern coastal regions of China, including lacustrine, fluvial, and marine facies, there are widely distributed organic-rich silt and clay layers. These formations constitute shallow Quaternary gas reservoirs at depths ranging from several tens to over a hundred meters. These shallow hazardous gases, commonly referred to as "biogas" or "methane", are generated in anaerobic conditions where organic matter within the soil is decomposed by anaerobic bacteria through reduction processes. The gases migrate and accumulate in nearby sand lens bodies or at the tops of sand layers, forming typical biogenic gas reservoirs [1, 2]. Different scholars illustrate examples of shield tunneling projects in which hazardous gases were discovered during recent surveys or construction periods [2–7]. The tunnels may traverse soft soil strata rich in methane and other hazardous gases, which pose significant challenges to construction. These effects primarily include excessive deformation of the tunnel, leakage or migration of hazardous gases causing soil deformation and settlement, thereby negatively impacting shield tunneling operations; and the ingress of hazardous gases into the tunnel, which could lead to fires or explosions upon encountering an ignition source, as well as endangering the health of construction personnel within the tunnel. The presence of these gases can lead to soil deformation and settlement, adversely affecting the shield tunneling process.

The total length of the river-crossing tunnel project for the Shanghai Natural Gas Main Pipeline Network, from Chongming Island to Changxing Island to the No. 5 Gully LNG Station in Pudong New Area, is approximately 15.2 km. The tunnel is constructed using the shield tunneling method with an internal diameter of Ø3.4 m. This project focuses on the B Line tunnel, with the river-crossing section extending approximately 6931 m. The construction scope mainly includes the river-crossing shafts at Pudong Caolu and South Changxing Island, the tunnel itself (including segment fabrication), installation of concrete platforms within the tunnel, anchor piers and connecting pipelines and supports, anti-corrosion measures, installation of communication optical cable brackets, related facilities, and tunnel flooding. The primary geological challenge is the construction through methane-rich strata [8, 9]. The geotechnical investigation report for this project indicates that shallow methane in Shanghai can be found at depths ranging from 8 to approximately 30 m. There are two main layers of shallow methane: the first layer, found above 20 m, is located within the marine facies sediments formed during the maximum transgression period in geological history, typically appearing as lenticular bodies, with shell and shell sand layers serving as the main gas reservoirs, constituting the shallowest gas reservoirs in the city; the second layer, around 25 m deep, is in the upper marine facies sediments, controlled by the undulating top of the central terrestrial facies layer, with the main gas reservoirs being sand layers, generally appearing as lens or pinched-out bodies. According to the geophysical survey report, there are several areas in the North Channel of the Yangtze River with abnormal shallow gas distributions. During construction, portable combustible gas detectors revealed methane leakage in some boreholes. The presence of hazardous gases poses significant safety risks to the construction and operation of the project. During the shield tunneling process, encountering high-pressure gas-bearing soil layers could result in the sudden influx of gas and mud into the tunnel, causing collapse and deformation. When the construction reaches the gas-bearing sand layers, shallow gas could breach the tunnel floor under pressure. Disturbance of the strata during construction may lead to the migration of hazardous gases into the tunnel, and in the presence of poor ventilation and open flames, explosions could occur, severely threatening construction quality and the safety of personnel [7–9].

This project focuses on the key technical issues in the construction of the B Line tunnel in the presence of hazardous gases. By combining numerical analysis, the research aims to develop critical safety control technologies for tunnel construction in hazardous gas-rich soft soil strata. Specifically, the study investigates the uneven settlement and reinforcement techniques following methane leakage and establishes an emergency response plan for shield tunneling through hazardous gas-bearing strata. A reasonable finite element model is established based on the geological report. Appropriate constitutive models and geological conditions are selected, and suitable model parameters are chosen. Reasonable monitoring points are set within the model to simulate the deformation and displacement patterns of the strata following methane gas leakage in the tunnel crossing area. The strength of the strata is enhanced through foundation reinforcement methods. The deformation and displacement patterns of the strata are then simulated again following methane gas leakage in the tunnel crossing area, and a comparative analysis is conducted with the previous model. The outcomes will provide a scientific basis and technical support for similar construction projects.

2 FEM Model and Parameters

Figure 1 shows the two-dimensional model of a tunnel with a gas-bearing formation according to detailed geological surveys and hydrological reports, shown in Table 1. A fully finite element modeling model is used, and detailed information can be found in the reference [5]. For pore water, the initial pressure is assumed to be the hydrostatic pressure below the free water surface. Methane exists only within the trapezoidal gas reservoir, and it is assumed that gas pressure is uniformly connected throughout the reservoir, resulting in a linear variation in gas pressure along the vertical direction due to gas density, which is much lower than that of the water phase. Therefore, gas pressure within the gas reservoir is essentially constant. At the saturated–unsaturated interface, water phase saturation is 1, matric suction is 0, and gas pressure equals water pressure. Five vertical monitoring points (D20–D24) are set in the segment gas-bearing formation to explore the distribution patterns of soil displacement, as shown in Fig. 1.

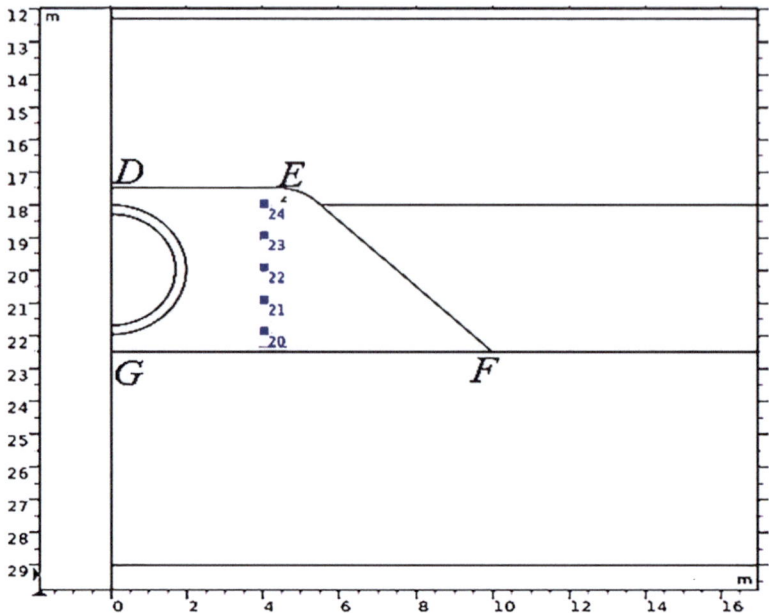

Fig. 1 Two-dimensional stratigraphic model of the gas-bearing formation and displacement monitoring points

Table 1 Stratum physical parameters

Geological parameter	Permeability κ (cm/s)	Density ρ (g/cm³)	Compression Modulus E_s(MPa)	Cohesion c(kPa)	Internal friction angle φ (°)	Porosity ratio
①2 Mud	2.0e−7	1.79	3.61	10.8	10.3	1.174
④ Silty clay layer	1.14e−7	1.69	2.6	14	12.0	1.373
⑤₁₋₁ Clay layer	1.15e−7	1.73	3.22	16	13.5	1.228
⑤₃₋₂ Interlayer of silty clay and clayey silt	6.03e−6	1.80	5.91	15	23.0	0.982
⑦₁ Medium sand interlayered with silty clay layer	5.90e−4	1.81	10.85	5	30.0	0.915

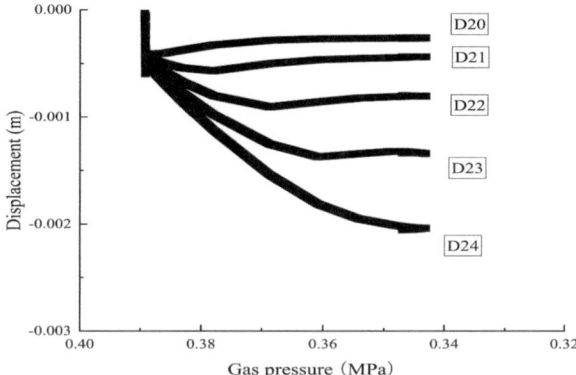

Fig. 2 Displacement changes at different monitoring points under methane leakage conditions

3 Influence of Gas Reservoir Leakage on Stratum Deformation

According to the initial conditions of the previous model, setting the height to 17.2 m, the impact of gas reservoir leakage is investigated on the displacement of the tunnel soil. The initial gas content of the reservoir is preliminarily set to 0.389 MPa. Based on the original gas-bearing reservoir model, over time, the gas in the reservoir leaks. The gas pressure after leakage is 0.347 MPa. In the gas-bearing formation of the segment, five vertical monitoring points (D20–D24) are uniformly set to explore the distribution patterns of soil displacement and other results. The displacement of the soil around the tunnel increases as the gas content in the reservoir leaks, leading to an overall subsidence trend in the formation. Details of the changes in stratum displacement are shown in Fig. 2. The displacement at different monitoring points in the gas-bearing formation shows a downward trend as the gas pressure in the gas reservoir leaks. Among these, the displacement change at monitoring point D24 is more significant. When the gas content in the reservoir decreases from the initial 0.389 MPa to 0.347 MPa, the displacement change at monitoring point D24 is approximately 2.0 mm.

4 Shield Tunneling Construction Monitoring

The displacement of the soil around the tunnel increases as the gas content in the reservoir leaks, leading to an overall subsidence trend in the formation. Details of the changes in stratum displacement are shown in Fig. 3. The displacement at different monitoring points in the gas-bearing formation shows a downward trend as the methane gas pressure in the reservoir leaks. Among these, the displacement change at monitoring point D24 is more significant. The displacement change at

Fig. 3 Displacement changes at different monitoring points under methane leakage conditions

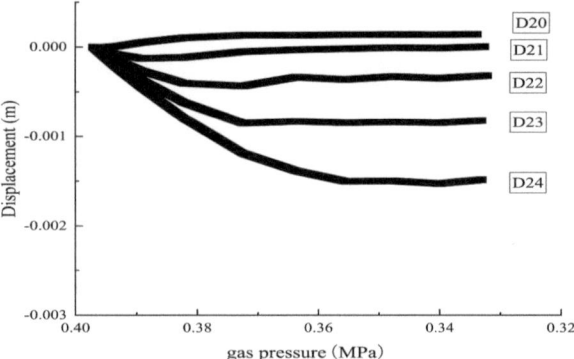

monitoring point D24 is approximately 1.5 mm. This indicates that when reinforcement measures are adopted, strengthening the formation, the maximum displacement of the tunnel formation under the same leakage conditions will be reduced by 25%.

5 Conclusion

The findings offer valuable insights for future urban tunneling projects crossing soft soil strata containing hazardous gases; emphasizing the importance of integrating geological considerations, the following conclusions are drawn:

(1) A reasonable finite element model of the tunnel is established according to the detailed geological investigation report.
(2) The initial gas pressure in the reservoir is set at 0.389 MPa, and the pressure after leakage is 0.347 MPa. The displacement variation at monitoring point D24 is approximately 2.0 mm.
(3) When reinforcement measures are implemented and the strata are reinforced, under the same leakage conditions, the maximum displacement of the tunnel strata decreases from the original 2.0 mm to 1.5 mm, reducing the overall strata displacement by 25%.

This study addresses the geological conditions of this tunnel in Shanghai. Future work can adapt specific constitutive models based on the actual engineering region, thereby making the simulation results more realistic.

References

1. Ayhan M, Aydın D, İmamoğlu MS et al (2019) Investigation of a methane flare during the excavation of the Silvan irrigation tunnel, Turkey. Bull Eng Geol Environ 78:2641–2652
2. Feng X, Ye B, Zhang X et al (2021) Analysis of the effects of shallow gas on a shield tunnel during leakage: a case study from the Sutong river-crossing gil utility tunnel project in China. KSCE J Civ Eng 25:2285–2299
3. Zhang F, Jia P, Wang Y et al (2023) Geological occurrence characteristics and engineering hazards of ultra shallow -buried gases in ningming basin. J Eng Geol 31:1105–1115
4. He J, Zhu H, Wei X et al (2023) Numerical and experimental analyses of methane leakage in shield tunnel. Front Struct Civ Eng 17:1011–1020
5. He J, Wei X, Yin M (2023) Study on the long-term performance of shield tunnel passing through the gas-bearing strata. Undergr Space 9:1–19
6. Guan Z, Wang Y, Cao Z et al (2020) Smart sampling strategy for investigating spatial distribution of subsurface shallow gas pressure in Hangzhou Bay area of China. Eng Geol 274:105711
7. Bilgin N, Balci C, Aslanbas A (2021) Case studies leading to the management of tunnel fire risks during TBM drives in an old coalfield. Tunn Undergr Space Technol 112:103902
8. Jiang X, Zhang Y, Zhang Z et al (2021) Study on risks and countermeasures of shallow biogas during construction of metro tunnels by shield boring machine. Transp Res Rec 2675: 105–116
9. Hu QJ, Tang S, He LP et al (2021) Novel approach for dynamic safety analysis of natural gas leakage in utility tunnel. J Pipeline Syst Eng Pract 12:06020002

Open Access This chapter is licensed under the terms of the Creative Commons Attribution 4.0 International License (http://creativecommons.org/licenses/by/4.0/), which permits use, sharing, adaptation, distribution and reproduction in any medium or format, as long as you give appropriate credit to the original author(s) and the source, provide a link to the Creative Commons license and indicate if changes were made.

The images or other third party material in this chapter are included in the chapter's Creative Commons license, unless indicated otherwise in a credit line to the material. If material is not included in the chapter's Creative Commons license and your intended use is not permitted by statutory regulation or exceeds the permitted use, you will need to obtain permission directly from the copyright holder.

Analysis of Deformation Influence of Gravel Shield Tunneling Under Existing Tunnel

Hongtao Pei, Rimei Han, Zezhan Shao, Shenqin Sun, Tianlu Xu, and Junmin Pan

Abstract In recent years, the rapid development of urban rail transit in China has led to an increasingly dense subway network. The construction of new subway tunnels frequently requires crossing many buildings in the city and existing infrastructure such as tunnels. Especially in the construction of Wumi Road Station of Kunming Metro Line 5, the construction phenomenon of crossing the existing structure is particularly significant. Based on this background, this paper uses the combination of model test and numerical simulation to explore the construction deformation law of shield tunneling under the existing tunnel in sandy cobble stratum of Kunming subway. After in-depth research on the three key factors of crossing angle, tunnel clearance, and formation loss rate, we have drawn the following conclusion: whether using a 60°oblique crossing method or a 90° straight crossing method when passing through the existing tunnel, their final impact on the surface and the existing tunnel is roughly the same. Further analysis reveals a significant negative correlation between the final deformation of the surface and existing tunnels, as well as the net distance between tunnels. That is, the larger the net distance between tunnels, the smaller the deformation; there is a positive correlation with the formation loss rate; that is, the higher the formation loss rate, the greater the deformation. With the increase of distance, the range of vertical deformation and surface settlement will also increase. In addition, there is a linear relationship between the formation loss rate and the final deformation value.

Keywords Existing tunnel · Gravel stratum · Model test · Numerical simulation · Construction deformation

H. Pei (✉) · R. Han · T. Xu · J. Pan
Xi'an Rail Transit Group Company Limited, Xi'an, Shaanxi, China
e-mail: 59241971@qq.com

Z. Shao · S. Sun
China Railway Xi'an Group Co., Ltd., Xi'an, Shaanxi, China

1 Introduction

With the continuous increase of infrastructure area, the contradiction between urban transportation resources and the speed of urban development has gradually become prominent, which has become a major problem in municipal construction. As the surface space is compressed due to the rapid development, which cannot meet the growing demand of urban road construction, more and more cities turn their eyes to underground and take the development of underground rail transit as the primary choice to relieve the urban traffic pressure [1–7].

In the actual situation of the shield tunnel underpass construction, the core mechanism that causes a series of deformation problems is due to the interaction between the new tunnel, the soil, and the existing structure. The soil bears the disturbance caused by the construction and transmits the displacement field generated by this disturbance to the existing structure [8]. As a common research method, model test plays an important role in exploring the deformation law of structure and formation in the construction process. For example, Zhu [9] et al. used the scale model test to explore the law of formation settlement with time and explain the effect of double-line tunnel superposition. Chen Renpeng et al. [10] used the self-made shield machine model to systematically study the potential impact of different buried depths on the boundary bearing force and the surface settlement of dry sand formation. Fang Qian et al. [11], based on the model test, deeply studied the influence mechanism of particle grading and other factors in the construction of sand formation and found that the change of particle size has a significant direct impact on the settlement value. Liu Xinjun et al. [12] combined with the engineering background of Nanjing Metro Line 5 through the existing Line 1, comprehensively elaborated the 3D spatial deformation characteristics of the orthogonal area by using model test and numerical simulation methods.

This paper takes the engineering practice of the shield under the existing Line 3 tunnel of Kunming Metro Line 5 as the background and carries out systematic research and analysis on the deformation of the shield under construction combined with model test and numerical simulation, hoping to provide reference for subsequent and similar engineering cases.

2 Research on Puncture Deformation Law Based on Model Test

2.1 Introduction of the Test Support Project

This chapter studies the potential impact on the existing line 3 tunnel and the surface subsidence of Kunming Metro Line 5. The simulation study analyzes the specific influence law of shield construction in the left and right lines of Line 5, especially when the tunnel of Line 5 (the buried depth of the vault is 23.5 m) intersects the existing tunnel of Line 3 (the buried depth of the vault is 14.9 m). In addition, the spacing between the left and right tunnels of Line 5 is precisely designed to be 8.8 m, while the left and right tunnels of Line 3 is 8.6 m. Note that the net distance between the two tunnels is 2.1 m and 2.2 m, respectively. When the shield machine of Line 5 tunnel crosses the tunnel of Line 3, the crossing angle between them is 62.8°, which brings some technical challenges to the construction process. In the actual construction, we use the "from left to right" order to ensure the safety and efficiency of the construction.

In order to simplify the test process and enhance the feasibility and efficiency of the study, we unified some parameters in the simulation test, setting the spacing between Line 5 and Line 3 to 8.7 m. At the same time, the spacing between the upper and lower tunnels was also simplified to 2.2 m. In addition, the crossing angle between the two tunnels was simplified and set to 60° to simulate the analysis. These adjustments are designed to ensure that the core objectives of the study are unaffected, while improving the convenience and efficiency of trials.

2.2 Determination of the Model Similarity Ratio

In the design of this model test, we fully referred to the research results of many scholars [12–15] and the similar theories mentioned above, aiming to comprehensively and comprehensively explore the influence law of multiple key indicators on the construction of shield tunneling through the method of dimensional analysis. These key indicators include geometric size (L), material bulk density (γ), compression modulus (E_s), and Poisson's ratio (μ). Based on these considerations, we formulated the similar parameters of the model trials and listed them in detail in Table 1.

Using the exponential method to set the geometric similarity constant and substituting the geometric similarity constant into the expression of stress, we derive the following dimensional relation:

$$[\sigma] = [F^a, l^b, \gamma^c, E^d] \tag{1}$$

Table 1 Expression table for each similar parameter of the model test

Name	Representation	Name	Representation
Material capacity parameters	$C_\gamma = \frac{\gamma_p}{\gamma_m}$	Stress parameters	$C_\sigma = \frac{\sigma_p}{\sigma_m}$
Compression modulus parameters	$C_{E_s} = \frac{E_{sp}}{E_{sm}}$	The strain parameters	$C_\varepsilon = \frac{\varepsilon_p}{\varepsilon_m}$
Poisson's ratio parameter	$C\mu = \frac{\mu_p}{\mu_m}$	Displacement parameters	$C_\delta = \frac{\delta_p}{\delta_m}$
Geometric parameter	$C_l = \frac{l_p}{l_m}$		

In each parameter dimension, you can obtain:

$$= [F^a L^b (FL^{-3})^c (FL^{-2})^d] \\ = [F^{a+b+c} \cdot L^{b-3c-2d}] \quad (2)$$

A comparison of the index refers to:

$$\begin{cases} a + c + d = 1 \\ b - 3c - 2d = -2 \end{cases} \quad (3)$$

By the above Formula (3):

$$[\sigma] = [Fl^{-2} \cdot \left(\frac{\gamma l^3}{F}\right)^c \cdot \left(\frac{El^2}{F}\right)^d] \quad (4)$$

The criterion equation is:

$$\frac{\sigma l^2}{F} = \phi\left(\frac{\gamma l^3}{F}, \frac{El^2}{F}\right) \quad (5)$$

There are similar criteria for this purpose:

$$\pi_1 = \frac{El^2}{F} \quad (6)$$

$$\pi_2 = \frac{\gamma l^3}{F} \quad (7)$$

$$\pi_3 = \frac{\sigma l^2}{F} \quad (8)$$

Similarly, the parameters in Table 1 into the displacement expression to obtain the relationship between the similarity constants as follows:

$$C_E = C_\sigma = C_\gamma C_l \quad (9)$$

$$C_F = C_E C_l^2 \quad (10)$$

After fully considering the purpose and conditions and space in the test, we selected the following similar parameters: geometric similarity constant C_l is set to 60, and the compression modulus and stress similarity constant C_E and C_σ are set to 60, to ensure that the physical quantity in the same proportional relationship between the model and the prototype. Meanwhile, meanwhile, because the material density is less sensitive to the model scaling, we set the similarity constant of the

Table 2 Table of similar ratios of each parameter

Parameter	Displacement	Modulus of elasticity	Poisson ratio	Unit weight	Stress	Meet an emergency	How much
Ratio of similitude	60	60	1	1	60	60	60

material density C_γ to 1; that is, the model is consistent with the material density in the prototype. The parameters determined are similar as shown in Table 2.

2.3 Design and Manufacture of the Test Model Box

In order to further study the settlement law of the existing tunnel and the ground surface, we set the long side of the model box as a section perpendicular to the excavation direction of the tunnel, while the short side is consistent with the excavation direction of the tunnel. To minimize this effect, we are careful to limit the effect of tunneling on the surrounding strata to an area of 3 to 5 times the tunnel diameter (D). Such a design ensures that we can more accurately observe and analyze the dynamic changes of the settlement curve, thus providing a deep understanding of the formation response mechanism during tunneling. After careful planning and calculation, careful planning, and accurate measurement, we finally determined the size of this test model to be 1.80 m × 1.00 m × 1.20 m, ensuring its high degree of accuracy and reliability.

After a thorough study of the design documentation and reference drawings (as shown in Figs. 1 and 2), we found that a 250 mm spacing was used for the Line 3 tunnel and a 288 mm spacing for the new Line 5 tunnel. In the model simulation, the center of the existing Line 3 tunnel is 313 mm away from the top of the model, while the center of the new Line 5 tunnel is farther away from the top of the model, specifically 456 mm. These precise dimensional parameters are the key factors to ensure the safe and efficient construction of shield construction.

The overall design of the model box takes into account the quality and stiffness, ensures its strong carrying capacity and stability (Fig. 3), and thus effectively reduces the error caused by the deformation of the box. And a working window is specially left on the front side of the model box, as shown in Fig. 4.

2.4 Monitoring Items and Layout Scheme

In order to comprehensively explore the impact of surface subsidence in the construction process of double-line shield, this experiment was carefully planned and specially set up two measuring lines A and B. The two test lines are arranged in

Fig. 1 Front-side diagram

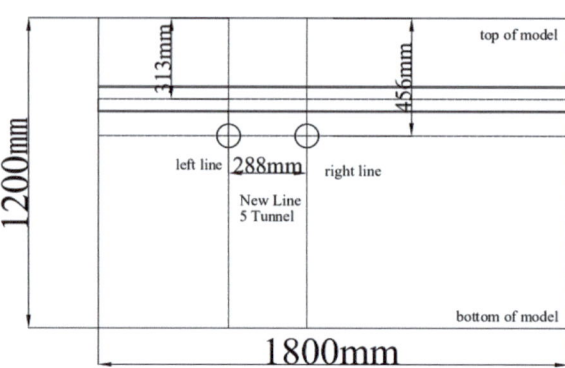

Fig. 2 Top surface

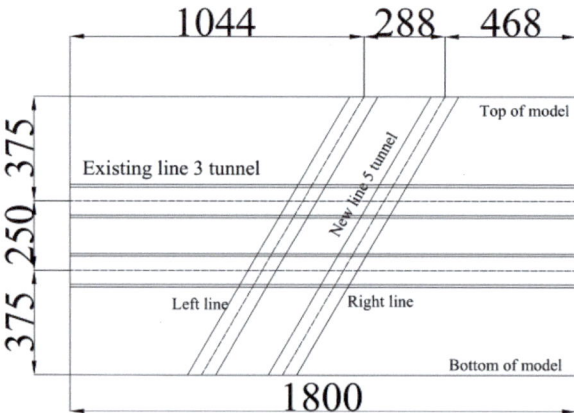

Fig. 3 Model test box (laboratory instrument, obtain a license)

Fig. 4 Replacement window for different working conditions (laboratory instrument, obtain a license)

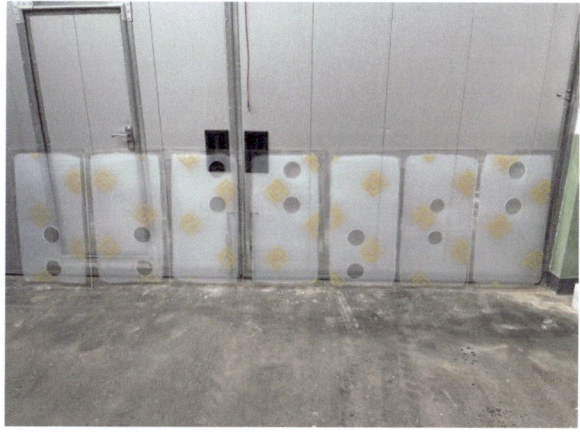

strict accordance with the criteria of being parallel to the tunnel of Line 3, aiming to ensure that the collected data is both accurate and highly representative. During the detailed layout process, line A is accurately set 218 mm from the long edge of the front edge of the model box. We place the measuring points numbered AD1 to AD11 from left to right. At the same time, the measuring line B is accurately placed directly above the left line of Line 3362 mm away from the long edge of the front of the model box. Similarly, we also set the measuring points numbered BD1 to BD11 on this line from left to right. This well-designed layout and numbering method not only provides a clear reference for the subsequent surface subsidence monitoring, but also provides an orderly reference basis for data analysis.

In terms of the measuring point layout, we followed the symmetrical layout principle based on the center line of the tunnel of Line 5. We specially put the no. 6 measuring point accurately in the central position, as the core reference point of the whole measuring line. Then, we used this central point as a benchmark and evenly distributed five measuring points on the left and right sides, with the aim of comprehensively and meticulously monitoring the subsidence of the surface around the tunnel. The distance between each adjacent measuring point was precisely set to 132 mm to ensure the accuracy and consistency of the data. In conclusion, each measuring line contains 11 sets of measuring points. The specific layout can be shown in Figs. 5 and 6.

In this test, we specially arranged two measuring lines, respectively, located at the left and right axis of Line 3. In order to accurately capture the settlement change, these two measuring lines are arranged at the vault position of Line 3, with the center line of Line 5 as the reference axis. The adjacent spacing between the measurement points is accurately controlled as 132 mm to ensure data continuity and accuracy. There are 11 measuring points set on each measuring line, shown in Figs. 7 and 8.

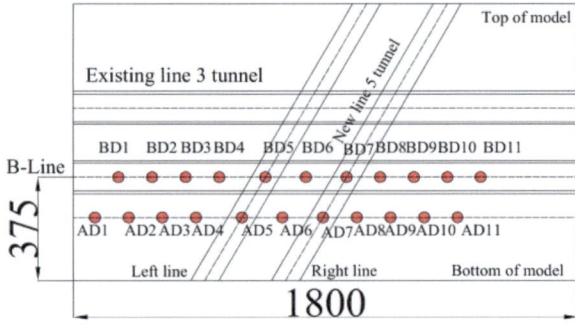

Fig. 5 Arrangement diagram of surface measuring line and measuring points

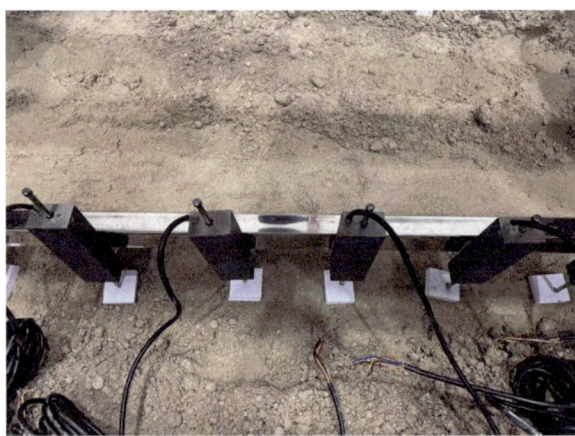

Fig. 6 Surface survey line layout diagram (laboratory instrument, obtain a license)

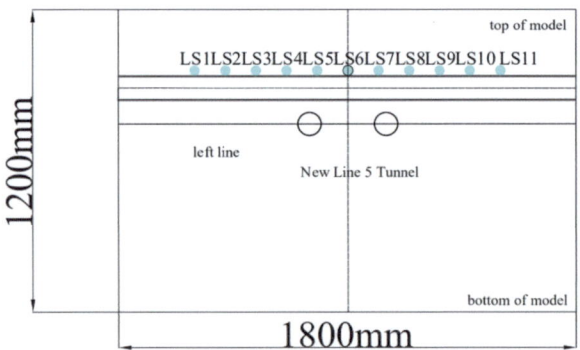

Fig. 7 Layout diagram of tunnel measuring line and measuring point of the existing left line

2.5 Model Test Protocol

In order to explore deeply the formation loss rate, the formation loss rate caused by the crossing angle, and how the net distance between the tunnels affects the surface

Fig. 8 Layout drawing of the existing left line tunnel measurement line (laboratory instrument, obtain a license)

Table 3 Model test conditions

Number	Through the angle	Formation loss rate (%)	Clear distance (D)
1	90°	0.75	0.4
2	60°	1.5	0.4
3	60°	2.0	0.4
4	60°	0.75	0.4
5	60°	0.75	0.8
6	60°	0.75	1.2

settlement and the deformation of the existing tunnels, we carefully planned six sets of test conditions. Detailed settings of each group working condition have been arranged in Table 3.

3 Analysis of the Model Test Results

3.1 Analysis of the Land Surface Settlement Results

Figures 9 and 10 intuitively shows A variety of test conditions, double tunnel mining work after the completion of A, B two line recorded by the surface subsidence data, these data reveals the specific situation of the settlement for us, help us to further analyze and understand the influence of tunneling on the surface, for we analyze the influence of tunneling on the surrounding environment provides an important data support.

Fig. 9 End of double-line excavation-A measuring line

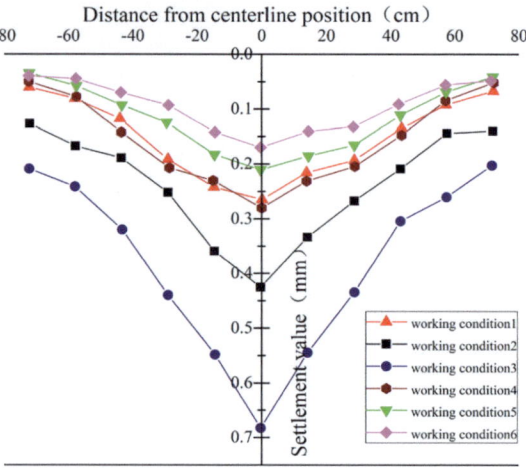

Fig. 10 End of double-line excavation-B measuring line

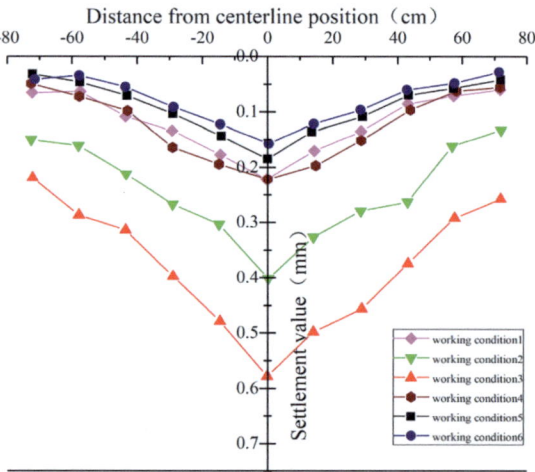

The following conclusions can be drawn from the drawing:

(1) When comparing the surface settlement data after excavation, the settlement amount of each point of measurement line A is generally significantly more than that of the corresponding point of measurement line B. This difference stems from the strong resistance of the existing tunnel to stratigraphic disturbances, which actually strengthens the original formation and effectively limits the diffusion of the displacement field, and thus has relatively little surface subsidence in the line B area.

(2) It can be seen from the figure that the settlement peak of measurement line A and measurement line B both converge at the measurement points near the

center line of the new tunnel. The settlement is negatively correlated with the distance from the center line; that is, the closer to the center line, the settlement and settlement speed increase. This surface settlement pattern presents a similar "V" groove, which intuitively reveals that in the process of tunnel excavation, the influence on the surrounding soil mainly focuses on the adjacent area of the excavation axis. That is, with the gradual narrowing of the distance from the excavation point, the disturbance degree of the soil is gradually enhanced. The effect is mainly limited to approximately 5 times the tunnel diameter (5D) from the excavation center.

(3) The test results show that under the condition of condition 3, the cumulative settlement of each measuring point of line A and B reaches the peak, while the trough is in condition 6. Comparing conditions 1 and 4, it is found that when the angle increases from 60° to 90°. The emergence of this settlement pattern is largely due to the parallel arrangement of the measuring lines A and B along the long side of the model box. In particular, at the crossing angle of 60°, the contact length between the measuring line and the tunnel is extended, so that the influence of the tunnel excavation on each measuring point becomes more significant and prominent.

(4) By comparing the data curves of working conditions 2, 3, and 4, we find that after fixing the crossing angle and the tunnel distance, the increase of the formation loss rate leads to the increase of the settlement value of the measurement points on the two measurement lines A and B. The mechanism behind this increase is that the increase of the formation loss rate leads to the expansion of the gap between the soil and the tunnel segment, which weakens the supporting conditions of the surrounding rock, thus aggravating the degree of surface subsidence.

(5) After comparing the data curves of working conditions 4, 5, and 6, we draw an obvious conclusion: when the formation loss rate and crossing angle are constant, the increase of net distance will lead to the decrease of the settlement of the measuring point near the center line. With the increase of the net distance of the tunnel, the position of the existing tunnel is relatively moved down, and the soil layer in the middle is thickened. This thickened soil layer can form a more obvious soil arch effect, which effectively strengthens the support ability of the stratum and then reduces the disturbance and subsidence of the tunnel excavation on the surface.

3.2 Analysis of the Vertical Displacement Results of the Existing Tunnel

Figure 11 clearly shows the cumulative vertical displacement changes of the existing tunnel vault on the left line after the completion of the double-line tunnel excavation under different test conditions. Accordingly, Fig. 12 intuitively reflects the cumulative vertical displacement trend of the existing tunnel vault on the right line under the same working conditions.

The following conclusions can be drawn from the figure above:

(1) After a detailed analysis of different construction conditions, we find that the final displacement values of the two tunnels are very close and within a similar numerical range. It is worth noting that in condition 6, the settlement curve is obviously "U" type, while in the rest of the construction conditions, the displacement curve is more "W" type.

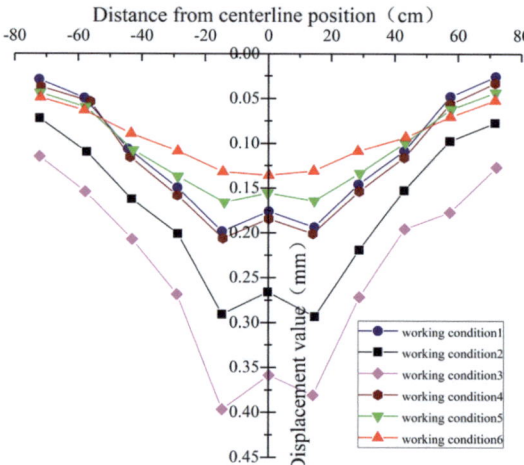

Fig. 11 Cumulative displacement curve of the left

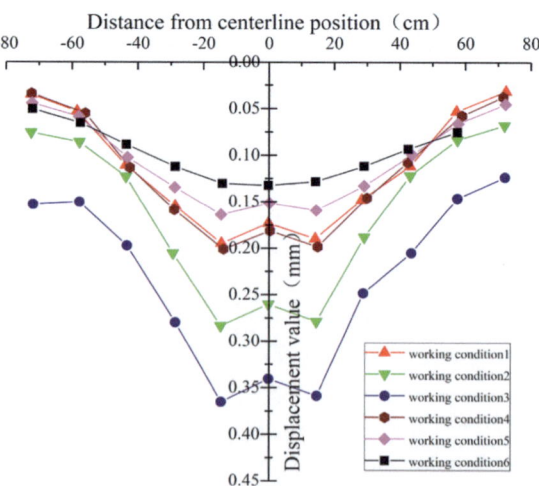

Fig. 12 Cumulative displacement curve of the right line

(2) After observing the similar settlement rules and similar settlement values of the existing tunnels on the left and right lines, we decided to focus on each working condition of the left line for detailed analysis. By comparing the displacement curves of working conditions 1 and 4, we observed two important phenomena: first, the maximum displacement value of both is concentrated in the LS5 measuring point. When the double-line tunnel excavation is completed, the cumulative vertical displacement change of the existing tunnel vault on the left line is 0.195 mm, while the right line is 0.200 mm.

(3) After comparing the curve data of working conditions 2, 3, and 4, it is found that the maximum displacement value of the vault is concentrated in the LS5 measurement point under all working conditions. Specifically, in condition 2, the LS5 displacement is 0.284 mm; in condition 3, the displacement increases to 0.388 mm; in condition 4, the LS5 displacement is 0.203 mm. It shows that if the factors are the same, the increase of the formation loss rate will directly lead to a significant increase of the cumulative settlement value of the existing tunnel.

(4) After the detailed curve comparative analysis of conditions 4, 5, and 6, we draw the following conclusions: in conditions 4 and 5, the maximum displacement of the vault appears in the measuring point of LS5, that is, the central axis of the left line of the new tunnel. Specifically, the displacement value of condition 4 is 0.202 mm, while the displacement value of condition 5 is 0.163 mm. This result shows that the displacement on the left central axis of the new tunnel is significant under similar construction conditions. However, in working condition 6, we observed a change in the position of the maximum displacement, transferred to the LS6 test site, the center line of the existing tunnel. The displacement value for this point is 0.132 mm. This change indicates that the center line of the existing tunnel may be more affected under specific construction conditions (such as the change of crossing angle) and requires special attention during the construction process.

3.3 Establishment of the 3D Numerical Simulation Model

Focusing specifically on the three core factors of tunnel net distance, crossing angle, and formation loss rate, we systematically analyze their key effects on the shield construction process and accurately evaluate the extent of these effects. This study provides an important basis for us to better understand the deformation mechanism during the shield construction process. In view of this, to address such problems more efficiently and comprehensively assess the construction impact, we plan to introduce numerical simulation methods as a validation and complementary means to the experimental results.

In order to ensure the accuracy of the simulation results and minimize the potential impact of the boundary conditions, based on the principle of Saint-Vernan, a relatively large simulation boundary range, that is, 3 to 5 times the outer diameter of the shield

Fig. 13 Numerical simulation calculated model figure

(a) Overall model diagram

(b) Tunnel, relative position diagram

machine as the boundary size of the simulation. At the same time, according to the size setting of the previous chapter model test, we carefully determined the size of the FLAC 3D 3 D model as 106 m 62 m 70 m to ensure that the simulation results can accurately reflect the deformation in the actual engineering.

In the model setting, we adopt the Mohr–Coulomb criterion to describe the strength characteristics of each soil layer. At the same time, the lining and isogeneration structures are regarded as isotropic ideal elastomers as shown in Fig. 13.

3.4 Layout Scheme of Surface Survey Line and Existing Tunnel Survey Line

On the basis of previous research, we realize that existing tunnels have some mitigation effect on surface subsidence. In order to further explore this influence mechanism and according to the research practice of selecting the most unfavorable conditions in engineering practice, this section will specifically study the surface subsidence deformation of the existing tunnel.

According to the results of the previous section, we observed that the left and right lines of the existing tunnel were similar in the final settlement, and the settlement change law during excavation also showed a high degree of consistency. In order to simplify the research content and avoid repeated analysis, we decided to study the left line of the existing tunnel as the main measurement line in the following discussion, so as to deeply analyze the influence of different factors on the vertical deformation of the existing tunnel.

4 Analysis of the Numerical Simulation Results

4.1 Simulation Analysis of the Surface Impact

As shown in Fig. 14, the adjustment of the tunnel net distance did not change the "V" pattern unique to the final settlement curve of the surface. It is worth noting that directly above the axis of the new shield tunnel is the most significant deformation of the existing tunnel. Through in-depth data analysis, we found that when the net distance is set to 0.2D, the excavation of the tunnel leads to be within 10.5D; when the

Analysis of Deformation Influence of Gravel Shield Tunneling Under … 197

net distance increases to 3D, this influence range is significantly expanded to about 12.3D. This obvious trend shows that as the net distance of the tunnel increases, the influence range of the surface transverse deformation will increase accordingly.

As shown in Fig. 15, when examining the influence of different net distance conditions on surface settlement, we observed that the maximum settlement value of the surface reached 17.64 mm when the net distance was 0.2D, while the minimum settlement value decreased to 6.78 mm when the net distance was expanded to 3D. The final surface settlement curve clearly shows a trend: the surface settlement decreases significantly as the net distance increases, but the rate of reduction gradually slows down. Importantly, even under the most unfavorable conditions, the surface settlement value does not exceed the specified 20 mm warning value. This finding shows that in the context of dependent engineering, the impact of changes in net distance conditions on surface settlement is in a controllable state and will not cause safety risks.

As shown in Fig. 16, for the underpass project discussed in this paper, we observed a remarkable phenomenon: the maximum settlement point in the surface survey line

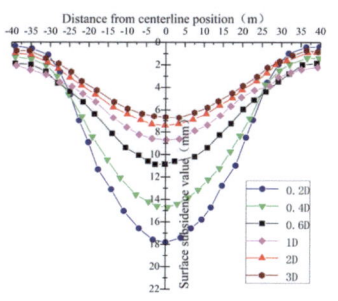
(a) Final settlement value curve of different net distance surface measurement line

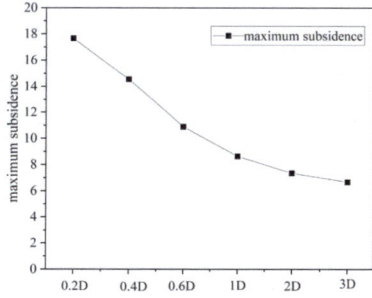
(b) Change curve of the maximum settlement value

Fig. 14 Decurve of surface line under different net distance

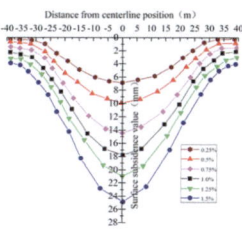
(a) Final settlement value curve of the surface measurement line for different formation loss rate

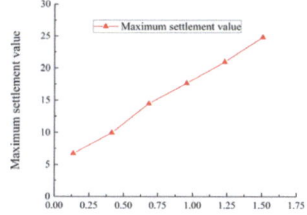
(b) Change curve of the maximum settlement value

Fig. 15 Decurve of surface line under different formation loss rates

is always located precisely directly above the center line of the two new tunnels. This observation shows that the "V" shape of the final surface settlement value curve remains unchanged regardless of how the formation loss rate is adjusted. Further analysis shows that there is an approximate linear growth relationship between the formation loss rate and the surface settlement. The detailed analysis shows that under the condition of strictly controlled formation loss rate of 0.25%, the ground surface transverse deformation caused by the excavation of the double-line shield tunnel is mainly concentrated in the range of 9.7D. However, as the formation loss rate climbs to 1.5%, the influence boundary of the ground surface transverse deformation extends outward to an area of approximately 12.1D. This data change intuitively reveals a rule: with the increase of the formation loss rate, the range of influence of the surface transverse deformation will gradually expand.

In the deep study of the influence of formation loss rate on surface subsidence, we reveal a remarkable law: with the increase of formation loss rate, the final surface subsidence also increases significantly. Specifically, when the formation loss rate climbs to 1.5%, the maximum surface subsidence value is 24.66 mm; otherwise, when the formation loss rate remains low at 0.25%, the minimum surface subsidence value is only 6.75 mm. It is particularly noteworthy that once the formation loss rate η reaches or exceeds the critical value of 1.25%, the formation settlement value will exceed the early warning range, which poses a potential threat to the safety and stability of the project. Therefore, in the actual engineering operation, the formation loss rate must be strictly controlled to ensure that it always remains within the safety threshold, so as to ensure the smooth progress and long-term safety of the project.

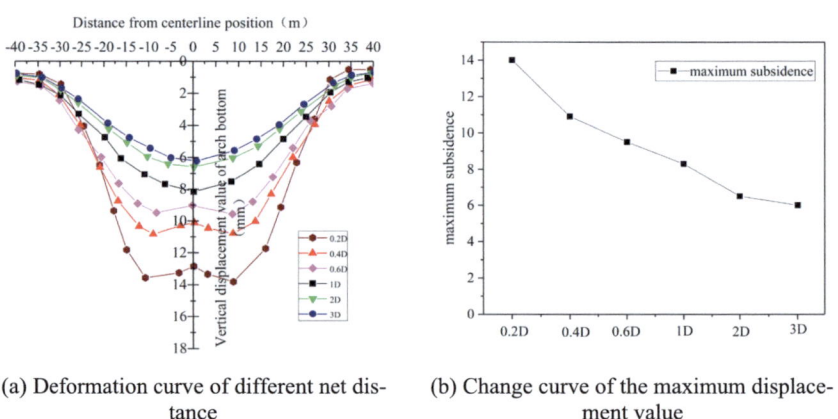

(a) Deformation curve of different net distance

(b) Change curve of the maximum displacement value

Fig. 16 Deformation of existing tunnels under different net distance conditions

4.2 Simulation Analysis of the Effects on Existing Tunnels

For the underpass engineering under specific formation conditions, the tunnel net distance has a significant influence on the deformation characteristics of the existing tunnel. Specifically, the deformation curve of the existing tunnel shows a typical "W" pattern when the net distance d is between 0.2D and 1D, and when the net distance d is greater than or equal to 1D. A deeper analysis shows that the adjustment of the tunnel net distance is directly related to the influence range of the transverse deformation of the existing tunnels. Especially in the net distance of 0.2D, the construction of the double-line shield tunnel will mainly concentrate the transverse deformation of the existing tunnel within a range of about 9.7D. However, as the tunnel net distance increases to 3D, this range of influence expands significantly extended to a region of about 11.2D. This obvious change trend clearly shows that the increase of the tunnel net distance will directly lead to the gradual increase in the influence range of the transverse deformation.

To study the specific effect of the net tunnel distance on the vertical displacement of the existing tunnel in detail, we focused on analyzing the data of the maximum vertical displacement data of the tunnel arch bottom under different working conditions, as shown in Figs. 17, 18 and 19. After careful analysis, we draw the following important conclusions: as the net distance of the tunnel gradually increases from 0.2D to 3.0D, the vertical displacement value of the characteristic point of the arch bottom above the left line axis shows a significant downward trend. Specifically, the maximum vertical displacement value was gradually reduced from the initial 13.91 mm, until reaching a minimum value of 5.62 mm. Of particular concern is that the final displacement values of the existing tunnel change significantly in the narrow range of net distance of 0.2D to 0.4D, and these values exceed the warning value of tunnel settlement control. This finding warns us that additional engineering control measures must be taken to effectively limit the vertical displacement of existing tunnels and ensure construction safety and engineering stability. Although the vertical displacement value decreases when the net distance exceeds 2.0D, there is still a significant displacement phenomenon. This result clearly shows that there is a limitation of relying solely on increasing the tunnel net distance to control the vertical displacement of existing tunnels, and that other and more refined engineering measures are needed to achieve more efficient displacement control.

From the detailed observation data in Fig. 17, we learn a key phenomenon: the change of the formation loss rate has little significant effect on the shape of the vertical displacement curve of the existing tunnel arch bottom, which always maintains the typical "W" type characteristics. Moreover, we note a remarkable rule that the maximum deformation of the existing tunnel is always precisely located directly above the axis of the new shield tunnel. Further analysis reveals a clear correlation between the formation loss rate and the influence range of the transverse deformation of the existing tunnel. Specifically, when we control the formation loss rate at 0.25%, the impact of the excavation of the double-line shield tunnel on the transverse deformation of the existing tunnel is roughly limited to the range of 9.8D. However,

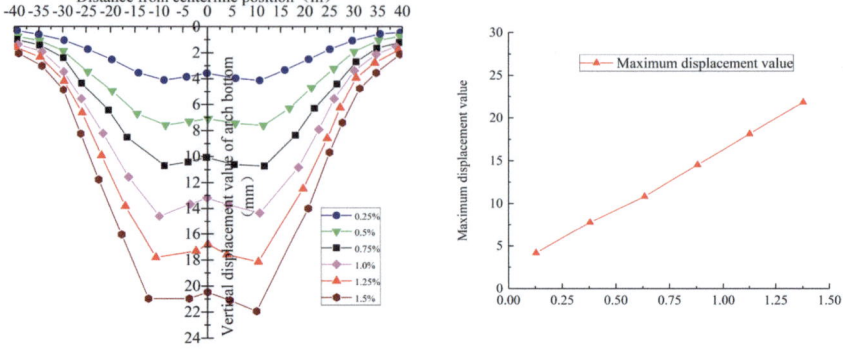

(a) The deformation curve of the loss rate of different strata

(b) Change curve of the maximum displacement value

Fig. 17 Deformation of existing tunnels under different formation loss rates

when the formation loss rate rises to 1.5%, this influence range increases significantly to about 12.1D. This data change directly proves that the increased formation loss rate will lead to a corresponding expansion of the influence range of the transverse deformation of the existing tunnel.

When studying the effect of the formation loss rate on the vertical displacement of the existing tunnel, we noticed that the vertical displacement value of the existing tunnel gradually increased from 3.98 mm to 10.56 mm in the range of 0.25% to 0.75%. This continuous and stable growth pattern indicates that the degree of deformation of the existing tunnel has not reached the predetermined safety threshold when the formation loss rate is controlled below 0.75%. Therefore, in this interval, the shield construction does not have a direct and significant impact on the safety of the existing tunnel.

However, the situation turns when the formation loss rate η touched or exceeded the limit of 0.75%. At this time, the maximum vertical displacement value of the existing tunnel has exceeded the predetermined safety warning value. In particular, when the formation loss rate climbs to 1.5%, the maximum displacement value of the existing tunnel is up to 21.68 mm, which significantly exceeds the control standard. When the parameters in shield construction are improperly selected, such as the soil bin pressure and grouting pressure is low, or the adjustment of shield posture is unreasonable, it may lead to a sharp increase in the formation loss rate, thus posing a serious threat to the safe operation of the existing tunnel.

5 Comparative Analysis of Numerical Simulation and Model Test

5.1 Comparative Analysis of Land Surface Subsidence

When evaluating the situation of surface settlement, we selected the data of model test line A to compare with the surface line data obtained by numerical simulation. To present these data more intuitively, we plotted them as shown in Fig. 18

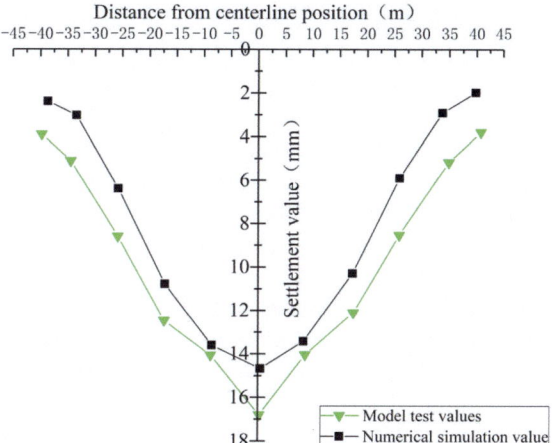

Fig. 18 Comparison of the final surface displacement values

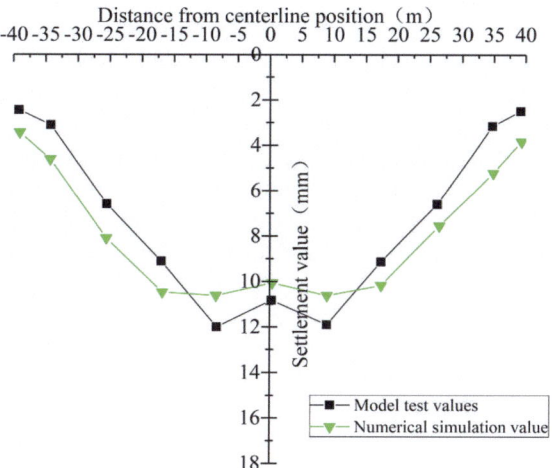

Fig. 19 Comparison of the final displacement value of the left line of the existing tunnel

By observing the data curves in the figure, we can clearly see that the model test line A is highly similar to the numerical simulated surface line in morphology, which reflects the consistency between the two. Specifically, the maximum settlement of the model test line A is 16.67 mm, while the maximum settlement of the numerical simulated surface line is 14.35 mm. Although the results of the model test are slightly smaller than those of the numerical simulation, the difference between the two is only small, less than 20%, which basically meets the relevant accuracy requirements.

From the diagram, the model test line A and the numerical simulated surface test line show similar characteristics in the settlement trend, but there is still a subtle difference between the two. This difference is mainly due to their different design considerations in the simulation methods. In the numerical simulation, to simplify the calculation, the soil is regarded as an elastic material, in the model test. Moreover, the operations involved in the test, such as repeated filling, digging, are difficult to ensure completely consistent conditions for each test, which inevitably introduces additional errors. In contrast, the numerical simulations can avoid the errors caused by such human factors. Comparative analysis of the existing tunnel deformation.

In view of the deformation of the existing tunnel, we specially selected the comparative analysis of the left vault data in the model test and the data of the left vault of the existing tunnel in the numerical simulation, and drew these data points as a curve, as shown in Fig. 19

According to the comparison of the data in Fig, the shape of the deformation curve of the left line vault in the model test and the existing tunnel arch of the left line vault in the numerical simulation is almost identical, showing a small difference. Specifically, the maximum displacement value measured by the model trial is 12.03 mm, while the numerical simulations give results of 10.60 mm. Despite the subtle difference between the two values, it is noteworthy that this difference represents only 11.95% of the maximum displacement value of the model trial, well below the threshold of 20%. Therefore, we can think that this result still meets the relevant requirements and further verify the validity and high consistency of the model test and the numerical simulations in predicting the tunnel deformation.

The difference in the results between the model test and the numerical simulation is mainly due to the difference in the excavation method and the selection of tunnel materials. In the model test, due to the influence of the actual pulling effect and the lack of real grouting process simulation, the maximum displacement value of the left line vault is slightly higher than the result of the numerical simulation. This difference reflects the possible dynamic effect and uncertainty under the real construction conditions.

6 Conclusion

In this paper, the three angles of crossing angle, net spacing, and bottom loss rate are studied, and the following rules of existing tunnel and surface settlement caused by shield tunnel crossing are obtained.

(1) When the formation loss rate gradually climbed from 0.25% to 1.5%, we observed that the vertical deformation of the existing tunnel and the influence of surface subsidence on the transverse and longitudinal sections were expanded. This increase is not only reflected in the expansion of the deformation range, but also in the significant increase of the maximum deformation value and the obvious acceleration of the deformation rate.
(2) With the gradual increase of the net distance from 0.2D to 3D, the vertical deformation of the existing tunnel and the surface settlement in the transverse and longitudinal sections have been expanded. During this change, the maximum deformation value decreases gradually, and the rate of deformation slows down.
(3) After the time-course curve analysis, we observed that in the left and right lines of the new tunnel, the first excavation caused significant disturbance effects on the surface and the existing tunnel.
(4) Comparing the data of numerical simulation and model test, we find that although there are slight differences between the two values, there is a high degree of consistency in the key data points and the changing trends shown.
(5) Under the condition of simulating the actual construction environment, we detected that the surface settlement value reached 14.24 mm and found that the settlement value of the existing tunnel was 10.53 mm, which has exceeded the warning standard but has not yet constituted a danger.

References

1. Yu L, Zheng X, Song Q et al (2017) Chinese mainland The evolution of traffic network accessibility. Geogr Res (12)/2
2. 2017–2018 Planning and construction projects in 2018. High-speed Railway and Rail Transit exclusive edition, 2017:20–24
3. Han B, Li Y, Lu F et al (2022) Summary of Operation Statistics and Analysis of World Urban Rail Transit in 2021. Urban Fast Rail Transit 35(1):7
4. Jia L, Wang J, Li X et al (2015) The control index of Beijing Metro Line 14 crossing the eastern suburb. J Beijing Univ Civ Eng Archit 31(04):24–27
5. Tan Y, Lu Y, Wang D (2021) Catastrophic failure of Shanghai Metro Line 4 in July, 2003: occurrence, emergency response, and disaster relief. J Perform Constr Facil 35(1)
6. Wayson (2019) Study on the influence of subway construction on the accessibility of transportation network in Xi'an City. Xi'an University of Technology
7. Bo L, Zhiwei Y, Ronghui Z et al (2021) Effects of undercrossing tunneling on existing shield tunnels. Int J Geomech 21(8)
8. Zhang Q (2017) Study on the response of soil and adjacent built tunnels under shield tunneling in soft soil area. Zhejiang University
9. Zhu X, Chen F, Xu M et al (2013) The experimental study of the formation movement law of shield excavation in Dalian Metro. Rock Soil Mech 34(S1):148–154
10. Chen R, Li J, Chen Y et al (2011) Experimental study on the stability model of the excavation face of dry sand shield tunneling. J Geotech Eng 33(01):117–122
11. Liu X, Tian J, Ye W et al (2020) Analysis on the influence of shield tunneling on the deformation of soft and plastic formations and existing tunnels. J Dis Prev Mitig 36(04):18–25

12. Jiang H, Li R, Yan R et al, Comparison of the difference and influence of the similarity degree of geometric scale in the tunnel model test. In: The 12th National conference of soil mechanics and geotechnical engineering of the chinese society of civil engineering
13. Wang S, Yu Q, Bloomberg et al (2016) Study on non-continuous contact model of two-layer lining of shield tunnel based on plastic damage. J Rock Mech Eng 35(2):17
14. Li W, Li X, Xue Y et al (2018) Stability model test of excavation surface of shield tunnel in sand pebble formation. J Geotech Eng 40(S2):199–203
15. Yu H, Zhang J, Ji Q et al (2017) Design and production of shield tunnel model based on 3D printing technology: Jane. J Railw Sci Eng 14(8):8

Open Access This chapter is licensed under the terms of the Creative Commons Attribution 4.0 International License (http://creativecommons.org/licenses/by/4.0/), which permits use, sharing, adaptation, distribution and reproduction in any medium or format, as long as you give appropriate credit to the original author(s) and the source, provide a link to the Creative Commons license and indicate if changes were made.

The images or other third party material in this chapter are included in the chapter's Creative Commons license, unless indicated otherwise in a credit line to the material. If material is not included in the chapter's Creative Commons license and your intended use is not permitted by statutory regulation or exceeds the permitted use, you will need to obtain permission directly from the copyright holder.

Damage Characteristics of Fractured Rock Under Freeze–Thaw Cycle

Yuanqiang Lv, Jingang Zhao, and Haibo Jiang

Abstract Infrastructure construction and resource exploitation in cold areas are of great significance to social development. The freeze–thaw cycle experiment, uniaxial compression, CT scanning and frost heave force monitoring were carried out on the precast fissure red sandstone of slope in Fugu area of northern Shaanxi Province. Utilising three-dimensional reconstruction techniques, we analysed the dynamic mechanical properties of rock samples, considering both frost heave force monitoring and the uniaxial compression and CT scanning. By introducing the fractal dimension, we established a freeze–thaw load damage model for fractured sandstone. Furthermore, we conducted a correlation analysis between the micro- and macro-scales, aiming to explore the effects of freeze–thaw process on rock samples. With the increasing number of freeze–thaw cycles, the quality and wave velocity of water-saturated sandstone decrease significantly, which is mainly due to the gradual penetration of internal defects, which makes the originally tight particle structure become loose, and the water-saturated sandstone is not able to be removed. The originally tight particle structure become loose, and at the same time, the internal pores and micro-cracks expand and connect with each other. By conducting frost heave force monitoring and performing dynamic mechanical testing on sandstone samples, it is observed that the fracture part of sandstone presents a positive microstrain due to the participation of water in the state of saturated water, and its peak strain gradually increases with the increase in freeze–thaw cycle.

Keywords Rock mechanics · Micro-defects · Freeze–thaw damage · Random strength · Statistical damage · Coupled damage · Coupled damage rate

---

Y. Lv (✉) · J. Zhao · H. Jiang
China Coal Xi'an Design Engineering Co., Ltd., Xi'an, Shaanxi, China
e-mail: 4109914@qq.com

1 Introduction

The northern Shaanxi region is known for its typical seasonal cold climate characteristics, where seasonal freeze–thaw phenomena are particularly significant. The freeze–thaw cycles not only cause the expansion and proliferation of internal cracks within the rock mass, but they also exacerbate the deterioration of the physical and mechanical attributes of the rock, ultimately resulting in a substantial reduction in the overall stability of the rock mass. Therefore, the impact of seasonal freeze–thaw cycles must be fully considered for the study and assessment of rock body stability in northern Shaanxi. After a certain number of freeze–thaw cycles, the internal damage of the rock mass occurs, accompanied by the expansion of primary fissures and the emergence, expansion and penetration of new fissures, and the deterioration of the mechanical properties of the rock mass is continuously straight [1].

Many scholars in China and abroad have done lots of research work on the mechanical properties and damage mechanism of rocks under freeze–thaw environment in both macroscopic and microscopic aspects and achieved many results. Fukuda [2], Chen et al. [3], Nicholson et al. [4] carried out a study on the effect of freeze–thaw cycles on rocks under various conditions and concluded that the main factors affecting the strength of freeze-thawed rocks are lithology, freeze–thaw temperature interval, the quantity of freeze–thaw cycles and the level of water content; both play significant roles. At the macroscopic level, Lu et al. [5] investigated the penetration mechanism of cracks in prefabricated single-fracture rocks subjected to freeze–thaw cycles under triaxial compression conditions. Ren et al. [6] investigated the strength deterioration patterns and damage forms of Luohe Formation sandstones after thawing by a complete freeze–thaw process at different freezing temperatures. Shen et al. [7] studied the fracture characteristics and fracture extension paths in the end region of single-fracture sandstones with different fracture inclinations under varying frequencies of freeze–thaw cycles. Zhao et al. [8] investigated the deformation and damage characteristics of vertically fissured rock samples of different lengths under the action of freeze–thaw cycles and found that the damage pattern tends to be complicated with the growth of the fissure length. In terms of fine-scale observation, Liu et al. [9] Rock samples that have undergone freeze–thaw cycles were carefully selected for exhaustive CT scanning tests. Subsequently, these scans were meticulously interpreted and professionally data processed, thus revealing in depth the internal structure and unique features of the rocks under different temperature conditions. This series of steps not only gave us a more comprehensive understanding of the freeze–thaw response of rocks, but also provided an important basis for their stability and durability assessment under extreme climatic conditions. Zhang et al. [10]. By combining the freeze–thaw cycle test and CT scanning technique, we used a binarisation method to analyse the rock samples in detail. This method successfully achieved accurate identification of the internal pores of the rock and revealed the effect of freeze–thaw cycles on rock damage. With the increasing number of freeze–thaw cycles, we observed that the damage variables showed a gradual increasing trend, which coincided with the development of the internal pores of the rocks, and further proved the significant effect of freeze–thaw cycles on the rock structure. Liu et al. [11] proposed a progressive damage zoning method for freeze–thaw sandstone layers dependent on CT scanning technology. Using advanced CT scanning technology, this method provides an in-depth analysis of the intrinsic connection between changes in elastic modulus, porosity and with regard to the number of freeze–thaw cycles, thus quantitatively revealing the development and evolution of progressive damage in freeze-thawed sandstone layers. The implementation of this method provides a more accurate and scientific means to understand and evaluate the impacts of freeze–thaw cycles on the properties of sandstone. This method provides a powerful tool for the in-depth understanding of the damage mechanism assess the changes in sandstone properties resulting from freeze–thaw action and helps to guide the damage assessment and risk control in related engineering practices. Yang et al. [12]. In order to deeply investigate the micro-structural changes of water-saturated sandstone under the action of freeze–thaw cycles, we designed and implemented a series of freeze–thaw cycle experiments covering different temperature intervals. The main goal of these experiments is to analyse in detail how key parameters such as fractal dimension, number of pores and volume mean specifically evolve and change with the increase in the number of freeze–thaw cycles. These experiments provide important insights into how freeze–thaw affects the structural characteristics of rocks. Through careful observation and analysis, we found that under the cumulative effect of freeze–thaw cycles, the original small pores gradually expanded and evolved into larger-sized pores, which was accompanied by a significant decrease in the rock physico-mechanical parameters. This finding not only provides a new perspective for us to understand the effect of freeze–thaw on gravels, but also provides an important theoretical basis for the stability assessment of gravels in related engineering practices. Park et al. [13] By utilising high-precision 3D CT reconstruction techniques, we have carefully observed the significant changes in the internal micro-structure of the rocks during the freezing and thawing process. These changes include the gradual separation of grains, initial crack initiation, further expansion and significant increase in porosity, which provide us with an intuitive and reliable basis for understanding the effects of freezing and thawing on the rock properties. Maji et al. [14] With the help of advanced CT technology, we were able to trace the complete dynamic process of micro-cracks from initial expansion, gradual growth, merging with each other, and finally evolving into macroscopic cracks in limestones and sandstones. These precise observations have greatly deepened our understanding of the internal structure and properties of rocks under the action of freeze–thaw, and we have accumulated a solid experimental foundation and rich data support in order to reveal the mechanism of freeze–thaw damage in rocks in a more in-depth manner. Although CT scanning tests have achieved remarkable, the study focuses on examining the mechanical properties of both intact and fissured rocks, revealing the outcomes thereof. We still face many challenges in understanding the mechanism of micro-fissure expansion after freeze–thaw cycles in macroscopically fissured sandstones. Particularly noteworthy are the comparative studies of intact and fractured sandstones under freeze–thaw and the quantitative analyses from the fine scale, which are still insufficient and need to be strengthened in order to fill the gaps in the existing body of knowledge. Research in these areas needs to be further strengthened and deepened. To this end, freeze–thaw cycling tests of single-gap sandstone bodies are carried out, on the basis of which uniaxial compression tests are used to study the mechanical characteristics, damage characteristics and damage modes of fissured sandstones, fine-scale parameter analyses of rock samples are carried out by using CT scanning experiments and monitoring of freezing and expansion forces, and dynamic mechanical properties are analysed. The freeze–thaw loading damage model of the fissured sandstone was constructed, through which the fine and macroscopic characteristics of the rock were analysed in an in-depth correlation. This analysis not only reveals the damage evolution of rocks under freeze–thaw cycles, but also further explores the intrinsic mechanism of their damage deterioration, providing a powerful tool for understanding and predicting the changes of rock properties in freeze–thaw environments. The research results are of great significance for the evaluation of engineering stability of fissured rocks in cold regions.

2 Test Content

2.1 Sample Preparation

The samples were taken from the red sandstone commonly found in the rocky slopes of Fugu County, northern Shaanxi Province, China. To ensure the accuracy and reliability of the experiments, we used cutting-edge CNC rock cutting technology to create cylindrical samples with dimensions up to 50 mm × 100 mm, in line with international standards. Any samples with significant cosmetic flaws are carefully rejected to ensure the accuracy and reliability of subsequent experimental results.

On this basis, the rock samples shown in Fig. 1 were prepared with a fissure length of 20 mm and a width of 1 mm. The fissure inclinations were 0°, 30°, 45°, 60° and 90°.

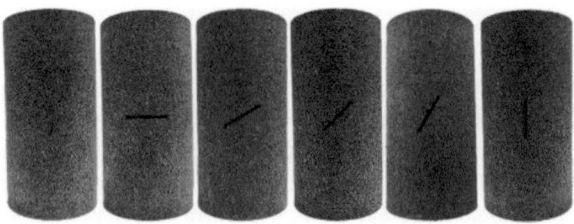

Fig. 1 Fracture arrangement of rock samples. From left to right the cleavage dip is for intact rock samples, 0°, 30°, 45°, 60°, 90° (test specimen)

2.2 Test Apparatus

For the freeze–thaw cycle test, we chose the TDR-28 Concrete Rapid Freeze–Thaw Tester, as shown in Fig. 2, which integrates a highly efficient heating and cooling system to accurately simulate temperature changes under extreme temperature conditions in cold regions. Its wide temperature control range can cover a wide interval from − 30 to 30 °C, ensuring the accuracy and reliability of the experiment. It also has a high-precision control capability of ± 0.5 °C, ensuring the accuracy and reliability of the test data.

A Phoenix Nanotom CT scanner was used for the CT scanning test, as shown in Fig. 3, which has an efficient non-destructive acquisition capability and is able to rapidly capture and generate greyscale images with a resolution of 30 μm, ensuring the accuracy and integrity of the data.

For the indoor freezing and expansion force characteristic test, we chose YBY-2001 strain gauge produced by Liyang Weihan Instrument Factory, as shown in Fig. 3. This strain gauge has excellent performance, the range covers ± 19,999 micro-strain (με), the resolution is as high as 1 micro-strain (με), and the measurement error is controlled within ± 3 micro-strain (με), which ensures the accuracy and reliability of the test data.

For uniaxial compression testing, the RTX-1500 low-temperature, high-pressure rock triaxial instrument was chosen, as shown in Fig. 4. The RTX-1500 has excellent performance, with a maximum axial pressure of 1500 kN and a peripheral pressure of 140 MPa. Even better, it can be operated in a wide range of temperatures from − 30 to 80 °C, which provides an ideal experimental environment for the study of rock mechanics under different conditions. In addition, RTX-1500 can be loaded with either displacement control or stress control, and this flexibility allows it to simulate triaxial mechanical states under a wide range of complex conditions, thus providing us with accurate and comprehensive data on rock mechanical properties.

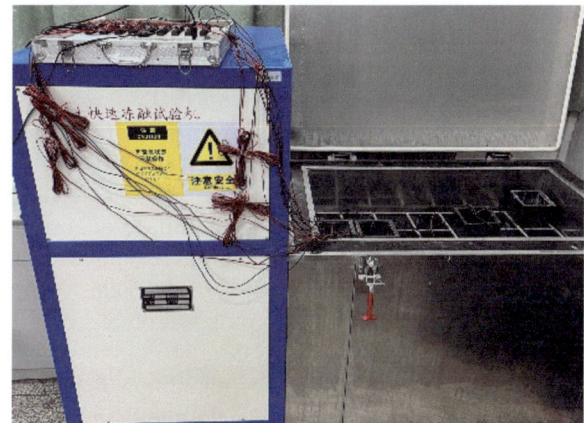

Fig. 2 Rapid freeze–thaw tester for concrete (laboratory instrument, obtain a licence)

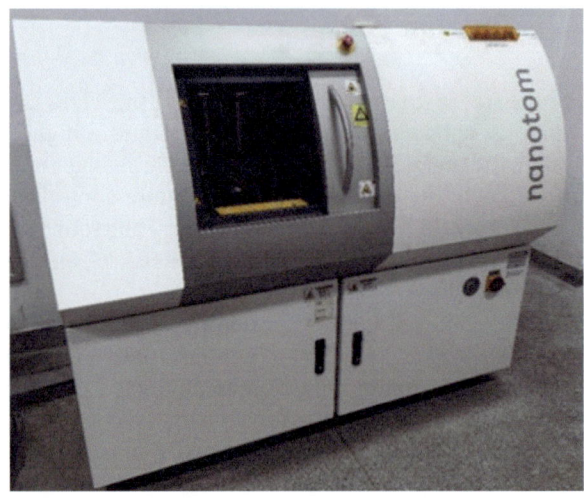

Fig. 3 High-precision CT scanning instrument (laboratory instrument, obtain a licence)

2.3 Pilot Programme

The monofissile sandstones with different dips were screened grouped and numbered with the following main test programme.

Freeze–Thaw Test Programme. To study the effect of freeze–thaw on the quality, longitudinal wave velocity and strength of the fissured red sandstone, the freeze–thaw cycle test was carried out on the fissured sandstone with fissure inclination angles of 0°, 30°, 45°, 60° and 90° (one freeze–thaw cycle is 8 h, 4 h for freezing and 4 h for thawing). The freezing temperature was set at − 20 °C, the thawing temperature at 20 °C, and the frequency of freeze–thaw cycles was 0, 7, 15, 30, 60, 90 and 120. And, the specimens before and after the freeze–thaw cycle were measured for various physical parameters.

CT Scanning Test Programme. For the water-saturated rock samples with 45° fissure inclination as well as the intact rock samples, we selected the CT group specimens experiencing different numbers of freezing and thawing (0, 30, 60, 90, 120) as the research objects and carried out a series of CT scanning tests. Before formally carrying out the tests, we first carried out fine parameter tuning of the scanning equipment to ensure the accuracy and reliability of the scanning results. During the CT scanning process, we set the scanning voltage to 180 kV and the exposure time to 0.3 s to ensure the image quality, and set the voxel resolution to 30 μm to capture the details. Each rock sample was scanned five times repeatedly to ensure the accuracy and reliability of the data. A complete dataset containing 1800 2D greyscale image slices per group was obtained by a top-down scanning approach. Regarding the images captured through computed CT scans, after careful study and analysis, we have thoroughly explored how the two core elements, the quantity or frequency of freeze–thaw cycles and the fissure inclination, specifically contribute to the fine-scale damage evolution characteristics of sandstones.

Fig. 4 GCTS low-temperature high-pressure geotechnical triaxial test system

Indoor Freezing and Expansion Force Characterisation Test study. Using a static strain gauge connected with strain gauges, we carried out tests of freeze expansion strain, tracking and monitoring the dynamics of the fissure freeze expansion force with temperature change in real time. After meticulous data analysis and careful comparative study of the freezing expansion mechanism under different fissure inclinations, we have deeply explored the damage characteristics of the fissured rock mass under the action of freeze–thaw cycles and successfully revealed the core mechanism of action behind it.

Uniaxial Compression Tests. A sequence of systematic conventional uniaxial compression tests were planned and carried out in order to fully analyse the specific effects of different freeze–thaw numbers and fracture inclinations on the properties of rock samples. These tests were designed to reveal how these factors affect the mechanical properties and damage modes of the rock samples. In the uniaxial compression tests, the rock samples were loaded at a constant rate of 0.06 mm/min until the samples lost their load-bearing capacity due to rupture. Upon successful completion of the test, the damage patterns of the specimens were accurately recorded and characterised in detail.

3 Effect of Freeze–Thaw on the Physical Properties of Fissured Red Sandstone

3.1 Surface Deterioration Analysis

To deeply investigate the damage deterioration mechanism of the fissured sandstone under the gradually rising number of freeze–thaw cycles, we recorded the surface freeze–thaw damage phenomenon of the water-saturated fissured sandstone and displayed it in Fig. 5. As can be seen from the figure, with the number of freeze–thaw cycles reaching 60, the surface of the fissured sandstone gradually showed abrasive particles, especially at the upper and lower edges where a limited quantity of particles were detached. At the same time, the rock samples produced a linear particle loss along the direction of the prefabricated fissures, as shown in Fig. 5a.

As the freeze–thaw cycle continues, as shown in Fig. 5b, significant spalling begins to appear at the ends of the sample and in the peripheral area around the fissure, which starts at the fissure and spreads radially to the surrounding area. The precise analysis of the freeze–thaw test clearly shows the effect of the damage on the rock samples in terms of the significant deposition of gravel and an augmentation in the count of pores on the surface of the rock samples. These detailed observations provide strong evidence of the significant effects of the freeze–thaw process on the micro-structure of the rock samples.

In Fig. 5c–e, we present exhaustively the gradual increase in the number of pores on the surface of the rock samples during the freeze–thaw process, which is accompanied by the remarkable appearance of frost crunching. Particularly, striking is our observation that a series of new tiny cracks gradually sprouted along the direction of the prefabricated fissures throughout the evolution of the freeze–thaw cycle. These nascent cracks undoubtedly provide conclusive evidence of the profound effects of the freeze–thaw process on the micro-structure of the rock samples and at the same

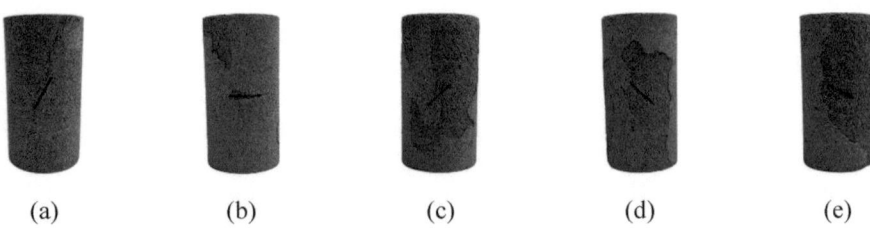

Fig. 5 Surface degradation of fissured rock samples under freeze–thaw action

time foreshadow that the rock samples may face more serious structural damage as the freeze–thaw cycle continues.

3.2 Impact on Quality

We conducted detailed freeze–thaw cycle tests on water-saturated fissured sandstone under the effect of freeze–thaw cycles. At the end of the tests, we carefully weighed the mass of the rock samples and recorded the mass changes in detail. The aim of this step was to analyse the potential effects of freeze–thaw action on the sandstone mass and thus to investigate the change in sandstone properties during freeze–thaw cycling, with a rate of change in the mass of the rock samples:

$$\emptyset_{An} = \frac{m_{A0} m_{An}}{m_{A0}} \quad (1)$$

In our experiments, we defined two key parameters: m_{A0} and ϕ_{An}. m_{A0} represents the weight or mass of the rock specimen when it is unfrozen and thawed (i.e. 0 freezing and thawing), whilst it represents the mass of the rock sample after n freezing and thawing cycles. In addition, ϕ_{An} represents the rate of change in regard to the mass of the rock sample after a freeze–thaw cycles.

A freeze–thaw cycle test was carefully planned and executed in order to investigate in depth the extent of damage to water-saturated sandstones with different angular fissures after undergoing freeze–thaw cycles. After the tests, we observed that individual rock samples showed significant variability in damage at the fissures. In order to more accurately assess these differences, the weight or mass of each individual rock sample was finely weighed, and the results were recorded in detail.

In order to ensure the accuracy and reliability of the results, and in order to obtain accurate data on the rate of change of sandstone mass, we carefully designed three sets of control group experiments and rigorously averaged the results of each set of experiments. Subsequently, we accurately and precisely substituted these averages into Eq. (1) for calculation to ensure that the rate of change of sandstone mass under varying frequencies of freeze–thaw cycles was accurately derived. This step provided an important basis for us to further analyse the effect of freeze–thaw cycles on sandstone properties.

The average mass of sandstones with different cleavage inclinations under water-saturated condition decreased by 0.06%, 0.14%, 0.29%, 0.51%, 0.64%, 0.89%, respectively, during the freeze–thaw process, and the sandstones with different inclinations showed similar trends of the mass loss rate changes in the freeze–thaw cycle under the condition of controlling the constant cleavage lengths and widths. In the initial stage, the mass loss rate of sandstone was relatively low, but it began to climb significantly with time, especially after the thirtieth freeze–thaw cycle. It is worth noting that the fissure-bearing sandstones showed a more pronounced variation in mass loss compared to the intact sandstones. This is mainly due to the fact that during freeze–thaw cycling, the water within the fissures undergoes repeated freezing and thawing processes, which results in a large freezing and expansion force that accelerates the sandstone mass loss. This freezing and swelling force leads to the detachment of some crystals at the edges of the fissures and makes the pores on the surface of the rock samples significantly larger. These phenomena fully illustrate that freeze–thaw cycles have a particularly pronounced and deeper damage effect on fissured sandstones.

After analysing the experimental data, we found that there is an obvious exponential relationship between the average mass change rate and the number of freeze–thaw cycles of the rock samples with different angles of fissures. In order to demonstrate this relationship more intuitively, we carried out the fitting of the exponential relationship and plotted the fitted curves together with the average mass change curves in Fig. 6a, b. The specific fitted relationship equation is as follows:

$$\begin{cases} \emptyset_{0°} = e^{(-3.725+1.052n-0.076n^2)} & (R^2 = 0.9868) \\ \emptyset_{30°} = e^{(-3.71+1.038n-0.073n^2)} & (R^2 = 0.9936) \\ \emptyset_{45°} = e^{(-3.563+0.975n-0.067n^2)} & (R^2 = 0.9910) \\ \emptyset_{60°} = e^{(-3.425+0.846n-0.049n^2)} & (R^2 = 0.9948) \\ \emptyset_{90°} = e^{(-3.605+0.992n-0.069n^2)} & (R^2 = 0.9877) \end{cases} \quad (2)$$

3.3 Impact on Wave Velocity

To investigate the effect of freeze–thaw cycles on the wave velocity of water-saturated fissured sandstones, the samples were first dried after a specific number of freeze–thaw cycles. The key objective of this step is to remove excess water from the samples until they reach a constant weight state, ensuring that the internal structure and properties of the rock can be accurately assessed in subsequent longitudinal wave velocity tests. After drying, we performed longitudinal wave velocity tests and calculated the rate of change of longitudinal wave velocity during freezing and thawing using Eq. (3)r_{An} to quantitatively analyse the specific effects of freezing and thawing on the properties of the rock samples.

$$r_{An} = \frac{r_{A0} - r_{An}}{r_{A0}} \quad (3)$$

In our experiments, we recorded the wave velocity of the rock samples before the freeze–thaw cycle (i.e. 0 freeze–thaw) r_{A0} and the wave velocity after n freeze–thaw cycles r_{An} with units of metres per second (m/s). Further, we defined r_{An} as the velocity of wave variation of the rock sample at n times of freezing and thawing.

To ensure the accuracy and reliability of the experimental results, we carefully set up three groups of control group experiments and meticulously calculated the average

Fig. 6 Mean mass and rate of change of rock samples with different angular fissures under different numbers of freeze–thaw cycles

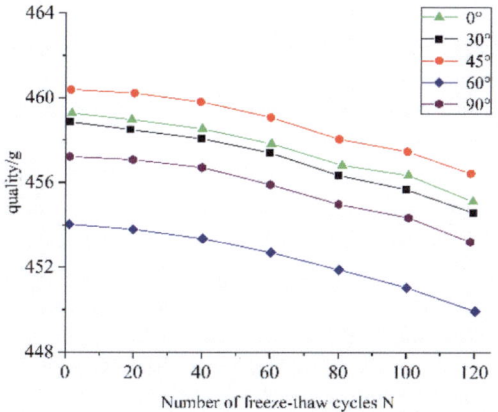

(a) Average mass of rock samples

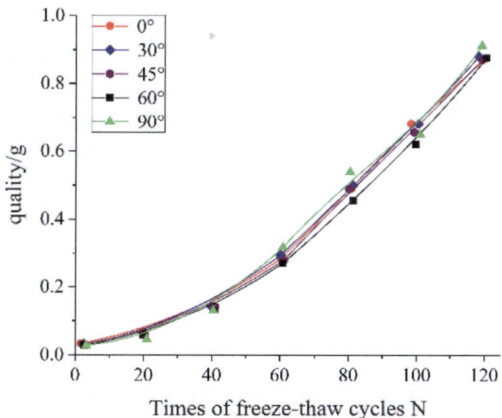

(b) Rate of change in the average mass of rock samples

values of each group of data. Immediately after that, we accurately substituted these averages into Eq. (3) and successfully calculated the rate of change of wave velocity of sandstone across varying frequencies of freeze–thaw cycles. This step provides an important basis for us to gain a deeper understanding of the effects of freeze–thaw cycles on the internal structure and properties of sandstone.

After reaching the saturated state, the sandstones treated by the freeze–thaw process, regardless of their fissure inclination, showed a clear trend of decreasing wave velocity, with specific decreases of 1.31%, 2.78%, 6.19%, 11.22%, 20.69% and 34.02%, respectively. It was observed that there was a clear positive correlation between the wave velocity loss rate and the number of freeze–thaw cycles of these sandstones, which was manifested by a gradual increase in the wave velocity loss rate with the increment of the number of freeze–thaw cycles. In particular, after the 15th freeze–thaw cycle, the tendency or pattern of increasing wave velocity loss rate becomes more significant.

Although the fracture lengths and widths of these sandstone samples remain consistent, their fracture dips vary. However, during freeze–thaw cycling, we noticed that the trends of their wave velocity loss rates exhibited remarkable similarities. This observation suggests that when analysing wave velocity variations in sandstones, the effect of cleavage inclination on them is somewhat universal. More significantly, the accumulation of the frequency or count of freeze–thaw cycles becomes the main determining factor leading to a significant decrease in the wave velocity of sandstones. After analysis, we found that when examining the average rate of change of wave velocity in rock samples with different angles of cleavage, we found a significant exponential relationship between it and the frequency or count of freeze–thaw cycles. By fitting the data, we obtained the corresponding curvilinear relational plots (shown in Fig. 7a), as well as the variation curves of the mean wave velocity (shown in Fig. 7b). The specific fitted relational equation is expressed as:

$$\begin{cases} \emptyset_{0°} = e^{(-1.415+0.903n-0.016n^2)} & (R^2 = 0.9971) \\ \emptyset_{30°} = e^{(-0.764+1.114n-0.068n^2)} & (R^2 = 0.9934) \\ \emptyset_{45°} = e^{(-0.808+1.051n-0.05n^2)} & (R^2 = 0.9970) \\ \emptyset_{60°} = e^{(-3.565+0.813n-0.028n^2)} & (R^2 = 0.9904) \\ \emptyset_{90°} = e^{(-0.808+1.041n-0.053n^2)} & (R^2 = 0.9980) \end{cases} \quad (4)$$

4 Macro- and Micro-damage Characterisation

4.1 Fine-Scale Characterisation of Fissured Red Sandstone Under Freeze–Thaw Action

In view of the insignificant influence of different fissure inclination and drying conditions on the damage degree of rock samples during freeze–thaw cycles, and considering the economy of the test cost, sandstone samples with 45° fissure inclination and intact rock samples were selected as the main research objects in this study. Through carefully planned freeze–thaw cycle CT scanning tests, combined with CT 3D reconstruction technology and fractal theory, we thoroughly investigated the dynamic evolution of micro-structures, such as porosity and pore parameters, of intact rock samples and fissured sandstones during freeze–thaw process. The analytical method has the ability to quantitatively assess the degree of fine-scale damage and non-homogeneity coefficients of sandstones with different saturations during the freeze–thaw process, and after in-depth study, we precisely reveal the exact influence of the number of freeze–thaw cycles and different fracture characteristics on the

Fig. 7 Mean wave velocity and rate of change of rock samples with different angular fissures under different numbers of freeze–thaw cycles

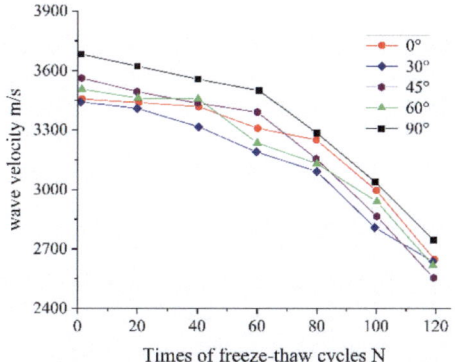

(a) Average wave velocity of rock samples

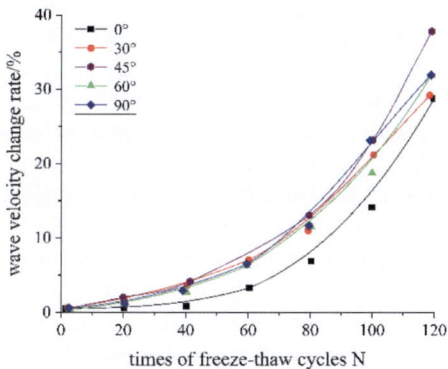

(b) Rate of change of average wave velocity in rock samples

internal micro-structural changes and damage evolution mechanisms of the rocks. These findings provide an important basis for our in-depth understanding of the mechanical properties and stability of rocks under freeze–thaw environments.

With the help of advanced 3D reconstruction software, we can accurately calculate the pore content, which is a key parameter in the sandstone centre cube pore equivalent model. When the pore volume and the total volume of sandstone are known, we can calculate the porosity of sandstone by the following mathematical formula:

$$W_p = \frac{V_{\text{pore}}}{V_{\text{voxel}}} \tag{5}$$

Here, we define W_p as the porosity of the rock sample, which is derived by calculating the ratio of the total volume of pores and fissures in the rock sample, V_{pore}, to the total volume of the rock sample, V_{voxel}. In order to visualise the effect of freeze–thaw action on the total pore volume V, the volumetric porosity W_p, the equivalent maximum radius of pores D_{eq} and the number of pores in the rock samples, we plotted the fitted curves of these parameters as a function of the number of freezes and thaws N (shown in Fig. 8). The fitting equations for these curves are detailed in Eqs. (6–9).

$$\begin{cases} V_{\text{fracture}} = 2.278 - 0.025x + 7.389x^2 & (R^2 = 9710) \\ V_{\text{integrity}} = 2.143 - 0.017x + 5.770x^2 & (R^2 = 9889) \end{cases} \quad (6)$$

$$\begin{cases} W_{p(\text{fracture})} = 3.511 + 0.081x + 1.841x^2 & (R^2 = 9943) \\ W_{p(\text{integrity})} = 3.540 + 0.018x + 1.754x^2 & (R^2 = 9804) \end{cases} \quad (7)$$

$$\begin{cases} D_{\text{eq(fracture)}} = 285.836 + 12.237x - 0.376x^2 + 0.004x^3 & (R^2 = 9993) \\ D_{\text{eq(integrity)}} = 276.288 + 11.806x - 0.348x^2 + 0.003x^3 & (R^2 = 9999) \end{cases} \quad (8)$$

$$\begin{cases} N_{p(\text{fracture})} = 104{,}495 + 119.177x & (R^2 = 9984) \\ N_{p(\text{integrity})} = 104{,}462 + 40.355x + 0.459x^2 & (R^2 = 9875) \end{cases} \quad (9)$$

As shown in Fig. 8, with the increment of the number of freeze–thaw cycles, the porosity, average pore volume, equivalent average radius of pores and the number of pores of the fissured sandstone rock samples showed different degrees of increasing trends. In detail, after 120 freeze–thaw cycles, the porosity of the fissured rock samples increased significantly by as much as 193.99%. In addition, the average volume of pores also increased tremendously, by 428%. Meanwhile, the equivalent mean radius of the pores also increased by 20.97%, whilst the number of pores also increased, specifically by 17.12%. It is noteworthy that the intact sandstone rock samples showed a similar trend after the same freeze–thaw cycles. After 120 freeze–thaw cycles, the porosity increased by 149.1%, the average pore volume increased by 323.27%, the pore equivalent average radius increased by 18.95%, and the number of pores increased by 16.4% compared to the untreated intact rock samples. These results indicate that the freeze–thaw cycle significantly affected the fine structural parameters of the sandstone, leading to a gradual increase. Further comparison of the changes between the fissured sandstone and the intact sandstone shows that the magnitude of the changes in the fine structural parameters of the fissured sandstone is even more significant, which is mainly attributed to the phenomenon of laminar damage expansion that the prefabricated fissured portion exhibits during freeze–thaw cycling [15]. This phenomenon highlights the sensitivity of fissured sandstones in freeze–thaw environments and the complexity of their structural changes [15].

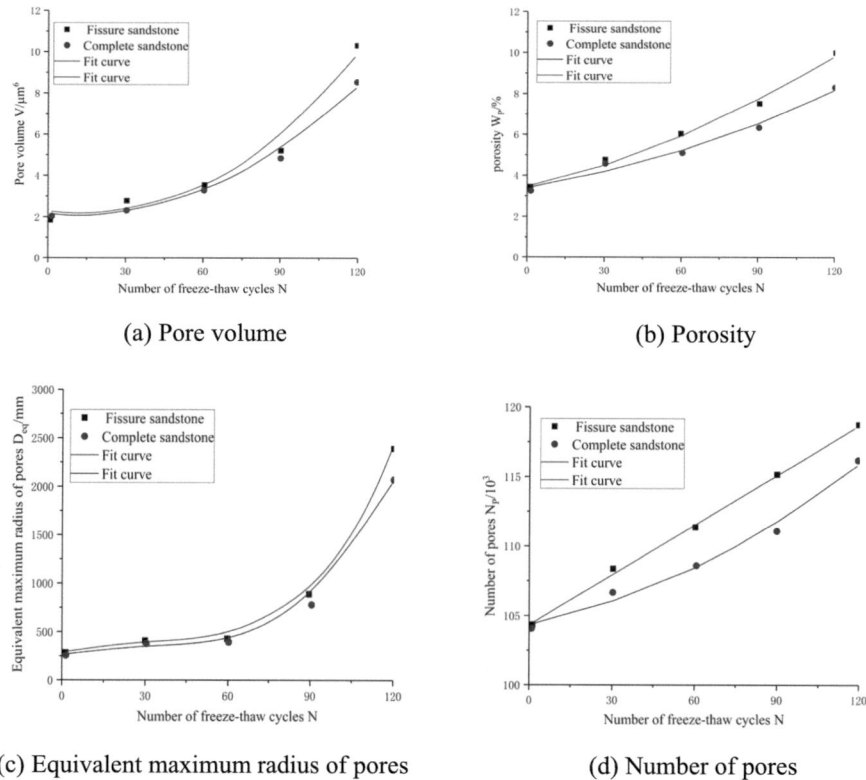

Fig. 8 Variation of pore parameters of intact and fissured sandstone with the number of freeze–thaw cycles

4.2 Mechanical Properties of Fractured Rock Under Freeze–Thaw Action

This study mainly focuses on the damage evolution of single-fracture rock samples under saturated moisture conditions during freeze-up. With the state-of-the-art technology of static strain gauges and strain gauges, we have successfully realised high-precision real-time monitoring of freezing strain, which enables us to accurately capture and continuously track the subtle changes in the freezing force of the fissure as a function of temperature fluctuations. Further, we have analysed the effects of different fissure inclination angles on the expansion mechanism of freezing and swelling, thus revealing the core mechanism of the damage characteristics of fissured rocks throughout the freeze–thaw cycles, and providing valuable scientific evidence for understanding and predicting the long-term effects of freeze–thaw environment on the rock structure.

Based on the data obtained from the experiments, we have plotted Fig. 9a–e, which present in detail the dynamic process of micro-strain changes experienced by rock samples with different cleavage angles in the cleavage region under water-saturated conditions during a complete freeze–thaw cycle. These diagrams provide us with a visual understanding that helps us to analyse the effects of the freeze–thaw process on the rock micro-structure.

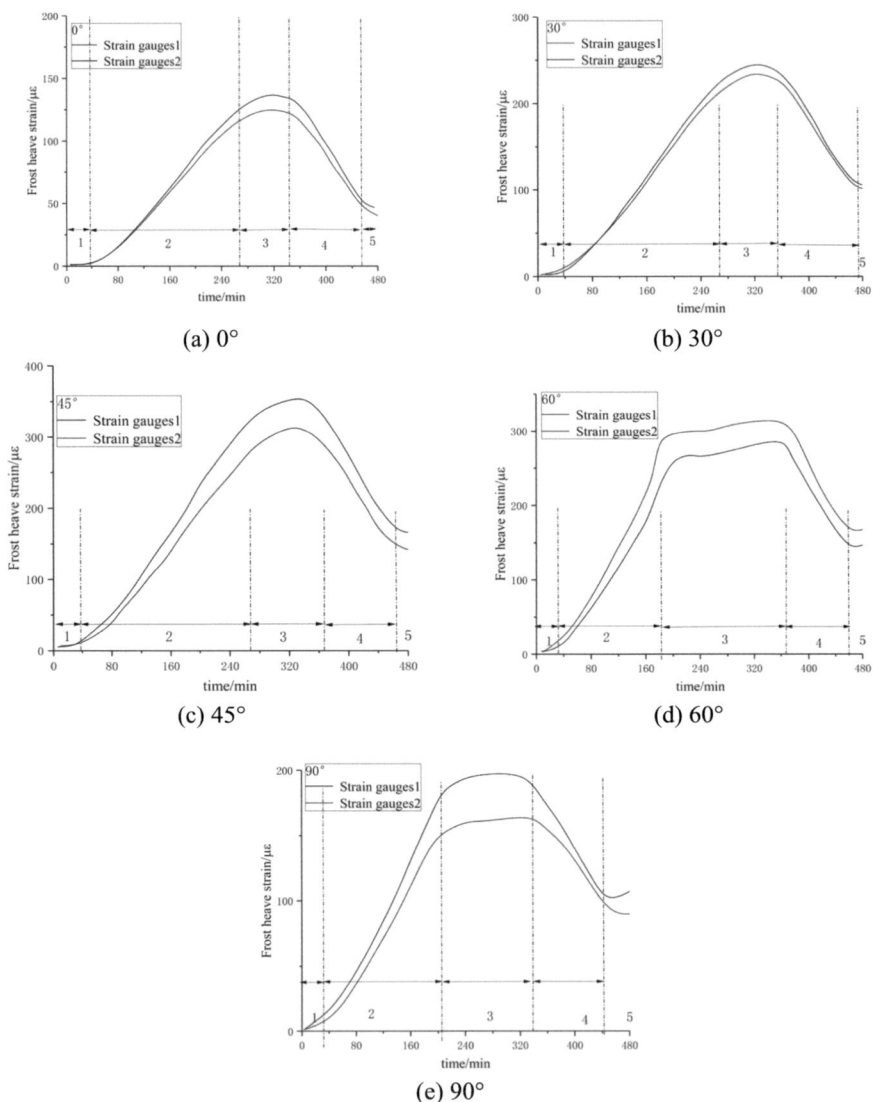

Fig. 9 Freeze-up strain curves of sandstone with different angles of cleavage in water-saturated state

From the figure, it can be observed that the rock samples with different fracture angles show roughly similar patterns in micro-strain changes. When the temperature starts to decrease, the micro-strain of the rock samples first shows a small contraction trend. As the temperature continues to drop, the micro-strain begins to increase dramatically and reaches a maximum value at some point in time, ranging from 147 to 358 μm. In particular, the change in micro-strain levelled off during the 240- to 320-min period when the temperature was stabilised at − 20 °C. Subsequently, as the temperature gradually increased, the micro-strain began to gradually decrease, eventually falling to between 38 and 173 μs. Finally, when the temperature was stabilised at 20 °C, the micro-strain correspondingly remained at a relatively stable level.

After an in-depth analysis of the freeze-up strain curves of sandstones with different angular fractures in the water-saturated state, we observe that the evolution of the freeze-up force can be clearly classified into the following five stages:

Cold Shrinkage Stage: When the temperature of the rock body starts to drop gradually from the initial 20 °C, it will undergo a short, relatively small degree of contraction due to cold. The duration of this contraction phenomenon is relatively short.

Freezing and Expansion Phase: With the gradual decrease in temperature, the water inside the fissure starts to undergo a water–ice phase transition and gradually transforms into solid ice. During this process, the not-yet-frozen fissure water and pore water will be affected by the driving force of ice crystal formation and gradually migrate to the fissure part and eventually freeze. This migration and freezing process results in a significant reduction of water content in the fractures. At the same time, with the formation and expansion of ice, the volume of the ice wedge within the fissure gradually increases, which triggers the stress concentration phenomenon. Under the combined effect of these factors, the freezing and expansion forces can suddenly increase to peak levels.

Warming-Up Hysteresis Stage: When the ambient temperature plummeted to − 20 °C, the ice wedge inside the crack began to produce a significant extrusion effect, and this change directly led to the obvious expansion of local cracks that we observed at specific locations in the crack. As these localised deformations occur, stresses are gradually released and relaxed. It is worth noting, however, that the subtle variations in the inclination of the cracks during this phase are characterised by differences in the inclination of the cracks and in the accuracy of the measuring instruments. At the same time, the unfrozen water continues to migrate towards the already frozen water within the crack, creating a smaller freezing expansion force. The interaction between the freezing expansion force and the stress relaxation induced by crack expansion has a counteracting effect, which gradually slows down the rise of the micro-strain curve. When the water in the cracks is completely frozen, the freezing expansion force remains basically stable, which further leads to the cessation of further crack expansion.

Melting and Shrinking Stage: With the gradual rise in temperature, the ice wedge inside the rock body fissure starts to melt gradually. The melted water will gradually penetrate into the deeper pores of the rock body along the pores or capillary channels, thus triggering the phenomenon of contraction and deformation of the cracks. During this process, the micro-strain will decrease rapidly until the ice wedge inside the fissure completely melts away. At this point, the contraction and deformation of the fissure will reach a relatively stable state.

Melt Shrinkage Stabilisation Stage: When the temperature is maintained at a stable 20 °C, the deformation process of the rock body is over, and its micro-strain value also tends to stabilise, no longer occurring significant changes.

4.3 Uniaxial Compression of Fissured Red Sandstone Under Freeze–Thaw Action

After conducting uniaxial compression tests on intact sandstone and monofissile sandstone after 0, 7, 15, 30, 60, 90 and 120 freeze–thaw cycles, respectively, we successfully obtained their respective uniaxial compressive strength data under varying iterations of freeze–thaw cycles. Based on these data, we plotted the corresponding uniaxial stress–strain curves, as shown in Fig. 10, in order to more intuitively demonstrate and analyse the implications or consequences of freeze–thaw cycles on the mechanical characteristics of sandstones.

Based on Fig. 11, after in-depth analyses, we can clearly conclude that freeze–thaw cycles significantly impact the mechanical attributes of rocks. This effect is mainly manifested in the exacerbation of the internal damage of the rock, which in turn leads to a significant decrease in its strength. Therefore, the overall strength of

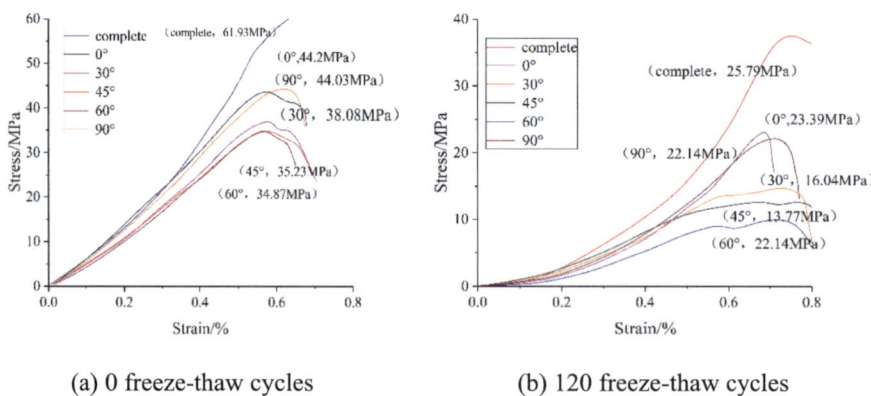

Fig. 10 Comparison of uniaxial stress–strain curves of sandstone under different numbers of freeze–thaw cycles

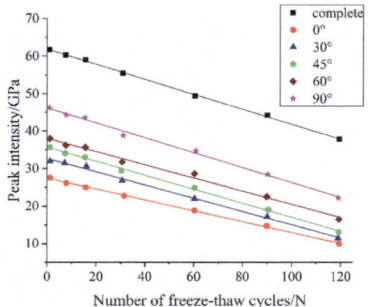
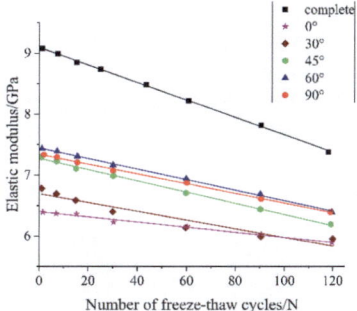

(a) Variation of uniaxial strength with the number of freezes and thaws

(b) Change in modulus of elasticity with number of freezes and thaws

Fig. 11 Variation of peak uniaxial compressive strength, modulus of elasticity with the number of freeze–thaw cycles

the rock samples shows a significant decreasing trend as the frequency of freeze–thaw cycles escalates.

When examining sandstones with different cleavage inclinations, we observe that they all exhibit a trend related to the number of freezing and thawing times, N. This trend is visualised in Fig. 11a, b, where it is accurately described by fitting curves as well as Eqs.

$$\begin{cases} \sigma_{\text{integrity}} = 61.62 - 0.20N\,(R^2 = 0.9963) \\ \sigma_{0°} = 46.8 - 0.20N\,(R^2 = 0.9889) \\ \sigma_{30°} = 36.21 - 0.19N\,(R^2 = 0.9904) \\ \sigma_{45°} = 35.59 - 0.19N\,(R^2 = 0.9963) \\ \sigma_{60°} = 32.56 - 0.18N\,(R^2 = 0.9960) \\ \sigma_{90°} = 27.37 - 0.14N\,(R^2 = 0.9810) \end{cases} \quad (10)$$

$$\begin{cases} \sigma_{\text{integrity}} = 9.13 - 0.01N\,(R^2 = 0.9750) \\ \sigma_{0°} = 6.42 - 0.01N\,(R^2 = 0.9806) \\ \sigma_{30°} = 6.82 - 0.15N + 6.7N^2\,(R^2 = 0.9853) \\ \sigma_{45°} = 7.30 - 0.01NR^2 = 0.9913) \\ \sigma_{60°} = 7.46 - 0.01N\,(R^2 = 0.9767) \\ \sigma_{90°} = 7.35 - 0.01N\,(R^2 = 0.9734) \end{cases} \quad (11)$$

Uniaxial Compression Damage Morphology Analysis

After completing the uniaxial compression damage tests on the fissured sandstone, we photographed and recorded the damaged rock samples. These photographs show the damage characteristics of the sandstone after 30 and 90 freeze–thaw cycles, as shown in Fig. 12. By comparing these photographs, we can clearly observe the influence of varying iterations of freeze–thaw cycles on the sandstone damage pattern.

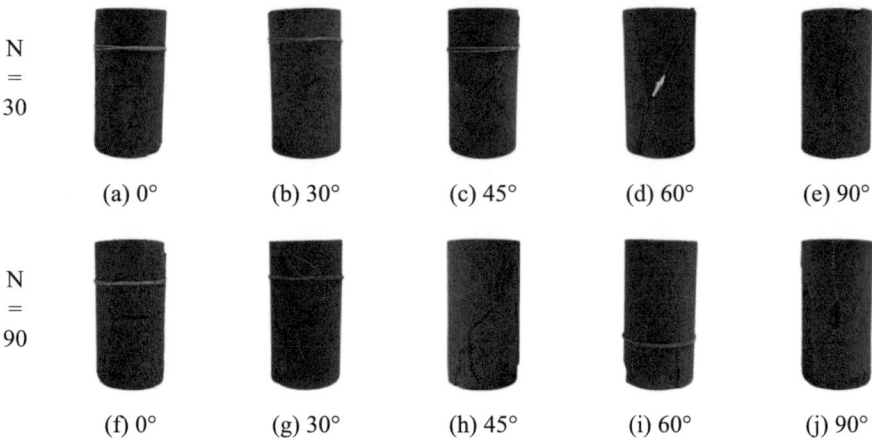

Fig. 12 Uniaxial compression damage in fissured sandstone

It can be observed from Fig. 12 that when the dip angle of the fractured sandstone is 0° or 90°, its main damage mode is significantly manifested as tensile or cleavage damage. In these cases, the initial formation of cracks tends to follow a path with a significant angle to the loading direction, followed by a gradual change in the damage mode to predominantly mixed tensile-shear damage under the dual action of tensile and shear forces. It is noteworthy that under the specific experimental conditions, the specimens mainly exhibited the characteristics of mixed tensile-shear damage even though the inclination angle of the prefabricated cracks was 30°. During the damage process, the tensile and shear forces intertwined with each other and jointly promoted the crack generation and expansion. Especially at the ends of the prefabricated cracks, these regions tend to become hot spots for crack development. In these hot spot areas, wing cracks and secondary cracks may appear successively under stress. With the increase in compressive stress, these cracks will continue to expand and extend, and eventually they will converge in the fragile region of the specimen, which will trigger the rupture and failure of the specimen. In addition, the damage process of the freeze–thaw cycle on the rock specimens has more intuitive manifestations, for example the blocky spalling phenomenon in localised areas of the specimens, as well as the fine cracks at the ends of the specimens, all of which significantly demonstrate the important influence of the freeze–thaw cycle on the damage process of the rock specimens. These phenomena are not only intuitively visible, but also profoundly reveal the weakening effect of freeze–thaw action on the mechanical properties of rock materials. In contrast, the damage pattern of the specimens at the prefabricated fissure inclination angles of 45° and 60° is similar to that at 30° inclination, and the main damage form is centred on shear damage. With the prolongation of the loading time and the increase in the load, the internal stress of the rock samples gradually accumulates until it reaches a certain degree, and the cracks begin to gradually expand and eventually lead to the destruction of the specimens. In this process, the equilibrium state at the crack tip suffered damage, which triggered the continuous growth of crack length and corner angle. At the same time, the deformation phenomenon of the cracks became more and more obvious, and the rock samples gradually approached the critical point of destruction, indicating that their stability was decreasing dramatically. This phenomenon profoundly reveals the dynamic relationship between crack extension and rock deformation, which is closely connected and interacts with each other, and the crack extension has a significant influence on the deformation characteristics of the rock.

On the other hand, as t the quantity or magnitude of freeze–thaw cycles increases, the characteristics of the fracture surfaces of the sandstone change significantly from a relatively smooth state to a rough and multi-faceted morphology. This transformation is manifested by the gradual differentiation of the original single-fracture surface and the formation of multiple fracture surfaces with rough surfaces, which highlights the cumulative effect of the freeze–thaw cycle on the structural damage of the rock. In the crack evolution process of fissured sandstone, the cracks tend to extend along their axial direction, and the formation and extension process of these cracks directly reflect the dynamic evolution of the fissure structure of the specimen under loading. When the specimen is finally damaged, the cracks are mainly divided into two categories: primary cracks and secondary cracks, which are intertwined with each other to form an intricate crack network. The in-depth observation and detailed analysis of the crack formation mode and its expansion path are of vital significance in revealing the mechanical properties and damage mechanism of the rock. Freezing and thawing significantly exacerbated the expansion and connection of micro-fractures, leading to a gradual weakening of the integrity of the rock samples, which was manifested by a significant increase in the number and volume of broken rock fragments. In the damage process of fissured sandstone, the through-going cracks of tension-shear composite were often formed between the upper and lower parts of the fissures, and this crack pattern played a dominant role in the damage of the specimens. At the same time, this crack pattern also promotes the nourishment of secondary cracks and exacerbates the dislodgment phenomenon of the local rock mass, which further reflects the complexity of the crack evolution mechanism of the fissured sandstone under loading conditions. The formation and expansion of cracks is an intertwined and interactive process, and they eventually converge into through-going tension-shear mixed cracks. When the through damage phenomenon occurs, the rapid increase in sub-cracks and the intensification of local falling block phenomenon are the visualisation of the continuous crack expansion and the damage of the internal structure of the rock under stress.

5 Conclusions

Macro- and fine-scale damage characteristics of freeze-thawed monofissile sandstones with different inclinations were investigated by indoor freezing and expansion force characteristic tests, CT fine-scale tests and uniaxial axial compression macro-tests, and the subsequent deductions were formulated.

(1) In the course of the freeze-up force monitoring tests on the fissured sandstone, we observed that in the water-saturated state, the micro-strain of the fissured sandstone showed a positive number, which was directly attributed to the involvement of moisture. It is worth noting that the peak maximum micro-strain induced by the freeze-up force showed a gradual increase with the repetition of the freeze–thaw cycle. This phenomenon suggests that moisture plays a key role in the freeze–thaw process, contributing to the accumulation and growth of micro-strain.

(2) After sophisticated analyses of uniaxial compression tests, we observed that the mechanical properties of intact sandstone as well as monofissile sandstone were affected by the number of freezing and thawing times. Specifically, the peak strength and modulus of elasticity of both sandstones show a gradual decrease with the increasing number of freeze–thaw events. It is particularly striking that the single-fissure sandstones show a higher degree of damage during freezing and thawing compared with the intact sandstones, which provides important clues for us to understand the mechanical behaviours of the rocks under freezing and thawing environments. Specifically, when the cleavage inclination angle was kept consistent, the peak strength and modulus of elasticity of the sandstone showed a significant decreasing trend with the increasing number of freeze-thawing, whilst the damage phenomenon of the cleaved sandstone became more obvious. On the other hand, at a fixed number of freezing and thawing times, the fissured sandstones with different inclinations exhibited a multi-peak pattern in the stress–strain curves compared to the intact sandstones. These sandstone samples have relatively long elastic phases. More detailed observations reveal that the peak strength and modulus of elasticity of the sandstones undergo a process of decreasing and then increasing with the gradual increase in the fissure inclination. In particular, the decrease is relatively small when the cleavage dip reaches 90°, whilst the decrease reaches a maximum at 60° dip.

(3) In the uniaxial compression test, the intact sandstone mainly shows splitting damage mode, and its crack distribution is relatively dense. However, when there are cracks in the sandstone, the damage pattern varies depending on the inclination angle of the cracks. Specifically, in the experiments, when we set the inclination angle of the fissure to 0° or 90°, the rock samples mainly experienced tensile or cleavage damage; however, it is interesting to note that, once the inclination angle of the fissure was adjusted to 30°, the pattern of deterioration in the rock samples significantly changed to a mixed tensile-shear damage, which was caused by the combined action of tensile and shear forces. On the other hand, with the cleavage inclination set to 45° and 60°, the rock samples are more likely to show shear damage characteristics dominated by shear slip. With the gradual increase in the number of freezing and thawing, the damage surface of the sandstone gradually shows more obvious roughening characteristics, showing the significant effect of freezing and thawing on the damage surface of the rock, and the direction of crack expansion is mainly along the direction of stress action. These cracks eventually penetrate into each other to form mixed tension-shear cracks, leading to more serious localised block loss in the rock samples.

These observations and analyses provide an important basis for our in-depth comprehension of the degradation process of sandstone.

(4) Based on the detailed analysis of experimental data, this paper successfully constructed a close relationship between the macro-mechanical parameters (including elastic modulus, compressive strength, etc.) and their micro-physical parameters (especially the change of pore structure) of fissured sandstone under the freeze–thaw environment. The findings indicate that not only causes the macro-mechanical parameters of the rocks to show a decreasing trend, but also their micro-structural characteristics exhibit significant changes. These changes indicate that the internal structure of the sandstone was damaged and reorganised during the freeze–thaw process, which had a profound effect on its overall mechanical properties. The mutual corroboration between macro- and micro-parameters provides a solid scientific basis for us to deeply reveal and understand the damage mechanism of fissured sandstone under freeze–thaw environment.

This paper mainly carries out the freeze–thaw cycle experimental research on the red sandstone with fissures in the cold zone, analysing the mechanical properties and micro-structure, and the results of the test can provide effective data support for the protection of the red sandstone rocky slopes in the cold zone, especially in the area of heavy rainfall according to the change of the water content of the red sandstone pre-treatment. However, the research in this paper did not consider the interaction of multiple factors in the natural environment, but only for the freeze–thaw cycle conditions for the study, in the future, research can be coupled under a variety of conditions under the red sandstone damage and deformation law to analyse.

References

1. Li C, Xiao Y, Wang Y et al (2019) Current status and trend of research on deformation and damage mechanism of rocky slopes in high-altitude cold areas. J Eng Process Sci 41(11):1374–1386
2. Fukuda M (1974) Rock weathering by freeze-thaw cycles. J Low Temp Phys 32:243–249
3. .Chen TC, Yeung MR, Mori N (2004) Effect of water saturation on deterioration of welded tuff due to freeze-thaw action. Cold RegionsScience Technol 38(2–3):127–136
4. Nicholson DT, Nicholson FH (2000) Physical deterioration of sedimentary rocks subjected to experimental freeze-thaw weathering. Earth Surf Proc Land 25:1295–1308
5. Lu Y, Li X, Wu X (2014) Mechanism of crack penetration in freeze-thawed single-fissure rock samples under triaxial compression conditions. Geotechnics 35(6):1579-1584
6. Ren J, Wang X, Chen X (2021) Research on physical and mechanical properties and damage characteristics of sandstone of Luohe Formation after thawing. Coal Eng 053(002):153–158
7. Shen Y, Yang G, Rong T et al (2017) Analysis of localised damage effects and end fracture characteristics of monoclinic sandstones under freeze-thaw cycling. J Rock Mech Eng 36(3):562-570
8. Zhao J, Xie M, Yu J et al (2019) Experimental study on mechanical properties and damage evolution law of fracture-containing rocks under freeze-thaw. J Eng Geol 27(6):1199-1207
9. Liu H, Yang G, Ren J (2007) Numerical analysis method of temperature field of freeze-thaw shale based on digital image processing. J Rock Mech Eng 26(08):1678–1683

10. Zhang H, Yuan C, Mu N et al (2022) CT image processing and detailed characterisation of freeze-thawed rocks. J Xi'an Univ Sci Technol 42(02):219–226
11. Liu J, Zhang H, Wang R et al (2021) Study on progressive damage deterioration law of sandstone layer under freeze-thaw cycle. Geotechnics 42(05):1381–1394
12. Yang H, Liu P, Sun B et al (2021) Study on microstructural damage mechanism of Maijishan Grotto gravels by freeze-thaw cycle. J Rock Mech Eng 40(03):545–555
13. Park J, Hyun CU, Park HD (2015) Changes in microstructure and physical properties of rocks caused by artificial freeze-thaw action. Bull Eng Geol Env 74:555–565
14. Maji V, Murton JB (2020) Micro-computed tomography imaging and probabilistic modelling of rock fracture by freeze-thaw. Earth Surf Proc Land 45(3):666–680
15. Zhang H, Meng X, Peng C et al (2019) Rock damage modelling based on residual strength characteristics under freeze-thaw-loading. J Coal 44(11):3404-3411

Open Access This chapter is licensed under the terms of the Creative Commons Attribution 4.0 International License (http://creativecommons.org/licenses/by/4.0/), which permits use, sharing, adaptation, distribution and reproduction in any medium or format, as long as you give appropriate credit to the original author(s) and the source, provide a link to the Creative Commons license and indicate if changes were made.

The images or other third party material in this chapter are included in the chapter's Creative Commons license, unless indicated otherwise in a credit line to the material. If material is not included in the chapter's Creative Commons license and your intended use is not permitted by statutory regulation or exceeds the permitted use, you will need to obtain permission directly from the copyright holder.

Research on a Machine Learning-Based Subgrade Compaction Degree Prediction Model

Feng Li, Jianfei Zhao, Hongzhao Li, Bing Hui, Zhenkun Wang, Wenjun Zhang, and Guangbo Liu

Abstract The highway subgrade is the foundation for ensuring the safe operation of highways, and subgrade compaction degree is a crucial parameter in subgrade construction. This study employs field tests to investigate subgrade compaction degree, analyzing the effects of rolling passes, rolling speed, and moisture content of subgrade fill material on subgrade compaction degree. Subsequently, four machine learning models are used to establish prediction models for subgrade compaction degree: random forest model, sparrow search algorithm optimized random forest model (SSA-RF), AdaBoost model, and BP-AdaBoost model. The results indicate that: (1) Subgrade compaction degree increases with the number of rolling passes, decreases with rolling speed, and is greater when the moisture content of the subgrade fill material is closer to the optimum moisture content. (2) The AdaBoost model shows the poorest prediction performance and is not suitable for predicting subgrade compaction degree, whereas the SSA-RF model achieves the best training results, with correlation coefficients (R^2) of 0.98 and 0.94 for the training and testing sets, respectively, closely matching the actual results, validating the feasibility of this method for predicting highway subgrade compaction degree.

Keywords Highway roadbed · Compaction degree · Field test · Machine learning · Model optimization

---

F. Li
Jining Highway Administration Wenshang Highway Bureau, Jining, Shandong, China

J. Zhao · Z. Wang
Gezhouba Group Transportation Investment Co., Ltd, Wuhan, Hubei, China

H. Li (✉) · B. Hui · W. Zhang · G. Liu
Shandong Transportation Institute, Jinan, Shangdong, China
e-mail: 1152774484@qq.com

1 Introduction

Highways play a pivotal role in economic development and social interactions, forming an essential component of modern transportation infrastructure. The roadbed is fundamental to the safe operation of highways, with its compaction degree being the most critical parameter during construction. Traditionally, the measurement of roadbed compaction relies on in situ experiments, primarily through sand cone and ring knife methods [1, 2]. However, these field tests are time-consuming and labor-intensive. Based on traditional compaction testing methods, this study establishes a predictive model for roadbed compaction using in situ compaction test parameters, which is of significant importance for controlling the construction process of highways.

In recent years, numerous experts and scholars have studied the factors affecting roadbed compaction [3–8], which primarily include moisture content, number of roller passes, and rolling speed. Researchers both domestically and internationally have analyzed these parameters and proposed various predictive models for roadbed compaction. For instance, Yang et al. [9] utilized a vast dataset from roadbed rolling experiments to develop a neural network model predicting roadbed compaction, introducing a novel approach to such predictions. Imran et al. [10] used a neural network model for predicting compaction and implemented real-time positioning technology to develop a compaction cloud map system. He [11] established a grey prediction model to predict compaction degrees across different rolling passes. Liu et al. [12] selected natural density, actual moisture content, optimal moisture content, and compaction conditions as input vectors for an artificial neural network prediction model. Wu [11] developed a real-time evaluation model for compaction quality based on support vector machines, showing certain classification and recognition capabilities. Wang et al. [13, 14] conducted extensive compaction tests to assess the compaction quality of various fill materials. By integrating these laboratory data and literature, a dataset was developed, leading to the creation of a PSO-BP-NN model that automatically predicts the compaction parameters of different filling materials.

These studies have established predictive models using machine learning and statistical analysis to predict roadbed compaction under various environmental conditions. However, current research on predicting the compaction of highway subgrades still faces several challenges. Firstly, there is a lack of large-sample and long-term observational data, which limits the accuracy and reliability of the research. Secondly, parameters like the number of roller passes, rolling speed, and moisture content of the roadbed fill, which are critical during construction, are seldom used in predictive studies.

Furthermore, soil conditions and construction techniques vary by region, making it difficult to directly apply research findings to other areas. Additionally, existing prediction models need further refinement and validation regarding their accuracy and reliability.

This study initially employs comprehensive experimental methods to conduct in situ tests on roadbed compaction, analyzing the effects of factors such as the number of roller passes, rolling speed, and moisture content of the roadbed fill. Subsequently, it uses machine learning models better suited for small-sample data predictions, such as the random forest and AdaBoost models, to establish predictive models for roadbed compaction. These models are then evaluated using field test data. The predictive models developed in this research could provide reference and guidance for highway construction.

2 Field Test

2.1 Overview of the Project

Field experiments were conducted to measure the compaction density of the roadbed. The test site was located along a segment of a highway at K8 + 931 to K9 + 201, spanning 270 m. The roadbed fill soil exhibited a maximum dry density of 1.734 g/cm^3 and an optimum moisture content of 12.4%.

2.2 Test Methods

The objective of this study was to develop a predictive model for roadbed compaction density, necessitating a large dataset; hence, a comprehensive experimental approach was adopted. The study focused on the impacts of rolling passes, rolling speed, and moisture content of roadbed fill material on compaction density. Specifically, the rolling passes were set to 6, 8, and 10, labeled as "A1, A2, A3", respectively; rolling speeds were set at 3 km/h, 4.5 km/h, and 6 km/h, labeled as "B1, B2, B3", respectively; moisture contents of the fill material were set at 10.4%, 12.4%, and 14.4%, labeled as "C1, C2, C3", respectively. The specific experimental methods are presented in Table 1.

As shown in Fig. 1, in the construction site, every 10 m was designated as a test area, resulting in a total of 27 test areas for conducting experiments. The compaction tests were carried out according to the experimental plan outlined in Table 1. After the completion of the compaction tests, sand cone tests were performed to determine the compaction density of the roadbed. Each test area underwent three repetitions of the sand cone test to ensure the stability and accuracy of the measurement data.

2.3 Results and Discussion

Field data from each test section is shown in Table 2. Figure 2 presents a comparison of roadbed compaction densities under different operational conditions, where larger

Table 1 Subgrade compactness field test scheme

Serial number	Test combinations	Serial number	Test combinations	Serial number	Test combinations
1	A1B1C1	10	A2B1C1	19	A3B1C1
2	A1B1C2	11	A2B1C2	20	A3B1C2
3	A1B1C3	12	A2B1C3	21	A3B1C3
4	A1B2C1	13	A2B2C1	22	A3B2C1
5	A1B2C2	14	A2B2C2	23	A3B2C2
6	A1B2C3	15	A2B2C3	24	A3B2C3
7	A1B3C1	16	A2B3C1	25	A3B3C1
8	A1B3C2	17	A2B3C2	26	A3B3C2
9	A1B3C3	18	A2B3C3	27	A3B3C3

Fig. 1 Subgrade compression field test

spheres indicate higher compaction densities. Compaction density increases with the number of rolling passes and decreases with the rolling speed. Moreover, the compaction density increases as the moisture content of the roadbed fill approaches the optimum moisture content. The highest representative compaction density value was observed in the test area with a moisture content of 12.4%, ten rolling passes, and a rolling speed of 3 km/h, reaching 98.89%.

Exploring the reasons, when the moisture content of the fill is low, water fails to provide sufficient lubrication between soil particles, increasing the frictional resistance during compaction. Continuous rolling does not overcome this resistance, leading to no further reduction in the gaps between soil particles, and the compaction density fails to meet the expected standards. From the perspective of rolling passes, too few passes fail to adequately compact the lower layers of the roadbed, potentially leading to a loose roadbed structure; conversely, excessive rolling not only extends the project duration and reduces cost-effectiveness but might also damage the bonding

Table 2 Test results of the test section

Serial number	Compaction (%)	Serial number	Compaction (%)	Serial number	Compaction (%)
1	93.59	10	94.06	19	94.85
2	96.38	11	97.61	20	98.89
3	94.04	12	95.09	21	96.06
4	92.28	13	93.29	22	93.96
5	95.12	14	97.19	23	98.38
6	93.84	15	94.89	24	95.24
7	91.53	16	91.92	25	93.52
8	94.96	17	96.39	26	97.93
9	93.62	18	94.43	27	95.09

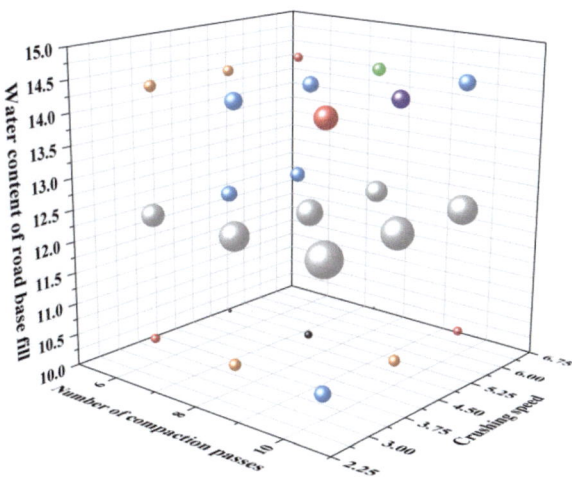

Fig. 2 Subgrade compactness comparison diagram

strength between soil layers, potentially causing future roadbed settlements or an inability to support the load of heavy vehicles.

Further, this study conducts a quantitative analysis on the relationship between subgrade compaction, number of rolling passes, rolling speed, and the moisture content of subgrade fill. Utilizing field trial data and compiled using Python 3.6 in the Anaconda distribution on the JupyterLab platform, correlation heat maps were created to illustrate the interdependencies between the number of rolling passes, rolling speed, moisture content of the fill, and the degree of compaction. The correlation coefficient, which quantifies the degree of association between two variables, is calculated as follows:

$$\rho_{A_1 A_2} = \frac{\mathrm{Cov}(A_1, A_2)}{\sqrt{DA_1, DA_2}} = \frac{EA_1 A_2 - EA_1 * EA_2}{\sqrt{DA_1 * DA_2}} \quad (1)$$

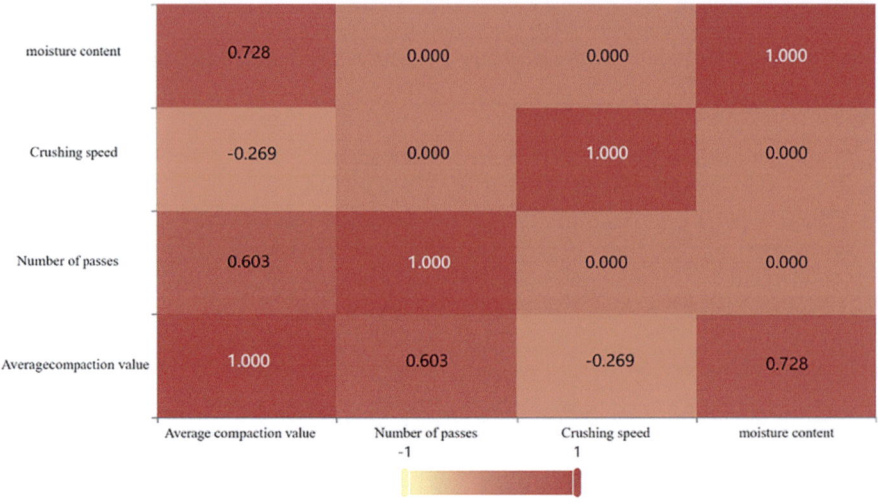

Fig. 3 Compactability-related factors thermal map

P denotes the correlation coefficient, Cov represents the covariance, and *E* denotes the expected value.

As illustrated in Fig. 3, the number of rolling passes is positively correlated with compaction degree, while rolling speed and compaction degree are negatively correlated. The absolute value of the correlation coefficient between moisture content and compaction degree is the largest, indicating the highest correlation between moisture content and compaction degree, followed by the number of rolling passes (0.603) and rolling speed (− 0.269).

3 Roadbed Compaction Prediction Model

3.1 Random Forest Model

In the 1980s, Breiman and others developed the classification tree algorithm, which significantly reduced computational demands through the repeated binary splitting of data for classification or regression purposes. In 2001, Breiman enhanced this approach by amalgamating multiple classification decision trees into a cohesive ensemble known as the random forest. This method constructs numerous decision trees and aggregates their predictive outcomes. It introduces a random process into the selection of input features (columns) and sample data (rows). This technique improves the model's predictive accuracy without notably increasing computational costs and remains relatively insensitive to high inter-variable correlation. Random forest algorithms exhibit robust resilience to missing values and unbalanced data, effectively predicting the impact of thousands of feature variables. In scenarios with limited sample sizes, random forests offer advantages against overfitting and excel in extracting significant features. Due to its exemplary performance, it is widely regarded as one of the finest algorithms available today. The specific algorithm is as follows:

(1) Utilizing information entropy to characterize the purity of a sample set, the information entropy of sample set B is computed by assuming that the proportion of the ith class samples in B is $Bi(i = 1, 2, \ldots, |y|)$, The information entropy, $\text{Ent}(B)$, is defined as:

$$\text{Ent}(B) = E[-\log B_i] = -\sum_{i=1}^{|y|} B_i \log_2 B_i \quad (2)$$

(2) Attribute b can take n values $(b_1, b_2, b_3, \ldots, b_n)$, resulting in the creation of n branch nodes. The Nth branch node encompasses all samples in D where attribute b is equal to b_n. These samples are denoted as B_n, for which the information entropy is calculated. Each B_n is assigned a node weight of $|B_n|/|B|$, where branches with a larger number of samples have a greater influence. The information gain $g(B, b)$ is computed by using b to partition B:

$$g(B, b) = \text{Ent}(B) - \sum_{n=1}^{N} \frac{|B_n|}{B} \text{Ent}(B_n) \quad (3)$$

(3) Select the variable that maximizes the information gain ratio and the Gini index as the split node variable, and continue until the entropy at the leaf node is reduced to zero. The information gain ratio is defined as:

$$gr(B, b) = \frac{g(B, b)}{f(b)} \quad (4)$$

where fb is the intrinsic value of attribute b.

$$f(b) = -\sum_{n=1}^{N} \frac{|B_n|}{|B|} \log 2 \frac{|B_n|}{|B|} \quad (5)$$

The formula for calculating the Gini index is:

$$\text{GINI}(B) = 1 - \sum_{i=1}^{|y|} Bi^2 \quad (6)$$

Based on the selected combinations of sample variable parameters, a substantial number of decision trees capable of parallel processing are generated. Each decision

Table 3 Random forest model parameter table

Parameter name	Parameter value
Min_samples_split	2
Min_samples_leaf	1
Max_depth	10
Max_leaf_nodes	50
Number of decision trees	100

tree yields a predictive outcome based on the data. The final regression result of the random forest is derived from aggregating the outcomes of all decision trees, selecting the result that appears most frequently. The parameters for the random forest model are presented in Table 3.

3.2 Optimize Random Forest Model Based on Sparrow Search Algorithm

The sparrow search algorithm (SSA) was introduced in 2020, primarily mimicking the foraging and anti-predatory behaviors of sparrows to optimize position and find local optima for some NP-hard problems [15]. Within the algorithm's framework, the sparrow population is divided into discoverers and followers, while also incorporating a danger warning mechanism that simulates real predation scenarios. Based on the sparrow search algorithm, an optimization of the number of decision trees and the minimum number of leaf nodes in the random forest model was conducted. The optimal configuration resulted in 96 decision trees and a minimum of one leaf node per tree. This model is hereafter referred to as the SSA-RF model. The specific workflow is illustrated in Fig. 4.

3.3 AdaBoost Model

AdaBoost is an iterative classification algorithm based on weak classifiers. Initially, each sample in the training set is assigned an equal specific initial weight, unless otherwise specified. The process begins by selecting the first classifier to train a weak classifier on the training data and calculating its error rate. The weights of the incorrectly classified samples are then increased, while those of the correctly classified samples are decreased. This adjustment causes the misclassified samples to be reclassified by the weak classifier within the same training set in subsequent iterations. This iterative process continues until the error rate meets the specified requirements. By iteratively changing the weights, the classifier increasingly focuses

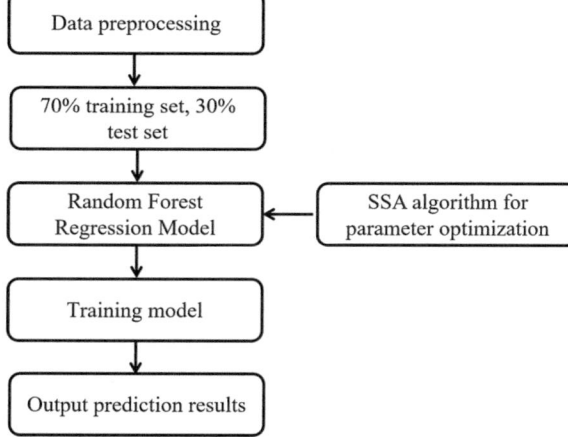

Fig. 4 Stochastic forest model flow based on SSA optimization

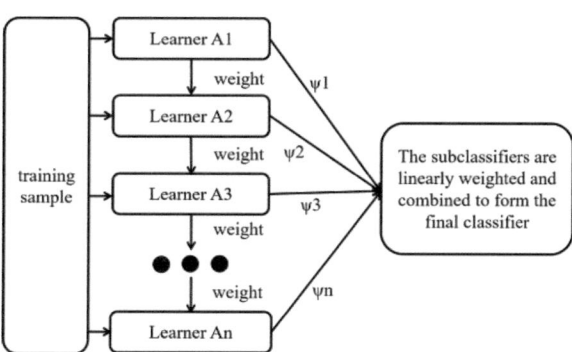

Fig. 5 AdaBoost flowchart

on samples that are prone to misclassification, thereby reducing the overall error rate. The AdaBoost model flowchart is shown in Fig. 5.

In the AdaBoost machine learning model employed in this study, the base learners are decision tree classifiers, with a total of 100 trees utilized. The loss function selected is the linear loss function.

3.4 BP-AdaBoost Neural Network Modeling

In this approach, the backpropagation (BP) neural network is employed as a weak regressor. By repeatedly training the BP neural network to predict sample outputs, the AdaBoost algorithm is utilized to aggregate multiple BP neural networks into a strong regressor. The flowchart of BP-AdaBoost network is shown in Fig. 6.

The detailed steps of the algorithm are as follows:

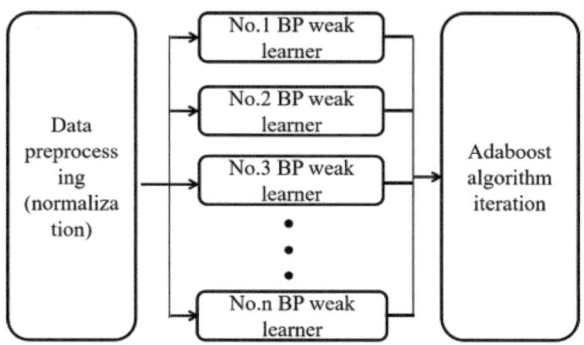

Fig. 6 BP-AdaBoost model flow

(1) Sample data selection and network initialization. Randomly select (t) groups of training data from the sample data, initialize the distribution weights of test data ($Di(i) = 1/t$), determine the structure of the neural network based on the input and output dimensions of the samples, and initialize the weights and thresholds of the neural network.
(2) Normalize the data;
(3) When training the (n)-th weak predictor, use a BP neural network to train the training data. Calculate the prediction error sequence (f_n) and the sum of prediction errors (e_n). The formula for the sum of prediction errors is:

$$e_n = \sum_n D_i(i); i = 1, 2, \ldots, t \tag{7}$$

(4) Calculate the prediction sequence weights. Based on the sum of prediction errors (e_n) from the prediction sequence (f_i), compute the weight (a_n) of the sequence:

$$a_n = \frac{1}{2} In\left(\frac{1-e_n}{e_n}\right) \tag{8}$$

(5) Adjusting weights of test data. Adjust the weights of new training samples based on the prediction sequence weight (a_n), using the adjustment formula shown as Eq. (9). In the equation, (B_n) is a normalization factor primarily used to ensure that the sum of distribution weights equals 1, while keeping the proportion of weights unchanged.

$$D_{n+1}(i) = \frac{D_n(i)}{B_n} \times \exp[-a_n y_i g_n(x_i)]; i = 1, 2, \ldots, t \tag{9}$$

(6) Output the predictor function. After (N) iterations, (N) sets of weak predictors, ($f(g_n, a_n)$), are obtained. These are then combined to form a strong predictor function, ($F(x)$), as shown in Eq. (10):

$$F(x) = \frac{a_n}{\sum_{n=1}^{N} a_n} f(x) \qquad (10)$$

4 Assessment of Projected Results

The sample data was divided into a training set and a test set, with the training set and test set accounting for 70% and 30% of the total data, respectively. Specifically, the training set consisted of 19 samples, while the test set included 8 samples. The allocation of samples to either the training or test set was determined using a random number approach. In this study, the experiment numbers for the training set were 1, 3, 4, 5, 6, 7, 8, 9, 10, 12, 13, 15, 16, 18, 19, 23, 25, 26, and 27. The experiment numbers for the test set were 2, 11, 14, 17, 20, 21, 22, and 24.

To mitigate the impact of significant disparities in the data scales of various input indicators on the algorithm's performance, it is essential to normalize each indicator. This study employed the Max–Min normalization technique, which involves a linear transformation of the raw data. Let Amax represent the maximum value of attribute A, and Amin denote the minimum value of attribute A. An original value (x) of attribute A is mapped to a value (x') in the range [0, 1] using Max–Min normalization, as illustrated by the following formula:

$$x' = \frac{x - A_{\min}}{A_{\max} - A_{\min}} \qquad (11)$$

To evaluate the performance of four different models, three statistical metrics are often used: the coefficient of determination (R^2), mean absolute error (MAE), and mean absolute percentage error (MAPE). Each of these metrics offers a different insight into the accuracy and performance of a predictive model. Here is a brief description of these metrics along with the formula for calculating MAE:

$$\text{MAE} = \frac{1}{n} \sum_{i=1}^{n} \left| Q_{e[i]} - Q_{m[i]} \right| \qquad (12)$$

where (n) is the total number of samples, $Q_{m[i]}$ represents the experimental measurement, and $Q_{e[i]}$ denotes the model-predicted value.

The mean absolute percentage error (MAPE) is defined as:

$$\text{MAPE} = \frac{1}{n} \sum_{i=1}^{n} \left| \frac{Q_{e[i]} - Q_{m[i]}}{Q_{m[i]}} \right| \times 100\% \qquad (13)$$

A MAPE value close to 0% indicates high predictive accuracy of the model, whereas a MAPE value close to 100% indicates very low predictive accuracy.

A larger correlation coefficient (R^2) indicates a higher modeling accuracy, which is calculated by the following equation:

$$R^2 = 1 - \frac{\sum_{i=1}^{n}(\hat{y}_i - y_i)^2}{\sum_{i=1}^{n}(\bar{y}_i - y_i)^2} \tag{14}$$

where y_i denotes the test value of roadbed compaction; $\hat{y}_i$ denotes the model-predicted value of roadbed compaction; $\bar{y}_i$ denotes the average test value of roadbed compaction; and n is the sample size.

Figures 7 and 8 illustrate the comparison between predicted and actual values for both the training and testing datasets, while Table 4 presents the performance metrics of four models across these datasets. As shown in Table 5, when using the number of rolling passes, rolling speed, and subgrade filler moisture content as inputs to predict the compaction degree of subgrade, the coefficients of determination (R^2) for the random forest, SSA-RF, and BP-AdaBoost models exceed 0.91 on the training set, with the MAE and MAPE both below 0.20. In contrast, the AdaBoost model alone exhibits a lower R^2 of 0.81 with both MAE and MAPE exceeding 0.30, indicating suboptimal training results and suggesting its unsuitability for predicting subgrade compaction.

For the four models, the training dataset consistently shows lower MAE and MAPE values compared to the testing dataset, and the values of R^2 are also higher for the training dataset. After optimization using the sparrow search algorithm, the SSA-optimized random forest model demonstrates lower MAE and MAPE values and a higher R^2 compared to the unoptimized version, signifying a significant improvement in parameter optimization. Further, the hybrid BP neural network and AdaBoost model, BP-AdaBoost, displays lower MAE and MAPE values and a higher R^2 than

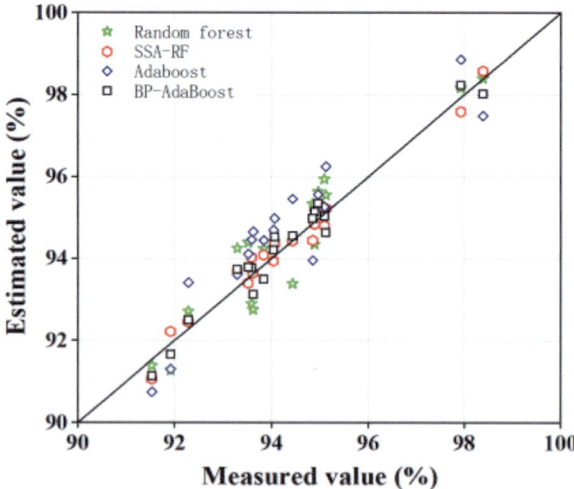

Fig. 7 Comparison of experimental and predicted values of the training set

Fig. 8 Comparison of experimental and predicted values of the test set

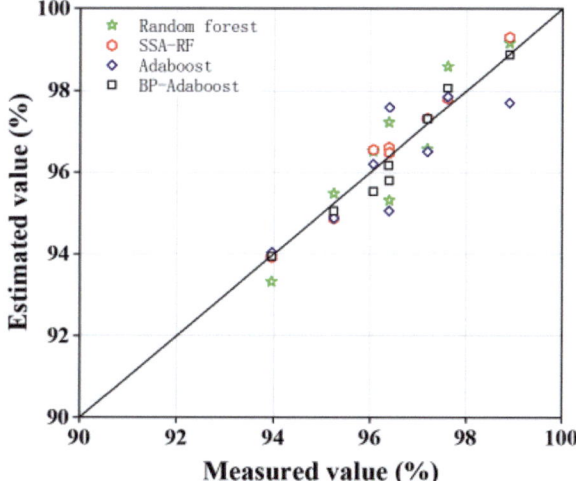

Table 4 AdaBoost parameter list

Parameter name	Parameter value
Base classifier	Decision tree
Number of base classifiers	100
Loss function	Linear

Table 5 Prediction model evaluation index

Model	Training set			Prediction set		
	MAE	MAPE	R^2	MAE	MAPE	R^2
Random forest	0.19	0.20	0.91	0.24	0.25	0.87
SSA-RF	0.11	0.12	0.98	0.17	0.18	0.94
AdaBoost	0.31	0.32	0.81	0.42	0.43	0.72
BP-AdaBoost	0.08	0.09	0.96	0.31	0.10	0.11

the standalone AdaBoost model, indicating its superior suitability for predicting subgrade compaction.

Among the four models, the SSA-RF model achieves the smallest MAE and MAPE and the highest R^2, indicating its superior predictive capability and precision.

In practical applications, given the rapid advancements in computer technology and the significant computational power of modern computers, this methodology holds extensive future potential for application.

5 Conclusion

This study first employs a comprehensive experimental approach to conduct field tests on subgrade compaction degree, analyzing the impact of environmental parameters on subgrade compaction degree. Subsequently, two machine learning models, the random forest model and the AdaBoost model, are used to establish prediction models for subgrade compaction degree. The conclusions are summarized as follows:

(1) Subgrade compaction degree increases with the number of rolling passes and decreases with rolling speed. The closer the moisture content of the subgrade material is to the optimum moisture content, the greater the subgrade compaction degree.

(2) The number of rolling passes is positively correlated with compaction degree, while rolling speed is negatively correlated with compaction degree. The absolute value of the correlation coefficient between moisture content and compaction degree is the highest, indicating the strongest correlation, followed by rolling passes (0.603) and rolling speed ($-$ 0.269).

(3) Four subgrade compaction degree prediction models are established: random forest, SSA-RF, AdaBoost, and BP-AdaBoost. In the testing set, the correlation coefficients (R^2) are 0.91, 0.98, 0.81, and 0.96, respectively. In the training set, the correlation coefficients (R^2) decrease. The AdaBoost model has the poorest training performance and is not suitable for predicting subgrade compaction degree, whereas the SSA-RF model has the best training performance and shows broad application prospects in predicting subgrade compaction degree.

(4) The experimental data of this study is derived from a single site, which may limit the generalizability of subgrade compaction degree predictions for different regions with varying soil types. However, with the continual expansion of data in the future, this method is expected to be applicable to different soil types across various regions.

References

1. Xu B (2021) Quality inspection method of layered compacted subgrade and engineering example analysis. E3S Web Conf 248:03068
2. Zheng M, Yang JM, Li ZL et al (2014) Plate load test applied to detect bearing capacity of the miscellaneous fill subgrade reinforced by the dynamic compaction. Adv Mater Res 1065–1069:778–782
3. Wang WM (2020) Variability of pavement compaction degree in transition section of highway and bridge and construction control technology. Highw Eng 45(02):128–132
4. Zhang D (2023) Influencing factors and control measures of highway subgrade compaction degree. Transpoworld (11):122–124
5. Zuo J, Gao X, Ma YJ et al (2022) Correlation and influencing factors of continuous test indexes of subgrade compaction quality. Railw Eng 62(10):17–21
6. Qiu YQ, Zhang LJ, Yin LH et al (2023) Research on compaction characteristics of Yunnan Red Clay subgrade filler. Highway 68(10):91–98

7. Hong L (2023) Study on factors and control technology of differential settlement of subgrade in expressway reconstruction and expansion. Transpoworld (24):44–46
8. Yang XC, He CP (2011) Prediction model of subgrade compaction based on BP artificial neural network. J Gansu Sci 23(03):132–135
9. Imran SA, Commuri S, Barman M et al (2017) Modeling the dynamics of asphalt–roller interaction during compaction. J Constr Eng Manag 143(7):04017015
10. He SY (2020) The GM(1, 1)model of Gray theory to forecast the subgrade compaction. Constr Technol 49(S01):1350–1353
11. Liu GG, Pei LY, Yang YM et al (2021) Compactness prediction of airport soil field based on artificial neural network. J Shenzhen Univ Sci Eng 38(01):54–60
12. Wu QL (2021) Real-time evaluation on compaction quality of pavement based on machine learning. Shandong University
13. Wang X, Dong X, Li J et al (2023) Developing an advanced ANN-based approach to estimate compaction characteristics of highway subgrade. Adv Eng Inform 56:102023
14. Wang X, Cheng C, Li J et al (2023) Automated monitoring and evaluation of highway subgrade compaction quality using artificial neural networks. Autom Constr 145:104663
15. Xue JK, Shen B (2020) A novel swarm intelligence optimization approach: sparrow search algorithm. Syst Sci Control Eng 8(1):22–34

Open Access This chapter is licensed under the terms of the Creative Commons Attribution 4.0 International License (http://creativecommons.org/licenses/by/4.0/), which permits use, sharing, adaptation, distribution and reproduction in any medium or format, as long as you give appropriate credit to the original author(s) and the source, provide a link to the Creative Commons license and indicate if changes were made.

The images or other third party material in this chapter are included in the chapter's Creative Commons license, unless indicated otherwise in a credit line to the material. If material is not included in the chapter's Creative Commons license and your intended use is not permitted by statutory regulation or exceeds the permitted use, you will need to obtain permission directly from the copyright holder.

Distribution Characteristics of Ground Stress Field in the Underground Caverns Under Complex Geological Conditions

Peiyang Yu, Jun He, Yang Qin, and Jianhua He

Abstract To explore the stress field characteristics of underground caverns under complex geological conditions, Yebatan hydropower station, located in Sichuan Province, is taken as the engineering background. A three-dimensional geological model of engineering region is established, which is based on the results of geological investigation. The regression analysis of ground stress characteristics is carried out in combination with the measured ground stress, and then, the evolution of the magnitude of ground stress in the whole engineering region is revealed. Finally, the stress orientation and magnitude characteristics of the stress field of the deep-buried underground caverns under complex geological conditions are studied, and the internal mechanism of stress field evolution is discussed.

Keyword Ground stress field · Underground caverns · Complex geological conditions · Multiple regression analysis

1 Introduction

The hydropower resources in Southwest China are extremely abundant. However, due to the influence of complex geological conditions and tectonic movements, the underground caverns of large hydropower stations in this region are often in a high ground stress environment, and the ground stress changes dramatically [1]. As the ground stress field is the main factor affecting the stability of underground caverns, it is necessary to study the stress field characteristics of underground caverns under

---

P. Yu · J. He (✉)
Key Laboratory of Geotechnical Mechanics and Engineering of the Ministry of Water Resources, Yangtze River Scientific Research Institute, Wuhan, China
e-mail: hejun@mail.crsri.cn

P. Yu
e-mail: yupeiyang18@mails.ucas.ac.cn

Y. Qin · J. He
Power China Chengdu Engineering Corporation Limited, Chengdu, Sichuan, China

complex geological conditions, which is of great significance to the site selection, design, construction, and operation of hydropower stations in complex geological engineering regions.

The ground stress testing and numerical analysis method are usually adopted to acquire the stress field characteristics of underground caverns. However, the measured ground stress magnitude only represents the stress state of isolated points in the engineering regions, and it is difficult to represent the distribution characteristics of ground stress field in the whole regions [2]. Therefore, according to the limited measured ground stress data, obtaining the stress field distribution of the engineering regions through numerical simulation has become the most common research method [3]. At present, the numerical analysis methods of in situ stress commonly used in engineering mainly include multiple regression analysis [4, 5], boundary loading adjustment method [6], stress/displacement function method [3], neural network, and genetic algorithm [7].

Therefore, many scholars have studied the stress field characteristics of underground caverns under complex geological conditions by means of ground stress testing and numerical simulation. The results show that in some regions with large topographic variations, complex lithology, and developed tectonic zones, such as deep valley slope, its peripheral stress field has obvious zonal phenomenon, which is mainly affected by various factors such as tectonic stress and unloading depth [8]. Meanwhile, the distribution of stress fields near faults is also affected by active faults, fault scale, fault geometry, fault parameters, and boundary stress ratio and other factors [9]. Although the above-mentioned research has achieved rich results, further exploration is still needed to understand the stress field characteristics of underground caverns under complex geological conditions.

On the basis of previous studies, this study takes the Yebatan hydropower station as an example and combines the measured ground stress data and geological investigation to explore the ground stress field characteristics of underground caverns. Furthermore, the influence of complex geological conditions on the distribution characteristics of stress field is obtained, which can provide reference for the construction of this project.

2 Regional Geological Environment

The engineering region is in the erosion plateau area in the south-east of the Qinghai–Tibet Plateau. Under the superposition of tectonic and elf-weight stress fields, the ground stress in the natural state in the engineering region is higher. The exposed bedrock in the engineering region is relatively simple, which is a simple composite strain of intermediate acid intrusion in Variscan age, and its lithology is quartz diorite and granodiorite. The lithology of the underground caverns is mainly quartz diorite, the rock mass is mainly classified as III_1 and III_2, and the rock mass around the fault is classified as IV and V.

Faults are relatively developed in the engineering region, amongst which there are mainly three faults with fracture zone widths greater than 1 m and extension lengths greater than 1000 m (i.e. F2, F3, and F4). And, 74 faults with fracture zone widths ranging from 0.1 to 1 m and extension lengths ranging from 100 to 1000 m are found. Hundreds of faults with width of less than 0.1 m and extension length of 10–100 m are revealed.

The measuring positions of ground stress are mainly selected in PD04 and PD08, and 13 groups of ground stress measurements are carried out by aperture deforming method. The maximum principal stress in underground caverns ranges from 16.51 to 37.57 MPa, with an average value of 24.31 MPa, and the orientation ranges from N82°E to N54°W, with an average orientation of N80.4°W, which is close to the direction of the regional tectonic principal compressive stress. The second principal stress is approximately 8.06–19.51 MPa, and the minimum principal stress is 3.85–15.03 MPa.

3 Inversion of Initial Ground Stress Field of Underground Caverns

3.1 Three-Dimensional Geological Model of Engineering Region

According to the geological investigation and the above analysis, the bedrock and main faults of the engineering region are generalized, and the three-dimensional geological model with the assistance of 3D discrete element numerical simulation method (i.e. 3DEC system) is established (Fig. 1). The influencing factors such as terrain and geomorphic features, rock mechanical properties, structural combination characteristics, and geological structure are considered comprehensively. Three large faults with fracture zone widths greater than 1 m and extension lengths greater than 1000 m (i.e. F2, F3, and F4), twenty-five small-scale faults with fracture zone widths of 0.1–1 m and extension lengths of 100–1000 m (i.e. f9, f21, f26, f27, f85, f87, f89, f82, f81, f28, f23, f71, f80, f88, f90, f91, f97, f31, fcF-14, fcF-10, fcF-9, fcF-7, fcF-13, fcF-6, and fcF-11), and two fracture dense zones, namely pd08-sc-lm2 and pd08-lm, are considered in the overall model. In addition, the unloading relaxation zone is set in the surrounding rock of main powerhouse, main transformer chamber, and tailrace surge chamber, respectively. Three large faults and two fracture dense zones are simulated by thin-layer solid elements in the model, whilst the other faults are simulated by contact surface elements.

In the three-dimensional geological model, the origin of the coordinate system is located at the centre point of Unit 1. The X-axis is parallel to the direction of water flow and points downstream, the Y-axis is parallel to the axis of the main plant and points from the centre point of Unit 1 to the centre point of Unit 6, and the Z-axis points vertically upward. The bottom elevation is 2400 m, and the model dimensions are

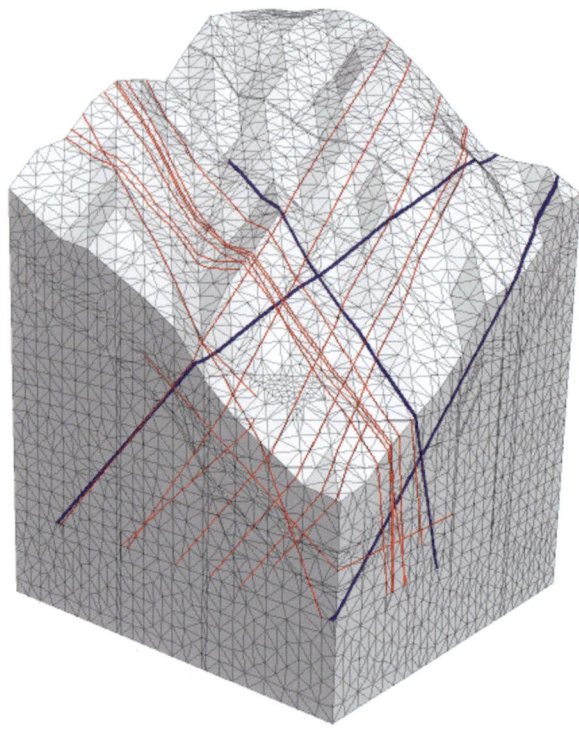

Fig. 1 Three-dimensional geological model of Yebatan hydropower station

660 m × 660 m × 1000 m ($X \times Y \times Z$). Tetrahedral elements are used for meshing, with 2,140,633 elements and 1,022,302 nodes. The rock mass constitutive model used is an ideal elastic–plastic model with the Mohr–Coulomb strength criterion with a tensile cut-off limit as the yield function. The fault constitutive model used is a regional contact elastic–plastic model under Coulomb sliding failure. Table 1 presents the value of rock mass mechanical parameters.

Table 1 Rock mass mechanical parameters

Materials	Young's modulus, E (GPa)	Poisson's ratio, n	Cohesions, (MPa)	Internal friction angle, (°)	Uniaxial tensile strength, σ_t (MPa)
Rock	11.00	0.26	1.10	47.20	1.00
F2, F3 and F4	1.00	0.35	0.30	28.81	0.165

3.2 Multiple Regression Analysis of Ground Stress Field

According to the geomechanical analysis, the inversion analysis of the ground stress field can consider the ground stress field in the computational domain as a linear superposition of the self-weight stress field and the boundary applied tectonic stress field. By decomposing and simulating the self-weight stress field and boundary load stress field, the initial stress field is finally formed by combining them. The self-weight stress field uses the measured weight of rock mass to generate the self-weight stress field under the action of gravity, with normal displacement constraints applied to the side and bottom surfaces of the model. The simulation of the tectonic stress field involves applying horizontal displacements on both sides to simulate the forces acting in the horizontal direction due to tectonic movement. The constraints on the non-loaded side boundaries and bottom boundaries are the same as those in the self-weight stress field simulation. The simulation of shear forces in the horizontal plane is achieved by applying boundary tangential horizontal displacements.

On the basis of the measured results of ground stress, the numerical simulation of stress field using multiple regression analysis method is as follows: the calculated value of ground stress regression is taken as the dependent variable, and the stress values corresponding to the measured points from simulation of the self-weight stress field and tectonic stress field are used as independent variables. The form of the regression equation is then determined:

$$\widehat{\sigma}_k = \sum_{i=1}^{n} L_i \sigma_k^i \tag{1}$$

Here, k is the serial number of observation points. σ_k^i is regression result of the k-th observation point. L_i is the multiple regression coefficient corresponding to the independent variable. Assuming that there are m observation points, the residual sum of squares of least squares is as follows:

$$S_r = \sum_{k=1}^{m} \sum_{j=1}^{6} \left(\sigma_{jk}^* - \sum_{i=1}^{n} L_i \sigma_{jk}^i \right)^2 \tag{2}$$

Here, σ_{jk}^* is observed value of the stress component at the k-th observation point. σ_{jk}^i is simulated value of the stress component at the k-th observation point. According to the principle of least squares, the formula that minimizes Sr is as follows:

$$\begin{bmatrix} \sum_{k=1}^{m}\sum_{j=1}^{6}\left(\sigma_{jk}^{1}\right)^{2} & \sum_{k=1}^{m}\sum_{j=1}^{6}\sigma_{jk}^{1}\sigma_{jk}^{2} & \sum_{k=1}^{m}\sum_{j=1}^{6}\sigma_{jk}^{1}\sigma_{jk}^{n} \\ \text{Symmetry} & \sum_{k=1}^{m}\sum_{j=1}^{6}\left(\sigma_{jk}^{2}\right)^{2} & \sum_{k=1}^{m}\sum_{j=1}^{6}\sigma_{jk}^{2}\sigma_{jk}^{n} \vdots \\ & \vdots & \vdots \\ \text{Symmetry} & \text{Symmetry} & \sum_{k=1}^{m}\sum_{j=1}^{6}\left(\sigma_{jk}^{n}\right)^{2} \end{bmatrix} \begin{Bmatrix} L_{1} \\ L_{2} \\ \vdots \\ L_{n} \end{Bmatrix} = \begin{Bmatrix} \sum_{k=1}^{m}\sum_{j=1}^{6}\sigma_{jk}^{*}\sigma_{jk}^{1} \\ \sum_{k=1}^{m}\sum_{j=1}^{6}\sigma_{jk}^{*}\sigma_{jk}^{2} \\ \vdots \\ \sum_{k=1}^{m}\sum_{j=1}^{6}\sigma_{jk}^{*}\sigma_{jk}^{n} \end{Bmatrix} \quad (3)$$

By solving the above equation, the undetermined regression coefficient $L = (L_1, L_2, \ldots, L_n)^T$ can be determined, and then, the regression value of initial stress at any point P in the calculation domain can be obtained by superimposing the calculated values of each boundary load condition at that point.

$$\sigma_{jp} = \sum_{i=1}^{n} L_i \sigma_{jp}^i \quad (4)$$

Here, $j = 1, 2, 3, 4, 5, 6$ corresponds to the six components of the initial stress.

3.3 Multiple Regression Results of Ground Stress Field

According to the above regression analysis, the stress distribution of rock mass in the engineering region is obtained, and the calculated stress value of the measuring points is compared with the measured value, as shown in Table 2. It can be observed that the regression calculated values of the initial stress are close to the measured values, indicating that the stress field in the engineering region obtained through multiple regression analysis is reasonable and reliable. Figure 2 shows the distribution of principal stresses in the engineering region.

4 Ground Stress Field Characteristics of Underground Caverns

4.1 Stress Orientation Characteristics of Ground Stress Field

Figure 3 shows the stress vector on the transverse profile of the centreline of Units 1 # and 4 #. The orientation of the maximum principal stress is predominantly NWW ~ EW, trending towards the valley, indicating that the inverted stress orientation is relatively consistent with the measured stress orientation. In the XZ plane, the maximum principal stress in Units 1# and 4# is deflected to the upstream, whilst

Table 2 Comparison of measured and calculated values

No.		σ_{xx} (MPa)	σ_{yy} (MPa)	σ_{zz} (MPa)	τ_{xy} (MPa)	τ_{xz} (MPa)	τ_{yz} (MPa)
σ_{PD08}-1	Measured value	−12.78	−13.28	−9.14	1.22	−0.44	4.41
	Calculated value	−19.05	−25.38	−17.52	5.97	−5.48	2.93
σ_{PD08}-2	Measured value	−23.04	−30.23	−17.31	9.03	−3.99	3.29
	Calculated value	−16.72	−22.43	−16.41	4.60	−6.12	4.73
σ_{PD08}-3	Measured value	−8.76	−17.82	−5.73	1.07	1.62	2.85
	Calculated value	−17.89	−25.89	−17.12	5.04	−6.58	4.76
σ_{PD08}-4	Measured value	−15.29	−22.36	−12.01	6.43	−6.17	−5.02
	Calculated value	−15.82	−21.66	−18.24	4.55	−3.69	3.10
σ_{PD08}-6	Measured value	−14.51	−12.01	−10.05	6.06	−4.52	5.69
	Calculated value	−15.69	−23.13	−18.89	5.75	−1.89	0.78
σ_{PD08}-7	Measured value	−14.30	−13.53	−17.53	2.52	−4.08	9.70
	Calculated value	−16.36	−23.59	−17.56	5.14	−5.42	3.14
σ_{PD08}-8	Measured value	−9.25	−22.11	−19.34	3.70	−5.58	7.46
	Calculated value	−16.11	−18.03	−17.18	4.39	−2.71	3.70
σ_{PD08}-9	Measured value	−12.87	−19.89	−12.47	6.68	−2.32	1.77
	Calculated value	−16.46	−20.24	−18.21	4.95	−3.39	5.00
σ_{PD08}-10	Measured value	−8.24	−19.61	−10.61	6.21	−1.14	−1.13
	Calculated value	−16.23	−20.58	−17.35	4.61	−5.09	5.39
σ_{PD08}-12	Measured value	−10.92	−7.50	−11.84	4.51	−3.45	0.97
	Calculated value	−14.64	−19.21	−17.55	4.49	−2.12	2.86

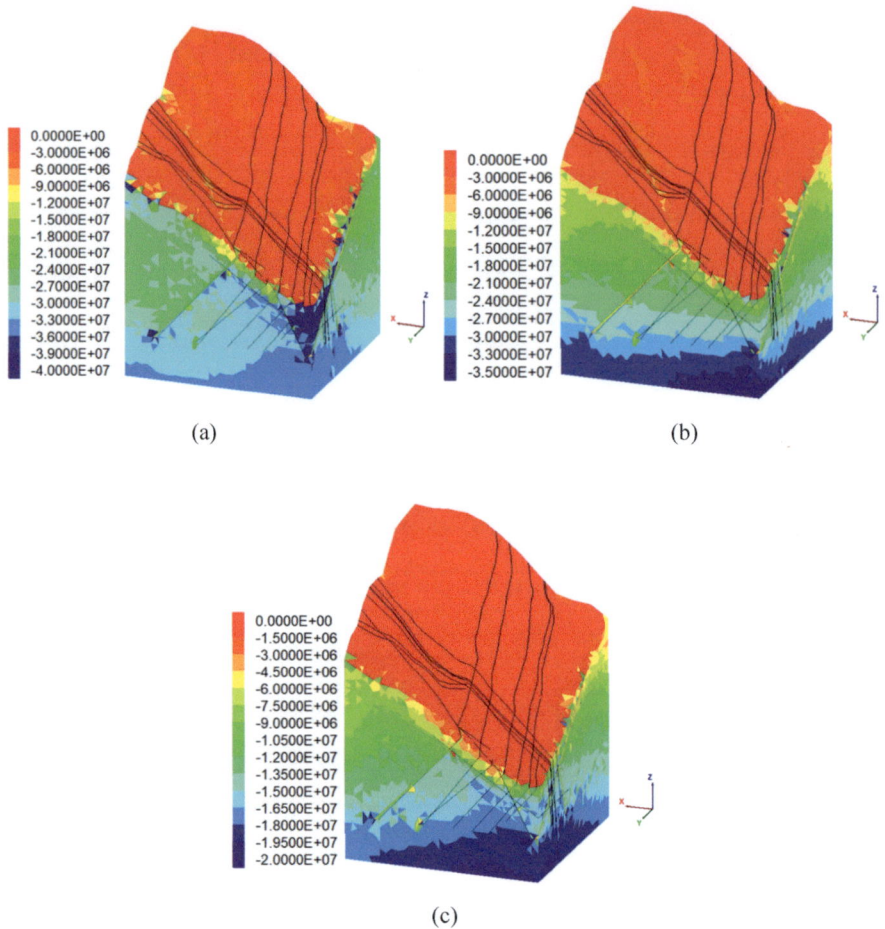

Fig. 2 Distribution of principal stresses in the engineering region: **a** maximum principal stress (σ_1), **b** intermediate principal stress (σ_2), and **c** minimum principal stress (σ_3)

the minimum principal stress tends to be near the valley side, which reflects the characteristics of the ground stress field in the valley region. The closer to the valley and the greater the depth, the more significant the horizontal tectonic action, resulting in a gentler maximum principal stress. Closer to the surface, the influence of surface topography on the maximum principal stress becomes more pronounced, leading to a more significant deflection towards the side near the valley.

The stress vectors on the longitudinal profile of the central axes of the main powerhouse, main transformer chamber, and tailrace surge chamber are presented in Fig. 4. Like the transverse profile, the principal stress directions on the three longitudinal profiles all deflect towards the side near the valley. The closer to the valley and the greater the depth, the gentler the maximum principal stress. Near the

Fig. 3 Stress vector on the transverse profile (i.e. *XZ* plane) of the centreline of typical unit section: **a** 1 # and **b** 4 #

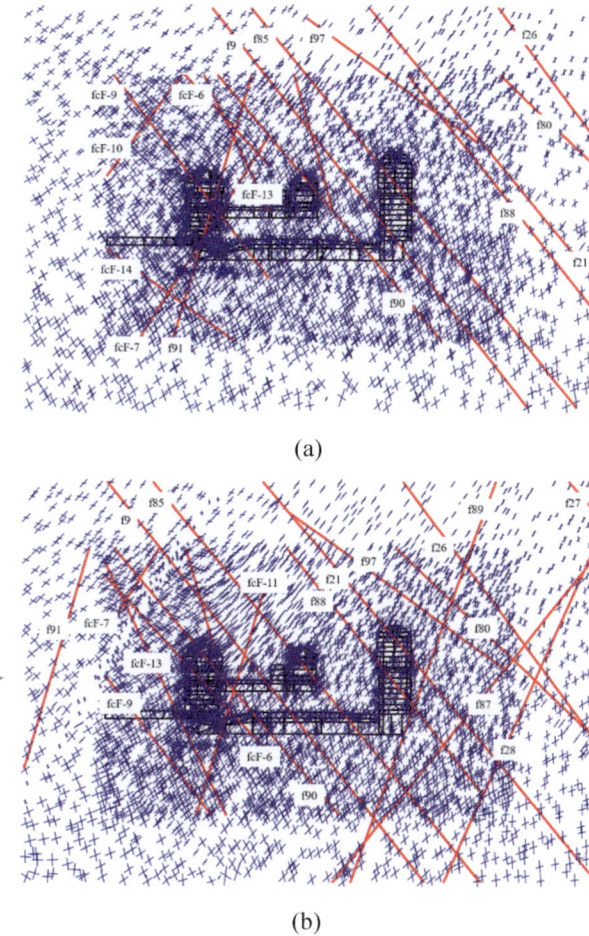

surface, the deflection of the maximum principal stress towards the side near the valley becomes more significant.

Figure 5 depicts the stress vector at the elevation of 2695 m. It can be observed that the maximum principal stress in the plane points towards the valley at a small angle of 30° ~ 40° relative to the axis of the powerhouse. Additionally, there is local variation in the stress field near the fault, indicating that the stress field in the underground caverns of the Yebatan project is significantly influenced by geological structures, further suggesting the complexity of the stress field in this engineering region.

Fig. 4 Stress vectors on the longitudinal profile (i.e. *YZ* plane) of the central axes of main caverns: **a** main powerhouse, **b** main transformer chamber, and **c** tailrace surge chamber

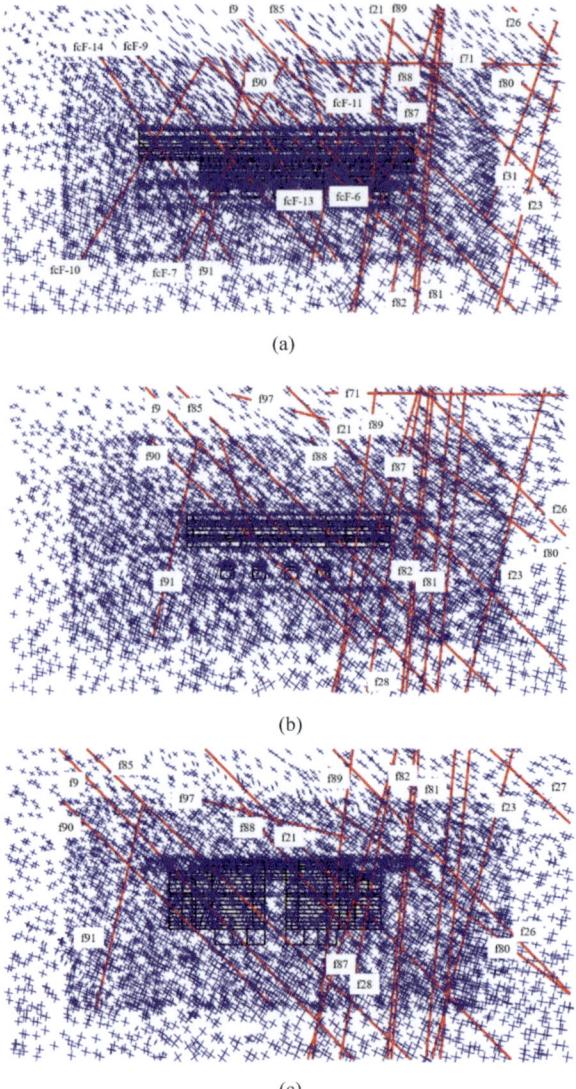

4.2 Stress Magnitude Characteristics of Ground Stress Field

The magnitude and distribution of the stress field in the engineering area are influenced by the evolution of the valley and the presence of faults, exhibiting a dual effect. The stress tends to concentrate towards the valley region, with the principal stresses being relatively high. Horizontal tectonic stresses become the main controlling factor, indicating the distribution characteristics of the stress field in the valley. Conversely, towards the interior of the mountains, the main controlling factor of

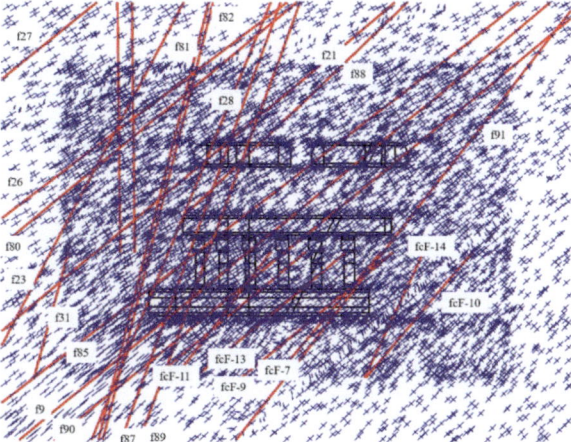

Fig. 5 Stress vector on the horizontal profile at the elevation of 2695 m

stress gradually transitions from tectonic stress to the self-weight of the mountains, leading to a gradual weakening of stress concentration and exhibiting characteristics of a self-weight stress field. Additionally, due to the presence of faults, there are significant variations in stress values near them, resulting in noticeable areas of stress relaxation and concentration (Figs. 6 and 7).

Therefore, the underground caverns of the Yebatan hydropower station are in a transitional zone from stress concentration zone to stress stable zone, locally influenced significantly by geological structures such as faults. The spatial distribution is uneven and exhibits considerable variability, indicating the complexity of the initial stress field in the caverns. Furthermore, in the main powerhouse, the magnitude of the maximum principal stress is approximately 14–37 MPa, the intermediate principal stress ranges from 6to 22 MPa, and the third principal stress varies from 4 to 16 MPa.

5 Conclusions

This study, based on the Yebatan hydropower station, investigates the distribution patterns of the measured ground stress in the engineering region. Through numerical analysis, the ground stress field is inverted, and distribution characteristics of the ground stress field in underground caverns under complex geological conditions are identified. The main conclusions are as follows:

(1) The stress field magnitude and distribution in the engineering area are influenced by the evolution of the valley and the presence of faults, exhibiting a dual effect. Towards the valley region, horizontal tectonic stresses become the main controlling factor. Conversely, towards the interior of the mountains, the main controlling factor transitions from tectonic stresses to the self-weight of the mountains, leading to a gradual weakening of stress concentration and exhibiting characteristics of a self-weight stress field. Additionally, due to the presence of faults, there are significant variations in stress values near them.

Fig. 6 Principal stress distribution of the central transverse section of Unit 1 #: **a** σ_1, **b** σ_2, and **c** σ_3

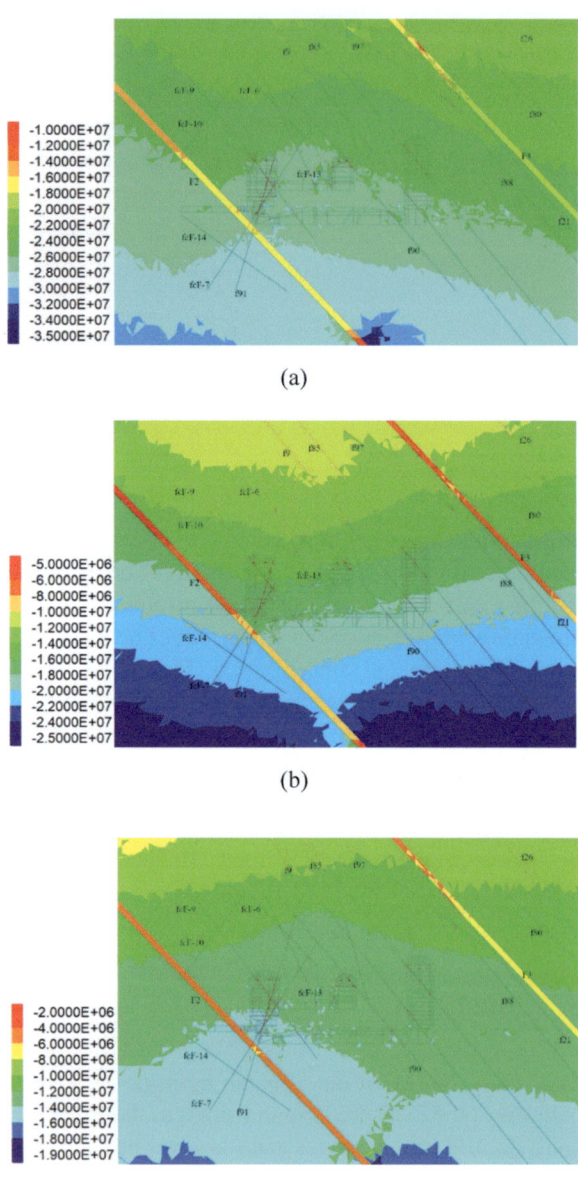

(a)

(b)

(c)

(2) The underground caverns of the Yebatan hydropower station are in a transitional zone from stress concentration zone to stress stable zone, locally influenced significantly by geological structures such as faults. The spatial distribution is uneven and exhibits considerable variability, indicating the complexity of the initial stress field in the caverns under complex geological conditions.

Fig. 7 Principal stress distribution of the central transverse section of Unit 4 #: **a** σ_1, **b** σ_2, and **c** σ_3

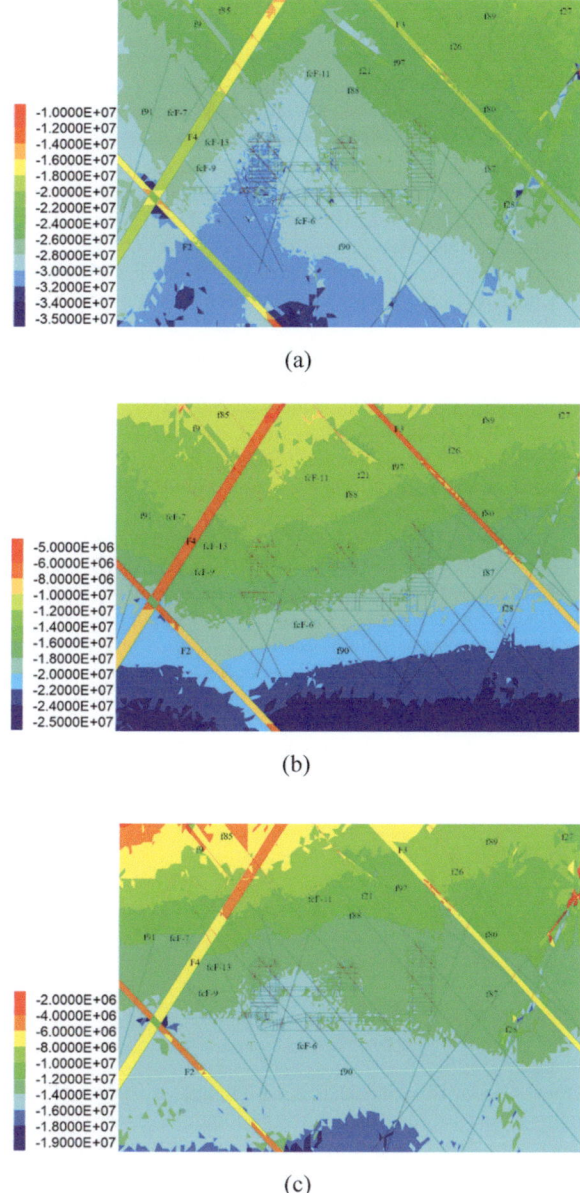

Acknowledgements This work was supported by the National Natural Science Foundation of China (Grant No. 52309122), the China Postdoctoral Science Foundation (Grant No.2023M730367), and the Fundamental Research Funds for Central Public Welfare Research Institutes of China (Grant No. CKSF2023323/YT).

References

1. Feng XT, Yang CX, Kong R, Zhao J, Zhou YY, Yao ZB, Hu L (2022) Excavation-induced deep hard rock fracturing: methodology and applications. J Rock Mech Geotech Commun 14:1–34. https://doi.org/10.1016/j.jrmge.2021.12.003
2. Wang C, Han ZQ, Wang YT, Wang CY, Wang JC, Hu S (2023) Rapid in-situ stress measurement in vertical borehole based on borehole diametrical deformation analysis. Rock Mech Rock Eng Commun 56:8289–8303. https://doi.org/10.1007/s00603-023-03472-3
3. Qin XH, Zhao XG, Zhang CH, Li PF, Chen Q, Wang J (2024) Measurement and assessment of the in-situ stress of the Shazaoyuan rock block, a candidate site for HLW disposal in Northwest China. Rock Mech Rock Eng Commun 57:4011–4031. https://doi.org/10.1007/s00603-024-03775-z
4. Li G, Mizuta Y, Ishida T, Li H, Nakama S, Sato T (2009) Stress field determination from local stress measurements by numerical modelling. Int J Rock Mech Min Commun 46:138–147. https://doi.org/10.1016/j.ijrmms.2008.07.009
5. Yong R, Wu JF, Huang HY, Xu E, Xu B (2022) Complex in situ stress states in a deep shale gas reservoir in the Southern Sichuan Basin, China: from field stress measurements to in situ stress modeling. Mar Petrol Geol Commun 141:105702. https://doi.org/10.1016/j.marpetgeo.2022.105702
6. Shang YQ (1999) The ways of determining boundary conditions in geomechanical numerical simulation. Chin J Rock Mech Eng Commun 18:201–204. https://doi.org/10.3321/j.issn:1000-6915.1999.02.019
7. Huang SL, Ding XL, Liao CG, Wu AQ, Yin J (2014) Initial 3D geostress field recognition of high geostress field at deep valley region and considerations on underground powerhouse layout. Chin J Rock Mech Eng Commun 33:2210–2224. https://doi.org/10.13722/j.cnki.jrme.2014.11.006
8. Pei SF, Zang DS, Li JH, He JH, Li GL, Chen BR (2023) Distribution characteristics and laws of ground stress field under deep valley Terrain. Northw Hydro, 8–14
9. Zhang XH, Yin JM, Ai K, Liu YK (2021) Study on stress field characteristics of deep-buried long tunnel under complex engineering geological conditions. Chin J Under Sp Eng Commun 17:421–429

Open Access This chapter is licensed under the terms of the Creative Commons Attribution 4.0 International License (http://creativecommons.org/licenses/by/4.0/), which permits use, sharing, adaptation, distribution and reproduction in any medium or format, as long as you give appropriate credit to the original author(s) and the source, provide a link to the Creative Commons license and indicate if changes were made.

The images or other third party material in this chapter are included in the chapter's Creative Commons license, unless indicated otherwise in a credit line to the material. If material is not included in the chapter's Creative Commons license and your intended use is not permitted by statutory regulation or exceeds the permitted use, you will need to obtain permission directly from the copyright holder.

Research and Application of Advance Bolt Support Mechanism of Highway Tunnel Under Complex Geological Conditions

Zhou Qiao

Abstract With the continuous expansion of the highway network, the relevant departments put forward stricter requirements for the construction of highways with tunnels. For highway tunnel engineering with complex engineering geology, during the excavation and construction of the tunnel face, anchor bolts and conduits are used to anchor the surrounding rock to ensure the safety of people and things during the excavation and construction of the highway tunnel. In this paper, a highway tunnel project in Shanxi Province is taken as a case. Through the analysis of the mechanism of advanced bolt support, scientific and reasonable advanced support construction design is carried out to ensure that this technology can play a full role. In addition, an ultra-high-strength aluminum alloy pipe advanced support is adopted to improve the safety and stability of tunnel construction.

Keywords Highway tunnel · Advance support · Bolt · Ultra-high-strength aluminum alloy conduit

1 Introduction

With the continuous expansion of the highway network, more and more complex engineering geological conditions are encountered. In the construction of highway tunnels, there have been multiple safety accidents related to excavation and support of the tunnel face, such as the "4.6" accident in the Shantouping Tunnel of Section A5 of Fuzhou Puyan Expressway in 2019, which resulted in fatalities [1]. In recent years, many measures for prereinforcement of rock layers have been proposed both domestically and internationally, mainly including advanced support construction technology. It is increasingly widely used in the fractured zone surrounding rock of high cut slopes and large section soft rock tunnels. For example, before excavation, the engineering hazard and stability are evaluated, and then scientific and reasonable

Z. Qiao (✉)
School of Engineering and Technology, Nanchang Vocational University, Nanchang, China
e-mail: kd_zq@qq.com

advanced support construction technology design and construction are carried out to improve the safety and stability of tunnel construction.

In advanced support construction, there are many advanced anchor rod support schemes, but of course, their specific application is somewhat difficult. This technology application scheme should be planned and implemented based on the actual situation of road tunnel engineering. Firstly, we should understand the mechanism of advanced anchor rod support as the theoretical basis for design, and then make a reasonable layout in construction technology, construction process, and other aspects to orderly construct and stabilize the tunnel. If encountering complex engineering geology, such as weak rock layers and fractured zone surrounding rocks, it is necessary to provide assistance and support, such as making reasonable use of metal support frames. Secondly, during the specific construction, not only should construction surveying and layout be done well but also a series of construction preparation work should be done well; Finally, according to the construction requirements and drawings, the excavation of the protective highway tunnel is carried out in an orderly manner. Based on the geological conditions and surrounding environment of the highway tunnel, reasonable planning and setting of support, such as brackets and anchor rod support, are carried out.

2 Analysis of Advanced Anchor Rod Support Mechanism

Since the 1970s, there have been numerous studies on the theory of rock and soil anchoring, mainly including Freeman's neutral theory [2] and Oreste's reinforcement theory [3]. Freeman's full-length bonded anchor rod test in the Kielder test tunnel summarized the stress process of the anchor rod and the distribution law of shear stress along the anchor rod, and proposed concepts such as "neutral point," "bonding length," and "anchoring length." Oreste's reinforcement theory proposes that the essence of anchor rod reinforcement is to change the stress state of the surrounding rock, improve the mechanical properties of the anchored rock mass, increase the number of joint cracks between the rock masses, and thus improve the mechanical parameters of the rock.

For the research on the reinforcement mechanism of advanced anchor rods, most people believe that the stress characteristics of advanced anchor rods are completely similar to those of fully bonded anchor rods in tunnels, and they also play a role in anchor rod reinforcement. The author believes that advanced anchor rod support plays a crucial role in the reinforcement of surrounding rock in engineering. By combining the fractured rock mass and forming an arched solidified thin shell under the action of arch effect, it supports the overlying fractured or loose rock layers. Then, using buckling theory, membrane theory, generalized variational principle, and virtual work principle, the theory of advanced anchor rod reinforcement arch for fractured zone engineering surrounding rock is proposed [4].

According to Figs. 1 and 2, the energy equation for internal buckling of the arch can be obtained by assuming that the second-order variation of the total energy of the system is equal to zero.

$$\frac{1}{2}\int_{-\alpha}^{\alpha}[EA(\delta\overline{w'}-\overline{v'})^2+\frac{EI_x}{R^2}(\delta\overline{v'}+\delta\overline{w'})^2+Q_s(\delta\overline{v'}+\delta\overline{w'})^2]Rd\theta=0 \quad (1)$$

Calculate the critical load

$$\sigma=\left[\overline{w'}-\overline{v}+\overline{v'D'}+\frac{1}{2}(\overline{v'})^2\right]E_m+\frac{yv''}{R}E_b \quad (2)$$

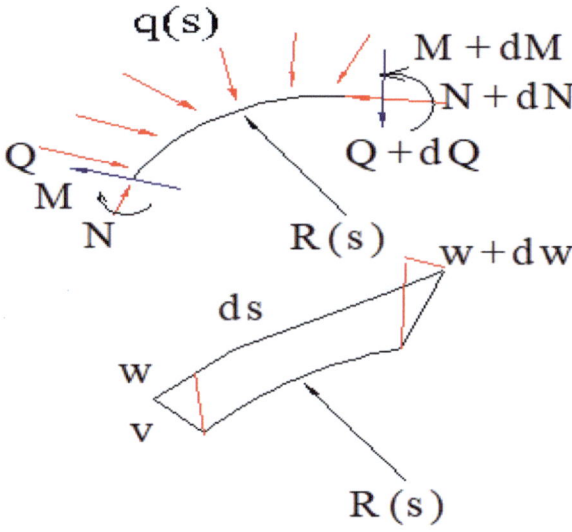

Fig. 1 The internal force and displacement of the anchorage arch in the fractured zone

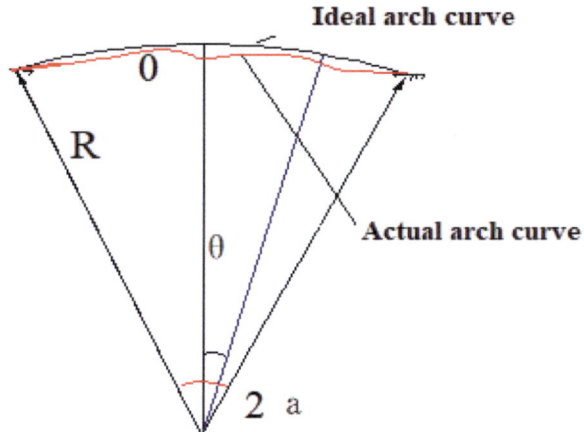

Fig. 2 Schematic diagram of the distribution of geometric defects in the anchorage arch of the fractured zone

The relationship between load and radial displacement is

$$A_3 \overline{q^2} + B_3 \overline{q} + C_3 = 0 \tag{3}$$

$$A_4 \overline{q_s^2} + B_4 \overline{q_s^2} + C_4 = 0 \tag{4}$$

where, $A_4 = 2A_3 + D_4$; $B_4 = 4A_3$; $C_4 = B_3 - C_3$;

$$D_4 = \frac{1}{2} - \frac{1}{\mu^2 \alpha^2} + \frac{\mu\alpha \cot(\mu\alpha)}{4} + \frac{3\cot^2(\mu\alpha)}{4} + \frac{\mu\alpha \cot^3(\mu\alpha)}{4} \tag{5}$$

The implicit function relationship $F(q, \mu) = 0$ between q and μ, Taking $dq/d\mu = 0$, the curve equation that determines the relationship between load q and displacement in dimension 1 is used to track the entire process of the advanced anchorage arch structure in the fractured zone from being loaded to buckling, thus obtaining the static buckling characteristics of the fixed anchorage arch under uniform load when considering the fractured rock mass (geometric imperfection).

Using this theory, the author predicted the displacement changes of the roof after on-site monitoring and theoretical analysis on the inclined slope of the fractured zone at the 230 level in Zijin Wuping Mining Company, as shown in Fig. 3.

By utilizing this pattern, the excavation of the working face is prevented from over excavation or under excavation, ensuring the safety and stability of the tunnel project.

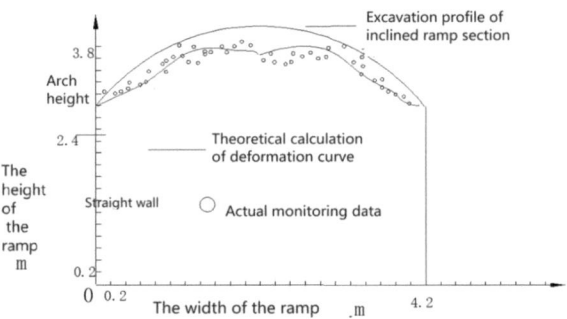

Fig. 3 Theoretical calculation and actual measurement of displacement at the arch part of the broken belt 230 slope anchor rod advanced reinforcement (Section B221)

3 Engineering Cases

3.1 Project Overview

A highway tunnel project in Shanxi Province has a steep natural slope and a loose geological structure. The soil type is loess. After meticulous and detailed engineering layout survey, the statistics of the surrounding rock of the tunnel are as follows: Level II 5642 m, Level III 478 m, Level IV 2432 m, and Level V 747 m, with a total length of 9299 m. The design of the tunnel is a double hole one-way driving double lane tunnel, with a net width of 9.75 m and a net height of 5.0 m. The starting and ending pile numbers of Tunnel 1 are K3 + 160 to K4 + 665, with a tunnel length of 1475 m and an uphill slope of 10.15‰; The maximum burial depth of the tunnel is about 161.0 m, and the excavation area of the tunnel section is 160 m². The starting and ending pile numbers of Tunnel 2 are K3 + 160 to K4 + 670, with a tunnel length of 1490 m and a slope of 11.45‰. The maximum burial depth is about 174.0 m.

The surrounding rock of the tunnel belongs to the loess collapsible rock and belongs to the fractured zone area, with a strength of $f = 7$. There is a weathering zone in some areas, which has poor stability and other characteristics. The maximum excavation width of the tunnel is 13.36 m, the maximum excavation height is 10.09 m, and the construction span reaches 18.9 m. Therefore, there is a significant risk of load displacement, and roof collapse may occur, making it difficult to effectively support the tunnel body. After conducting an engineering hazard and stability assessment by the project team, it is believed that advanced support construction technology should be used to stabilize the entrance and tunnel body, ensuring the safety and stability of tunnel construction.

3.2 Design of Advanced Anchor Rod Support

By collecting and organizing geological data and the shape characteristics of the highway tunnel engineering, advanced anchor rod support construction technology is adopted to ensure the robustness, stability, and durability of the entire project. Advanced anchor rod support usually includes ordinary advanced anchor rod reinforcement method, advanced anchor rod plus anchor cable reinforcement method, tunnel boring machine construction advanced anchor rod reinforcement method, advanced special anchor rod reinforcement method, excavation anchor rod groove advanced anchor rod reinforcement method, and arch bracket advanced anchor rod reinforcement method. To fully utilize the advanced anchor rod support construction technology, it is necessary to first determine based on the actual project. The tunnel surrounding rock belongs to loess collapsible rock, with low strength and many small faults in the surrounding area. Therefore, the arch bracket advanced anchor rod reinforcement method is adopted.

This design combines advanced anchor rods, arch supports, and small conduit support, with steel frame nets installed and equipped with drilling machines, inclinometers, and other equipment. The advanced anchor rod adopts R32N Mai type self-propelled anchor rod, injected with cement water glass double liquid slurry, with an external insertion angle of 10°–15° [5]. The small conduit adopts an ultra-high-strength aluminum alloy conduit, with an external insertion angle of 5° to 10°, and an arch support spacing of 30 cm for advanced support.

3.3 Construction Process of Advanced Anchor Rod Support

In advanced support construction, the organic combination of advanced anchor rods, arch supports, and small conduit support can achieve good results. When it is required to control the quality of support construction, such as drilling, grouting construction, etc., based on the actual situation of road and tunnel engineering, and to develop a design plan for advanced support construction, construction personnel should carry out support construction in sections, that is, the advanced support construction of the tunnel entrance and tunnel body.

Advance Support Construction of Tunnel Entrance Section

Before the advanced support construction of the tunnel entrance, surveying and layout, construction of the gutter and intercepting ditch, slope earthwork construction, and slope reinforcement treatment should be carried out. Excavators should be used to excavate the tunnel entrance longitudinally in sections, from top to bottom, and in layers. At the same time, side and front slopes should be repaired and anchor spraying protection should be carried out. Other related work shall be carried out in an orderly manner according to construction requirements and procedures to meet standard requirements. as shown in Fig. 4.

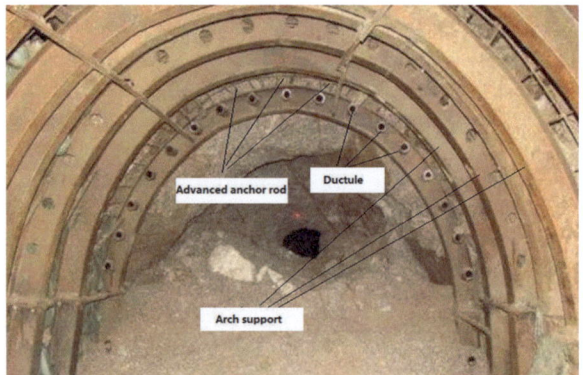

Fig. 4 Construction site for advanced support at tunnel entrance. *Source* Photo taken by myself at the construction site of the 230 ramp at Wuping Zijin Mining

Advance Support Construction of Tunnel Body (Main Tunnel)
Relatively speaking, the construction of the tunnel body is more difficult. Due to the adverse geological structure of the tunnel project, when carrying out excavation and advanced support construction of the tunnel body, on-site measurement and control are directly carried out after the support construction at the entrance, and the management of the construction technology of the fractured zone surrounding rock is done well. The new Olympic method is used to carry out the excavation and support construction organization design of the tunnel body. After engineering analogy and comprehensive consideration of safety, schedule, cost, and other related schemes, the step excavation method is adopted to excavate the tunnel.

Stability Construction of Surrounding Rock
The engineering overview clearly states that the surrounding rock strength of the project is poor and difficult to effectively support the tunnel body, so it is particularly necessary to stabilize the surrounding rock construction [6]. That is to say, during the excavation and construction of the tunnel body, attention should be paid to understanding the geological conditions and surrounding rock strength, and then reasonable planning and layout of pipe sheds and conduits should be carried out to set up support structures and stabilize the surrounding rock. After entering the core area, it is best to use weak blasting for excavation construction, and then arrange pipe sheds reasonably to form a strong and stable support structure, improving the robustness and reliability of the tunnel core area.

It should be noted that the pipe shed support is aimed at the surrounding rock of the tunnel, with the upper and lower parts jointly supported to form a circular support contour, providing all-round support to the surrounding rock and achieving the goal of stabilizing it [7]. For this road tunnel project, it is best to use a comprehensive support mode of I-beam, anchor, spray, and mesh for the upper part of the pipe shed support. At the same time, steel frames, anchor rods, and concrete should be used to finally set up a plum blossom shaped support.

Reinforcement Mesh Layout Construction
The purpose of laying steel frame mesh is to strengthen the support of surrounding rock working face again, and to create conditions for improving the construction quality of road tunnel engineering. In the specific construction of steel mesh layout, the first step is to select the most suitable steel according to the construction requirements and relevant standards, and cut the square pieces according to the specification of 2 m x 2 m; Secondly, appropriate methods should be adopted to remove rust and dirt from the steel bars, in order to improve the quality of their support [8], Once again, lay out the steel mesh tightly against the tunnel wall, and ensure that the length of the steel reinforcement is closely aligned with the concrete spraying surface. Finally, using a welding machine to weld the steel mesh, steel frame, and anchor rod can truly improve the stability and support of the steel mesh.

Tunnel Support Construction
During tunnel support construction, attention should be paid to the selection and use of waterproof membranes to divide the anchoring layer and lining layer, and blind

ditches should be used to drain the underground water in the tunnel and discharge it. Afterward, in the specific direction of the blind ditch, the spray anchor layer is reinforced, and the lining layer is laid using waterproof rolls, with attention paid to adding geotextiles in the middle.

Concrete Spraying Construction

For the sake of improving the safety of road tunnel construction, it is necessary to strictly follow the construction requirements and relevant standards to carry out concrete spraying construction, which means comprehensively and effectively cleaning the relatively loose rock blocks on the surrounding rock surface of the tunnel. Then, a standard setting is set every 1 m to determine the spraying direction and thickness. During the initial excavation of the surrounding rock working face, concrete spraying construction should be carried out according to the construction process flow; Afterward, the layout of steel frames, steel mesh, and anchor rods will be carried out, and concrete spraying construction will be carried out again after the above construction is completed, as shown in Fig. 5.

4 Conclusion

Based on the advanced anchor rod support mechanism, the application plan of advanced anchor rod support technology is planned and implemented according to the actual situation of road tunnel engineering, targeting the complex engineering geology of the fractured zone surrounding rock of high cut slopes and large cross section soft rock tunnels. This article takes a highway tunnel project in Shanxi as a case study, and applies advanced support construction technology to the tunnel entrance section and tunnel body (main tunnel). This further improved the theory of advanced anchor rod reinforcement arch for fractured zone engineering surrounding rock. At the same time, a high-strength aluminum alloy conduit is used for advanced support to improve the safety and stability of tunnel construction.

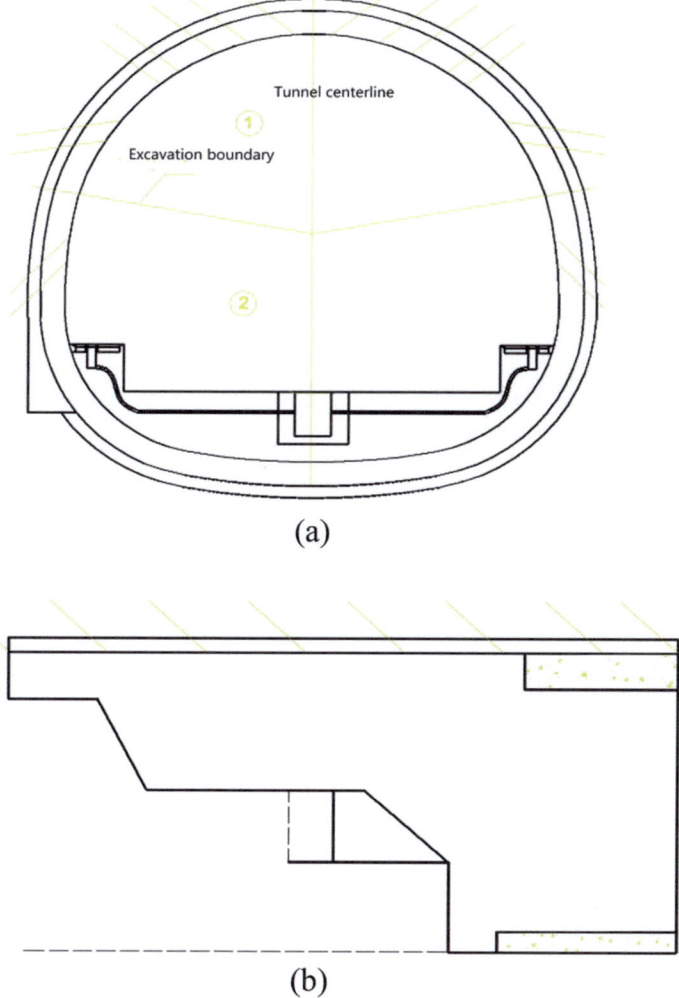

Fig. 5 Construction site of advanced anchor rod support for K420 section

Acknowledgements Project Fund: Key Project of Science and Technology Project of Jiangxi Provincial Department of Education (NO: GJJ206301)

References

1. Investigation report on the "4.6" death accident in Shantouping Tunnel of Section A5 of Fuzhou Puyan Expressway [EB/OL], website of Rongtai County People's Government in Fujian Province, 2019. http://www.yongtai.gov.cn/xjwz/zwgk/zfxxgkzdgz/aqsc/sgfxjyd/201912/t20191227_3156046.htm
2. Freeman TJ (1978) The behavior of fully-bonded rock bolts in the Kielder experimental Tunnel. Tunnles Tunnel J 10:37–40
3. Oreste PP, Peila D (1997) Modeling progressive hardening of shotcrete in convergence-confinement approach to tunnel design. J Tunnel Underground Space Technol 12(3):425-431
4. Zhou Q, Guo Q (2010) Arch structure of bolt forepoling in surrounding fractured rock. J Beijing Univ Sci Technol 6(32):697–701
5. Zhou Q, Guo Q, Xu HT (2009) Study on setting angle of bolt forepoling in surrounding fractured rock. J Coal Sci Technol 12:1594–1598
6. Zheng YR, Abi E (2022) On stability analysis and classification of surrounding rocks in rock tunnels, modern tunnelling technology. 59(1):1–13
7. Qiao JG, Peng B PENG, Qin JD (2022) Study on the optimization of three step excavation technology of large section soft rock tunnel. Highway Eng 47(1):75–78
8. Bao Q, Yan YX, Yao CK et al (2024) Analysis on reinforcement effect of advanced grouting for tunnel crossing fault fracture zone. J TransP Sci Eng 40(3):91–99. https://link.cnki.net/urlid/43.1494.U.20240527.1003.001

Open Access This chapter is licensed under the terms of the Creative Commons Attribution 4.0 International License (http://creativecommons.org/licenses/by/4.0/), which permits use, sharing, adaptation, distribution and reproduction in any medium or format, as long as you give appropriate credit to the original author(s) and the source, provide a link to the Creative Commons license and indicate if changes were made.

The images or other third party material in this chapter are included in the chapter's Creative Commons license, unless indicated otherwise in a credit line to the material. If material is not included in the chapter's Creative Commons license and your intended use is not permitted by statutory regulation or exceeds the permitted use, you will need to obtain permission directly from the copyright holder.

Experimental Study on Deformation Characteristics of Mud–Stone Mixture

Jian-bao Fu, Jian Yu, He-wen Liu, and Yu-bin Guo

Abstract This article focuses on the mud–stone mixture in Dalian New Airport Project and carried out large-scale consolidation tests of mud–stone mixture with different stone contents and scaled standard consolidation tests. The empirical formula for the compression modulus of the mud–stone mixture and the empirical formula for calculating the compression modulus of the on-site mud–stone mixture based on the results of the scaled standard consolidation test were obtained. The research results show that the compression modulus of the mud–stone mixture increases with the increase in the stone content, reaches the maximum value when the stone content is about 80% and then begins to decrease with the increase in the stone content as the mud between the crushed stones decreases. The compression modulus of the mud–stone mixture in this paper is 2–6 MPa. The maximum relative error between the compression modulus obtained by the empirical formula in this paper and the test value is 4.4%, indicating that the empirical formula in this article has high accuracy.

Keywords Mud–stone mixture · Compression modulus · Large-scale consolidation tests · Scale test · Standard consolidation test

---

J. Fu (✉) · J. Yu · H. Liu · Y. Guo
Tianjin Port Engineering Institute Co., Ltd. of CCCC First Harbor Engineering Co. Ltd., CCCC, Tianjin, China
e-mail: jian2203@foxmail.com

First Harbor Engineering Company Ltd., Tianjin, China

Key Laboratory of Port Geotechnical Engineering, Ministry of Communications, Tianjin, China

Key Laboratory of Tianjin Port Geotechnical Engineering, Tianjin, China

Y. Guo
e-mail: guoyubin@tpei.com.cn

1 Introduction

The Dalian New Airport project filled an artificial island in the sea as the airport site by using the riprap island-building method. This process has become a widely adopted method in the construction of artificial islands due to its simplicity and fast speed. During the riprap filling process, the mountain blasting crushed stones and seabed mud formed a layer of mud and stone mixture, which has unique engineering properties, neither like the riprap layer nor the mud layer, and has a significant impact on foundation settlement and stability. The study of its deformation characteristics is very necessary.

Currently, in the research literature, the mud–stone mixture is rarely mentioned, and generally, the soil and stone mixture is said. Many scholars have conducted a lot of research on the concept of the soil and stone mixture in the field of engineering geology, such as Wang et al. [1], Zhile et al. [2] believe that the soil and stone mixture is a loose body composed of soil and stone with significantly different properties and is a special geological body between the soil body and the rock body. The mud–stone mixture in this article refers to the mixture formed by the mixing of the mountain blasting crushed stones and the seabed mud under the impact force, which mostly exists at the junction of the riprap and the mud, and also exists near the surface of the mud bag caused by the riprap filling process.

Some scholars researched on the soil and stone mixture [3–11]. Through relevant experimental studies, Liu Liping et al. explored the influence of different gradations, different stone contents, and damage rates on the compaction performance of the soil and stone mixture. In view of the significant difference in the physical and mechanical properties between the soil and stone mixture and the fine-grained soil, Zhou Zhijun carried out a large-scale compaction test and obtained the influence law of factors such as the stone content, the maximum particle size, and the particle gradation on the compaction quality of the soil and stone mixture. For the soil and stone mixed slope in the southern part of Shaanxi Province, Yu Wei carried out the mechanical performance test of the soil and stone mixed slope, established the binary polynomial relationship between the internal friction angle of the soil and stone mixed system, the cohesion, the stone content, the water content, and the maximum dry density and obtained a method to quickly obtain the shear strength based on the maximum dry density on site.

Overall, the current research shows that the stone content has a significant impact on the engineering properties of the mud–stone mixture, but at present, there is basically no research on the deformation characteristics of the mud–stone mixture, especially the mud–stone mixture formed by the mud and riprap. In response to this situation, this article carried out large-scale consolidation tests of the mud–stone mixture with different stone contents and scaled standard consolidation tests after that, and obtained the relationship curves between the compression modulus and the stone content under two test conditions, providing an empirical formula for estimating the compression modulus of the on-site mud and stone mixture using the scaled standard consolidation test, which can provide a reference for the foundation settlement estimation and construction of the Dalian New Airport project.

2 Large Consolidation Test of Mud–Stone Mixture

Field sampling was carried out at the Dalian New Airport, and the properties were detected in the laboratory. The detection results of the mud are shown in Tables 1 and 2.

The test used the large-scale consolidation instrument modified by the research team itself, as shown in Fig. 1. The sample size is 60 cm in length, 40 cm in width, and the maximum height of the sample is 60 cm. In the test, the maximum particle diameter of the crushed stones is taken as 6 cm, and the crushed stones with a particle diameter larger than 6 cm on the site are removed. The apparent density of the crushed stones is 2.76 g/cm^3, and the particle size gradation of the crushed stones used in the test is shown in Fig. 2. The test process refers to "Coarse Particle Soil Consolidation Test" (SL237-058-1999), and a 5 kPa pre-pressure is first applied, and the pressure levels thereafter are 25, 50, 100, 150, and 200 kPa. Each level of pressure is consolidated for at least 24 h until the settlement basically does not increase anymore.

Six tests with stone contents of 100, 70, 60, 40, 30%, and 0 were carried out, and the relationship curve between the sample settlement and time was obtained. The settlement–time relationship curve with a 70% stone content is shown in Fig. 3. It

Table 1 Properties index of the mud

Particle composition			Physical properties of soil					
Fine sand 0.25–0.075	Powder particle 0.075–0.005	Clay particle < 0.005	Water content ω	Specific gravity of soil particles G_s	Wet density ρ	Dry density ρ_d	Saturation S_r	Porosity ratio e
%	%	%	%	–	g/cm^3		%	–
6.12	57.76	36.12	45.88	2.75	1.76	1.2	99.45	1.293

Table 2 Properties index of the mud

Limit water content			Compressibility		Engineering classification
Liquid limit ω_L	Plastic limit ω_P	Plastic index I_P	Compression coefficient $a_{v1\text{-}2}$	Compression modulus $E_{s1\text{-}2}$	
%	%	–	MPa^{-1}	MPa	
41.4	21.4	20	0.83	2.765	Silty clay

Fig. 1 Large-scale consolidation instrument for mud and stone mixture (owned by the research team)

Fig. 2 Particle size gradation curve of the crushed stones used in the test

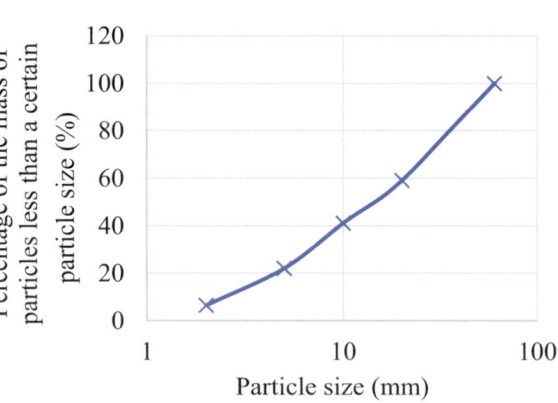

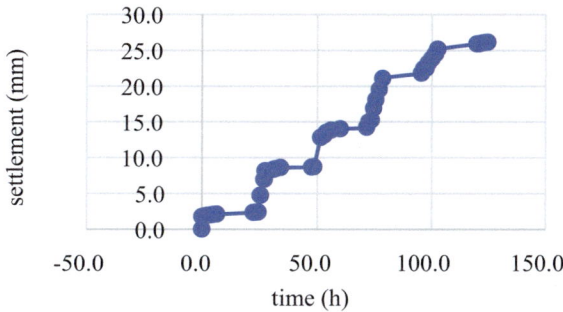

Fig. 3 The settlement–time relationship curve of the 70% stone content test

can be seen from the figure that as the time and load increase, the settlement amount of the sample continues to increase; when each level of load ends, the growth of the sample settlement has been very small, indicating that the sample has basically stabilized at this time.

According to the "Standard for Geotechnical Test Methods" (GBT50123-2019), the compression modulus of 100–200 kPa at different stone contents is shown in Fig. 6. It can be seen from the figure that the compression modulus of the mud–stone mixture increases with the increase in the stone content, and reaches the maximum value when the stone content is about 80%. Then, as the mud between the crushed stones decreases, the void increases, and the compression modulus begins to decrease with the increase in the stone content. In this paper, the compression modulus of the mud–stone mixture is 2–6 MPa.

According to the fitting of the test data, the empirical formula of the compression modulus of the mud–stone mixture with different stone contents can be obtained, as shown in Eq. (1).

$$E_s = -48.09R^4 + 74.94R^3 - 31.28R^2 + 5.70R + 2.50 \tag{1}$$

In the formula, R is the stone content of the mud–stone mixture.

3 Standard Consolidation Test After Scaling Down

Although the compression modulus obtained from the large-scale consolidation test can basically be considered as the compression modulus of the on-site mud–stone mixture, the large-scale consolidation test equipment is currently relatively rare, and many test and detection units do not have the test conditions. Therefore, this paper also carried out the standard consolidation test of the mud–stone mixture after scaling down to explore the method of determining the compression modulus of the mud–stone mixture using the conventional standard consolidation instrument as shown in Fig. 4. The area of the ring knife used in the test is 30 cm², and the height is 20 mm. The maximum particle diameter of the stones is taken as 5 mm, and the similar gradation method is used for scaling down. After the stones on the site are obtained and crushed, screening is carried out, and the particle size gradation of the stones before and after scaling down is shown in Fig. 5.

The crushed stones after scaling down and the mud are mixed evenly, and then the standard test is carried out. The test is divided into six groups, and the stone content

Fig. 4 Standard consolidation instrument (owned by the research team)

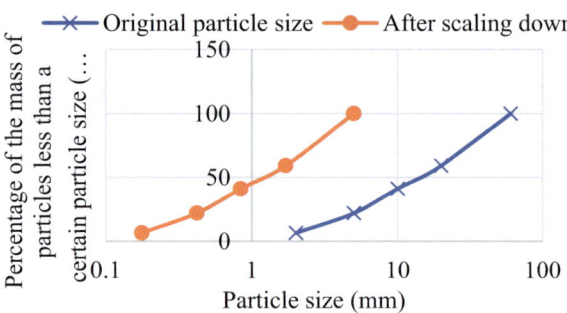

Fig. 5 Particle size gradation of the stones before and after scaling down

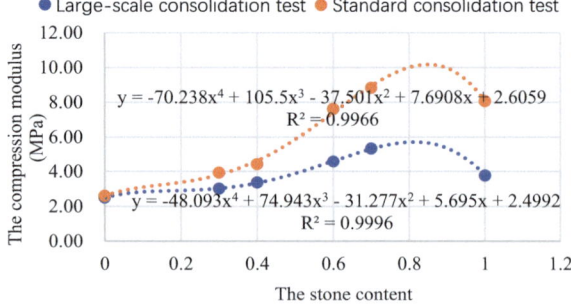

Fig. 6 The compression modulus obtained from the standard consolidation test and the large-scale consolidation test

is 0, 30%, 40%, 60%, 70%, and 100%, respectively. Each group conducts six parallel tests, and the obtained compression modulus is shown in Fig. 6.

Figure 6, respectively, shows the compression modulus and fitting curve obtained from the standard consolidation test and the large-scale consolidation test. It can be seen from the figure that the variation law of the compression modulus obtained by the two test methods with the stone content is similar.

This article used the data fitting function in the WPS spreadsheet software to fit the experimental data. The empirical formula of the compression modulus of the mud–stone mixture with different stone contents obtained from the standard consolidation test is shown in Eq. (2).

$$E_s = -70.24R^4 + 105.5R^3 - 37.50R^2 + 7.70R + 2.61 \tag{2}$$

4 Derivation of the Estimation Formula for the Compression Modulus of the Mud–Stone Mixture

Figure 7 shows the variation law of the ratio of the compression modulus of the large-scale consolidation test to the compression modulus of the standard consolidation test with the stone content and its fitting curve. The following fitting formula is obtained from the figure.

$$E_{sL}/E_{ss} = -0.49R + 0.94 \tag{3}$$

In the formula, E_{sL} is the compression modulus obtained from the large-scale consolidation test, and E_{ss} is the compression modulus obtained from the standard consolidation test after scaling down.

This paper considers that the compression modulus obtained from the large-scale consolidation test is the compression modulus of the on-site mud–stone mixture. Then the following formula is obtained.

Fig. 7 The variation law of the large-scale consolidation test compression modulus/standard consolidation test compression modulus with the stone content and the fitting curve

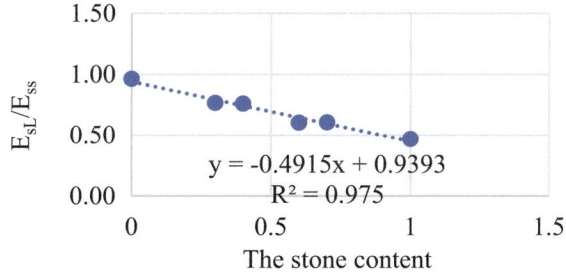

Table 3 The compression modulus obtained from the large-scale consolidation test and the values obtained from the empirical formula

Stone content (%)	Test value	Empirical formula value	Relative error between the empirical formula value and the test value (%)
100	3.77	3.63	−3.7
70	5.33	5.34	0.1
60	4.58	4.79	4.4
40	3.37	3.45	2.6
30	3.01	3.03	0.6
0	2.5	2.45	−1.9

$$E_s = (-0.49R + 0.94)(-70.24R^4 + 105.5R^3 - 37.50R^2 + 7.70R + 2.61) \quad (4)$$

The compression modulus obtained based on Eq. (4) and the test values can be seen in Table 3.

According to Table 3, it can be known that the maximum relative error between the test values and the compression modulus obtained from the empirical formula in this paper is 4.4%, indicating that the accuracy of the empirical formula in this paper is relatively high.

5 Conclusions

In this paper, the large-scale consolidation test of the mud–stone mixture with different stone contents in Dalian New Airport and the standard consolidation test after scaling down were carried out, and the following research results were obtained.

(1) The compression modulus ES100-200 of the mud–stone mixture increases with the increase of the stone content. Until the stone content is about 80%, the compression modulus reaches the maximum value. Then, as the mud in the gap between the crushed stones decreases, the gap increases, and the compression modulus begins to decrease with the increase in the stone content.
(2) The compression modulus ES100-200 of the mud–stone mixture in this paper is 2–6 MPa.
(3) The empirical formula for the compression modulus of the mud–stone mixture with different stone contents in Dalian New Airport and the empirical formula for calculating the compression modulus of the on-site mud–stone mixture based on the results of the standard consolidation test after scaling down are obtained.
(4) The maximum relative error of the compression modulus obtained from the empirical formula in this paper and the test values is 4.4%, indicating that the accuracy of the empirical formula in this paper is very high.

References

1. Wang Y, Li X, Li SD et al (2015) Cracking deformation characteristics for rock and soil aggregate under uniaxial compressive test. Chin J Rock Mech Eng 34(S1):3541–3552
2. Zhi-le S, Xin-rong L, Bao-xian L, Yue L (2010) Granule fractal properties of earth-rock aggregate and relationship between its gravel content and strength. J Central South Univ (Sci Technol) 41(3):1096–1101
3. Qiangqiang J, Yangqing X, Hao W (2020) Resarch on shear deformation characteristics acteristics of soil rock. J Eng Geol 28(5):951–958
4. Xiaohui Y, Yanpeng Z, Nan G (2017) Strength and deformation characteristic of soil-rock mixture and settlement prediction in high filled projects. Chin J Rock Mech Eng 36(7):1780–1790
5. Hanlong L, Minghua Z (2015) Review of ground improvement technical and its application in China. Chin Civil Eng J 49(1):96–115
6. Yanyi Z, Gan D, Yinqi Z et al (2020) Experimental study on the characteristics of consolidate—Wetting deformation of soil-aggregate mixture materials. SHUILI XUEBO 51(11):1393–1399
7. Teng W, Daisong H, Ga Z (2018) Experimental study of the behavior of soil-gravel mixtures. J Tsinghua Univ (Sci Technol) 58(5):456–460
8. Xu WJ, Zhang HY (2013) Research status and development trend of soil-rock mixture. Adv Sci Technol Water Resour 33(1):80–88
9. Jiaguo X, Ruilin H, Shengwen Q (2017) Large-scale triaxial shear testing of soil rock mixtures containing oversized particles. Chin J Rock Mech Eng 36(8):2031–2039
10. Kwan A, Mora CF, Chan HC (1999) Particle shape analysis of coarse aggregate using digital image processing. Cem Concr Res 29(9):1403–1410
11. Reid TR, Harrison JP (2000) A semi-automated methodology for discontinuity trace detection in digital images of rock mass exposures. Int J Rock Mech Min Sci 37(7):1073–1089

Open Access This chapter is licensed under the terms of the Creative Commons Attribution 4.0 International License (http://creativecommons.org/licenses/by/4.0/), which permits use, sharing, adaptation, distribution and reproduction in any medium or format, as long as you give appropriate credit to the original author(s) and the source, provide a link to the Creative Commons license and indicate if changes were made.

The images or other third party material in this chapter are included in the chapter's Creative Commons license, unless indicated otherwise in a credit line to the material. If material is not included in the chapter's Creative Commons license and your intended use is not permitted by statutory regulation or exceeds the permitted use, you will need to obtain permission directly from the copyright holder.

Statistical Characterization of the Geotechnical Properties of Changshou Rock

Jian-gong Chen and Bing Lu

Abstract This study presents a statistical characterization of rock properties in Changshou, Chongqing. The log data of boreholes are collected and used to develop a 3D geological model. The statistical analysis of rock properties is performed to describe the spatial variability. The lognormal distribution and Weibull distribution are adopted to describe probability density function of rock properties. The results of this research can help to establish random field considering rock spatial variability for reliability analysis in geotechnical engineering.

Keywords Rock properties · Geological model · Statistical characterization

1 Introduction

In recent decades, incorporating spatial correlation of soil properties into numerical analysis framework has drawn attention to geotechnical researchers and practitioners. Moreover, the soil uncertainty can be mainly classified into two categories: spatial variability of soil properties within one nominally homogeneous layer and geological uncertainty in the heterogeneous layer [1]. The spatial variability of geotechnical materials' properties refers to the inherent variability, which is the variation of geotechnical materials' properties from one point to another in space due to different deposition conditions and different loading histories [2, 3]. Due to the subsoils' natural origin, their characteristics render highly variable and rarely homogeneous. Considering the construction of buildings on the rocks, it is important to better quantify the behaviors of the rock, which is to characterize the density, strengths of the mudstone and sandstone. It has many statistical characterization researches of the subsoil [4], however, the physical and mechanical properties of rocks remain worth noting [5, 6]. According to the data of geological survey reports from the Changshou district in Chongqing, the main rocks of the area are the mudstone and sandstone. Thus, the empirical correlations of strength can be also established.

J. Chen (✉) · B. Lu
School of Civil Engineering, Chongqing University, Chongqing, China
e-mail: cjg77928@126.com

The 3D model of stratigraphy is shown in part 2. The statistical analysis of rock parameters is shown in part 3, information from the geological investigation reports including the physical and mechanical properties that can be employed to produce a geotechnical database in extra. The spatial variability of rock properties is shown in part 4.

2 3D Modelling of Stratigraphy

It is obvious that limited two-dimensional data in geotechnical engineering cannot indeed satisfy with designing of structures and analyzing of information. To create an all-around perspective of the subsurface structural geological stratigraphy, a 3D geological model using BIM is introduced in this section. Obtaining all the values in each location of different geological layers is not attainable, financially. There are six construction projects and over 1000 data from the reports in the Changshou district; therefore, this paper has shown a realistic picture of stratigraphy of one of six projects and physical and mechanical properties collected from all the projects. A collection of 283 drilled hole data points employed to build a 3D stratigraphic model of the place, as shown in Figs. 1 and 2. Solid modelling approaches have been investigated by several researchers [7].

The procedure for creating a solid model was implemented as demonstrated [8]. Firstly, the point groups of subsurface stratigraphy for each geological element were built up, followed by defining the triangled irregular networks (TIN) for the upper and lower surfaces of each geological element. Then, the TINs were extruded vertically to build primitive solids of different colours, the building model is shown in Fig. 2, finally. The 3D geological modelling in this paper was built up based on drilled hole data via a BIM software.

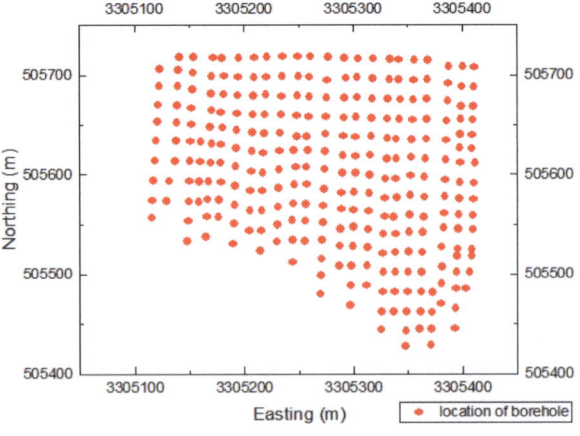

Fig. 1 Location of boreholes

Fig. 2 3D model of stratigraphy

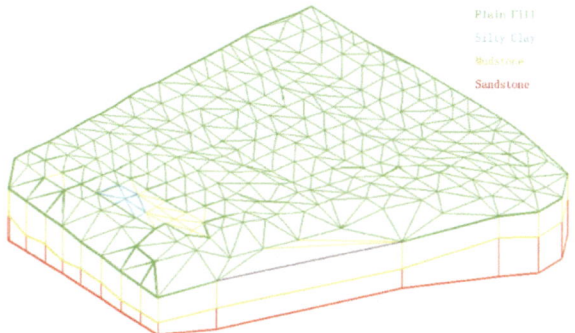

3 Statistical Analysis of Rock Parameters

In view of n rock samples, the traditional estimator of the mean is:

$$\mu = \frac{\sum_{i=1}^{n} x_i}{n} \quad (1)$$

where x_i is the value of the rock parameter of the ith sample.

The standard deviation (σ) and coefficient of variation (COV) are defined as follows:

$$\sigma = \sqrt{\frac{1}{n-1} \sum_{i=1}^{n} (x_i - \mu)^2} \quad (2)$$

$$\text{COV} = \sigma/\mu \quad (3)$$

The value of rock parameter is converted by the logarithm function, then the probability density function is shown as follows:

$$f(x) = \frac{1}{x \sigma_{\ln x} \sqrt{2\pi}} \exp\left(-\frac{(\ln x - \mu_{\ln x})^2}{2\sigma_{\ln x}^2}\right) \quad (4)$$

where $\mu_{\ln x}$, $\sigma_{\ln x}$ are the mean value and standard deviation value of $\ln x$.

The value of the rock parameter follows the Weibull distribution, then the probability density function is shown as:

$$f(x; \beta, \eta) = \frac{\beta}{\eta^\beta} x^{\beta-1} e^{-\left(\frac{x}{\eta}\right)^\beta} \quad (5)$$

where β is shape parameter of rock property, η is scale parameter of rock property.

3.1 Density (ρ)

On the basis of 150 data and 83 data for the mudstone and sandstone collected from six construction projects' investigation reports, the mean values of density were 2.53 g/cm³ ($n = 150$) and 2.48 g/cm³ ($n = 83$), respectively. The histogram shows that the connection between the rank of data and the frequency division of the data. In this research, the frequency polyline and the histograms are produced for both types of rock to recognize the number of times that range of value, as shown in Figs. 3a and 4a.

- **Mudstone**

A total of 150 statistics for mudstone, the density (ρ) differed from 2.41 g/cm³ to 2.65 g/cm³ with the standard deviation of 0.039 g/cm³ and coefficient of variation (COV) of 1.54% are summed up in Table 1. Figure 3a shows that the values of density for mudstone are all distributed between 2.4 and 2.65, and the values from 2.5 g/cm³ to 2.6 g/cm³ take the most part of data, where the frequency of 2.53 g/cm³, 2.54 g/cm³, 2.55 g/cm³ and 2.56 g/cm³ takes over 10% of the whole data, respectively.

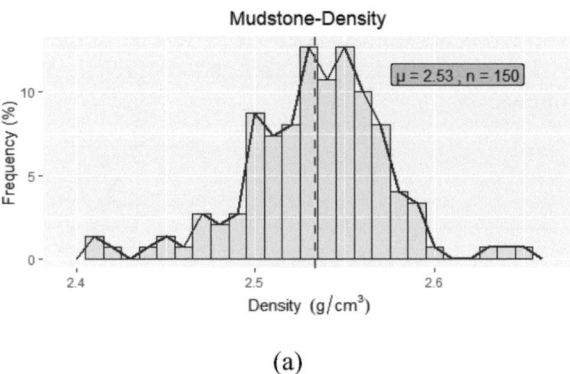

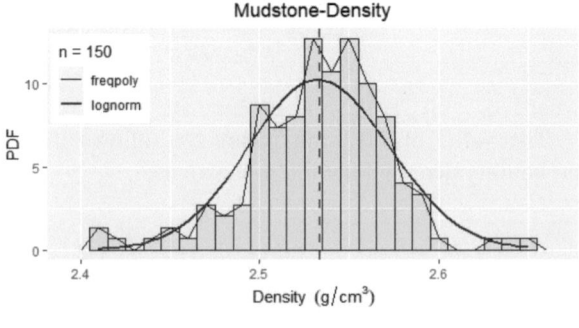

Fig. 3 **a** Frequency histogram of density for mudstone; **b** probability density function of density for mudstone

Fig. 4 **a** Frequency histogram of density for sandstone; **b** probability density function of density for sandstone

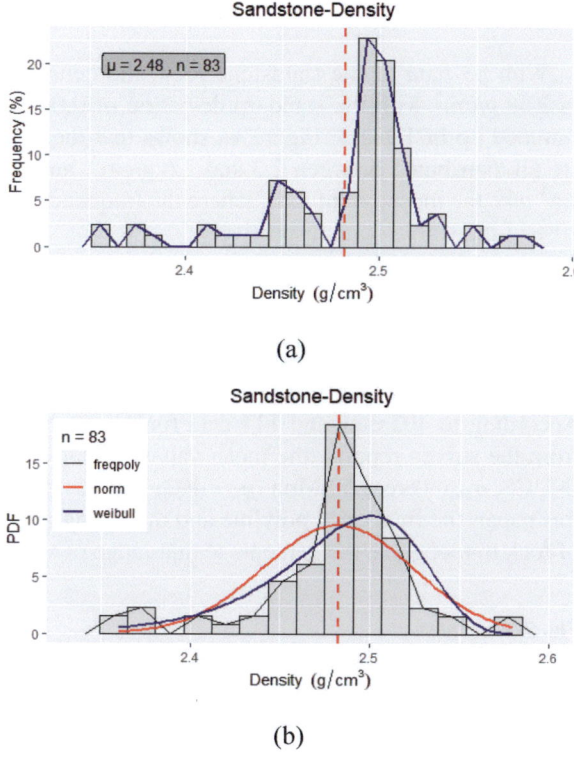

(a)

(b)

Table 1 Statistical variation of rock properties

Statistical parameter		Density	NUCS	SUCS
Mudstone	No. of data	150	402	402
	Range	2.41–2.65	2.16–13.4	1.12–8.66
	Mean (μ)	2.53	6.59	4.05
	Standard deviation (σ)	0.039	2.294	1.575
	COV	1.54%	34.82%	38.9%
Sandstone	No. of data	83	114	114
	Range	2.36–2.58	14.7–50.4	9.46–37.1
	Mean (μ)	2.48	35.14	26.2
	Standard deviation (σ)	3.83%	8.361	6.75
	COV	1.68%	23.8%	25.8%

- **Sandstone**

With 83 data of the sandstone rock, the density (ρ) differed from 2.36 g/cm^3 to 2.58 g/cm^3 with the standard deviation of 0.042 g/cm^3 and COV of 1.68% are summed up in Table 1. Figure 4a shows that the values of density with sandstone are all distributed between 2.3 and 2.6 g/cm^3, and the values from 2.45 to 2.52 g/cm^3 take the most part of data, where the frequency of 2.49 and 2.5 g/cm^3 takes over 20% of the whole data, respectively.

3.2 Uniaxial Compressive Strength (σ_c)

Natural Uniaxial Compressive Strength (σ_{cn}.)

According to 402 data and 114 data for the mudstone and sandstone accumulated from the survey reports, the mean values of natural uniaxial compressive strength (NUCS, σ_{cn}) were 6.59 Mpa ($n = 402$) and 35.14 Mpa ($n = 114$), respectively. In this paper, the frequency polyline and the histograms were produced for both types of rock to recognize the percentage that range of value, as shown in Figs. 5a and 6a.

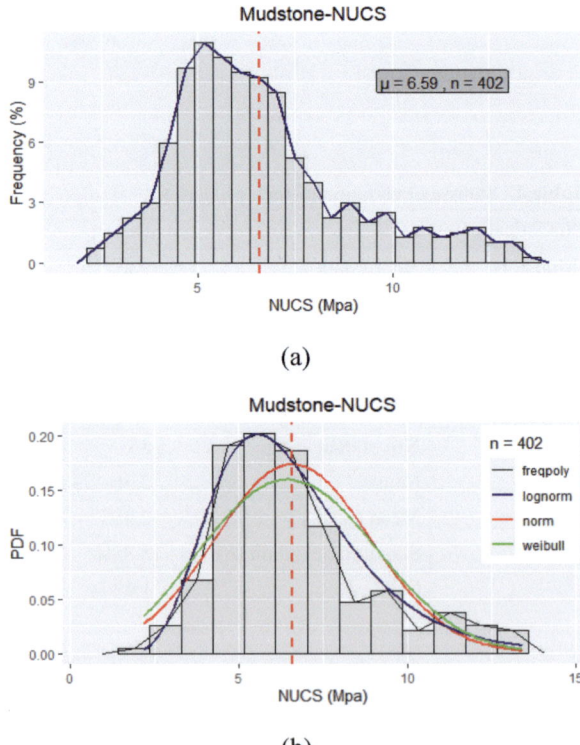

Fig. 5 **a** Frequency histogram of NUCS for mudstone; **b** probability density function of NUCS for mudstone

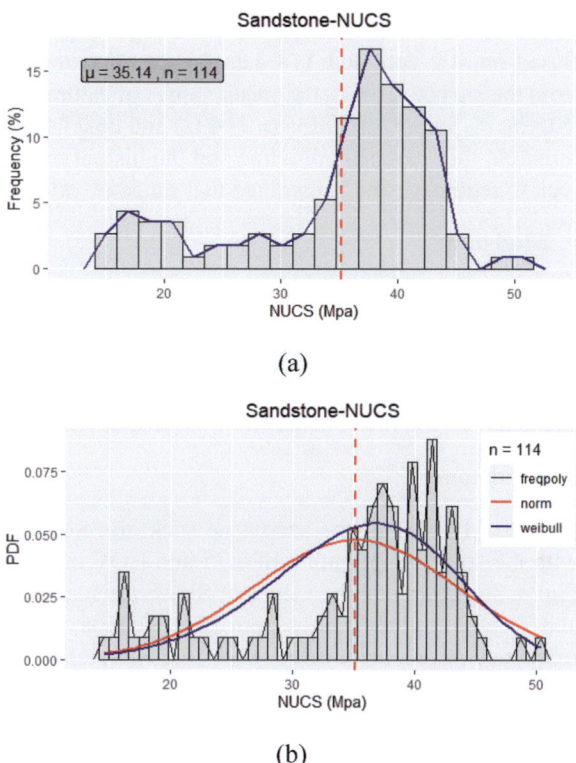

Fig. 6 **a** Frequency histogram of NUCS for sandstone; **b** probability density function of NUCS for sandstone

- **Mudstone**

A total of 402 data for mudstone, the NUCS varied from 2.16 to 13.4 Mpa with the standard deviation of 2.294 Mpa and COV of 34.82% are summed up in Table 1. Figure 5a shows that the values of NUCS for mudstone are all distributed between 2 and 14 Mpa, and the values from 5 to 10Mpa take the most part of data, where the frequency of the mean value $\mu = 6.59$ Mpa) takes over 9% of the whole data.

- **Sandstone**

With 114 data of the sandstone rock, the NUCS varied from 14.7 to 50.4 Mpa with the standard deviation of 8.361 Mpa and COV of 23.8% are summed up in Table 1. Figure 6a shows that the values of NUCS for sandstone are all distributed between 14 and 50 Mpa, and the values from 30 to 45 Mpa take the most part of data, where the frequency of the mean value ($\mu = 35.14$ Mpa) takes over 10% of the whole data.

Saturated Uniaxial Compressive Strength (σ_{cs}.)

Based on 402 data and 114 data for the mudstone and sandstone rocks collected from the survey reports, the mean values of saturated uniaxial compressive strength (SUCS, σ_{cs}) were 4.05 Mpa ($n = 402$) and 26.2 Mpa ($n = 114$), respectively. In this research, the frequency polyline and the histograms were produced for both type of rock to recognize the percentage that range of value, as shown in Figs. 7a and 8a.

- **Mudstone**

A total of 402 data for mudstone, the SUCS varied from 1.12 to 8.66 Mpa with the standard deviation of 1.58 Mpa and COV of 38.9% are summed up in Table 1. Figure 7a shows that the values of SUCS for mudstone are all distributed between 1 and 9 Mpa, and the values from 2.5 to 5 Mpa take the most part of data, where the frequency of the mean value ($\mu = 4.05$Mpa) takes over 6% of the whole data.

- **Sandstone**

With 114 data of the sandstone rock, the SUCS varied from 9.46 to 37.1 Mpa with the standard deviation of 6.75 and COV of 25.8% are summed up in Table 1. Figure 8a shows that the values of SUCS for sandstone are all distributed between 9 and 37 Mpa, and the values from 25 to 31 Mpa take the most part of data, where the frequency of the mean value ($\mu = 26.2$Mpa) takes nearly 10% of the whole data.

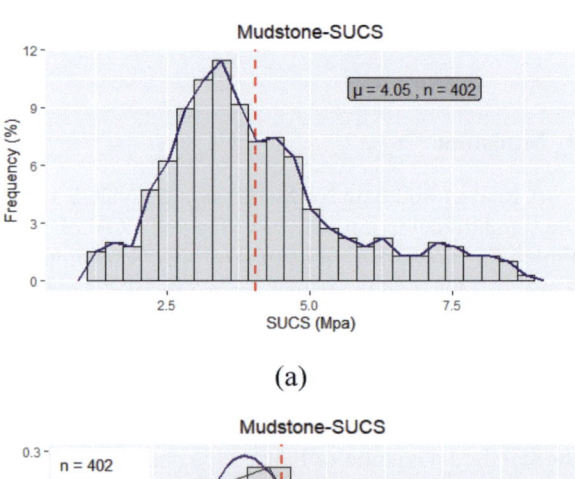
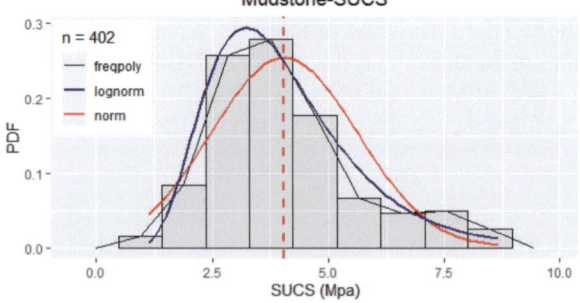

Fig. 7 **a** Frequency histogram of SUCS for mudstone; **b** probability density function of SUCS for mudstone

Fig. 8 **a** Frequency histogram of SUCS for sandstone; **b** probability density function of SUCS for sandstone

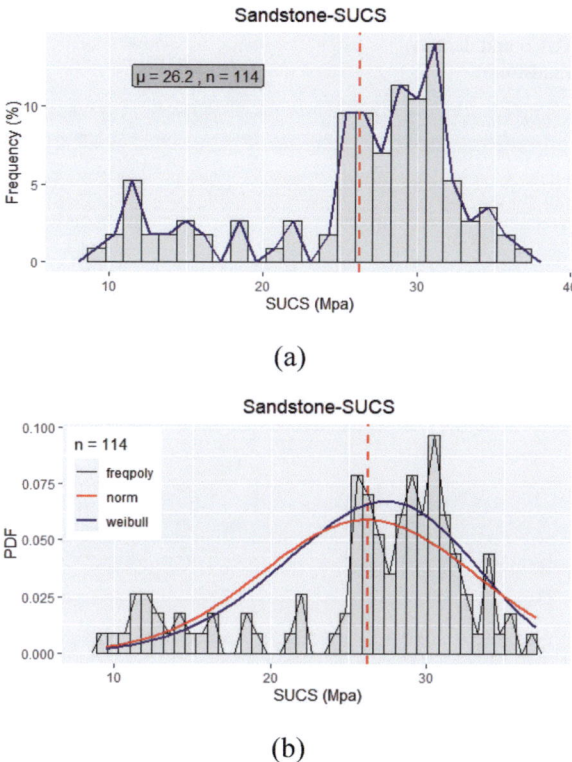

3.3 Property Correlations

The correlation of rock physical and mechanical characteristics has been investigated by several investigators [9–11].

- **Mudstone**

Total of 134 data were utilized to investigate the correlation between the NUCS and density (ρ), thus the regression line and point plot are shown in Fig. 9. It seems that no straightforward correlation is observed between the density and the NUCS for mudstone.

- **Sandstone**

Based on 79 data were utilized to investigate, the correlation between the NUCS and density (ρ), thus the regression line and point plot are shown in Fig. 10. The coefficient of determination (R2) is 0.304. It seems that low relevance is observed between the density and the NUCS for sandstone.

Fig. 9 Correlation between NUCS and density (mudstone)

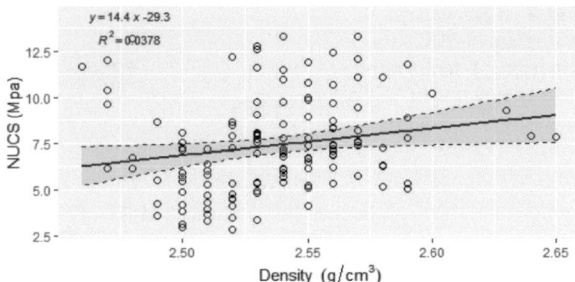

Fig. 10 Correlation between NUCS and density (sandstone)

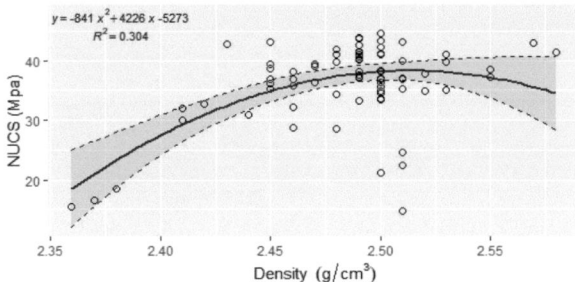

Considering SUCS and NUCS, the softening coefficient (S) is defined:

$$S = \frac{\sigma_{cs}}{\sigma_{cn}} \qquad (6)$$

where σ_{cs} and σ_{cn} represent the values of SUCS and NUCS.

Figure 11 shows that the SUCS is lower than the NUCS universally, which also shows that the SUCS takes the 66.7% of the NUCS ($S \approx 0.667$). The R2 is 0.933, which means the correlation of the SUCS and NUCS is strongly linear.

Figure 12 shows that the SUCS is lower than the NUCS universally, which also shows that the SUCS takes the 80% of the NUCS ($S \approx 0.8$). The R2 is 0.938, which means the correlation of the SUCS and NUCS is strongly linear.

According to Figs. 11 and 12, it implies that both two kinds of rock for UCS generally have a decrease in strength due to water absorption.

4 Results and Discussion

The system involves grouping of rock property COV into three major classes: (1) rock properties with COV values less than 20% are considered to have low variability, (2) those with COV values ranging from 20 to 50% are considered to have medium variability and (3) those with COV values greater than 50% are considered to have high variability [12]. The COV values of NUCS and SUCS for mudstone and

Fig. 11 Correlation between NUCS and SUCS for mudstone

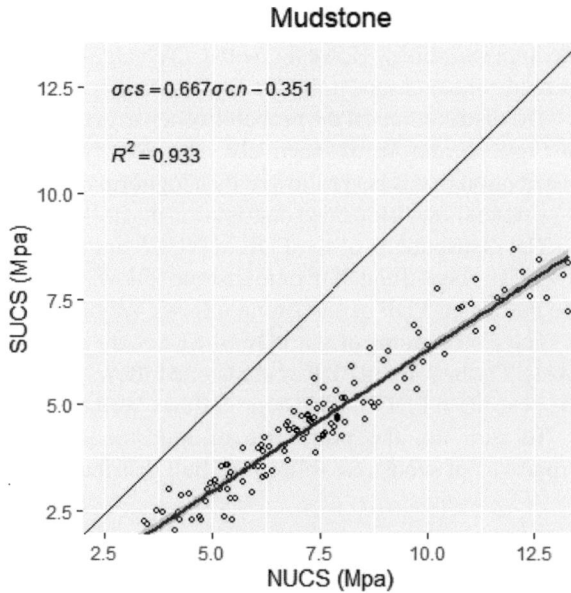

Fig. 12 Correlation between NUCS and SUCS for sandstone

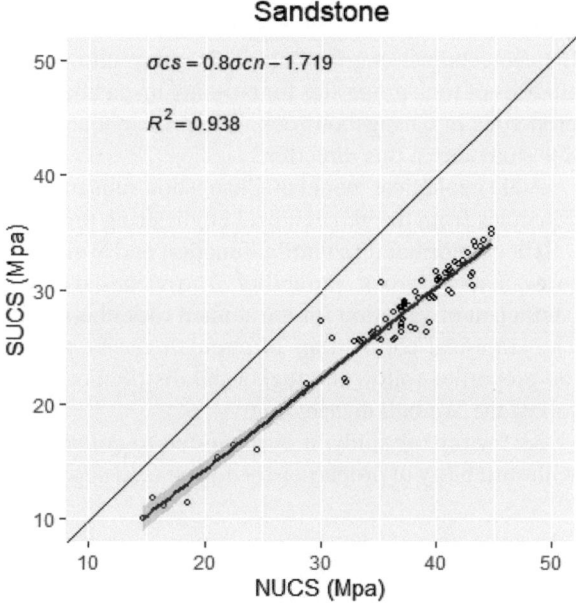

sandstone are more than 20%, indicating that the UCS values of these two rocks are medium variability. However, with COV value of 1.54 and 1.68%, the density values of mudstone and sandstone are low variability.

The distributions of the probability density function (PDF) of density are presented in Figs. 3b and 4b, respectively. The density of mudstone shows a symmetrical distribution that is best followed by a lognormal distribution. However, the density of sandstone shows a lopsided distribution that is best followed by a Weibull distribution.

The distributions of the PDF of NUCS are submitted in Figs. 5b and 6b, respectively. The best-fitted PDF of mudstone follows lognormal distribution. Nevertheless, the best-fitted PDF of sandstone follows Weibull distribution.

The distributions of the PDF of SUCS are submitted in Figs. 7b and 8b, respectively. The best-fitted PDF of mudstone follows lognormal distribution. Nevertheless, the best-fitted PDF of sandstone follows Weibull distribution.

To sum up, the properties of mudstone follow lognormal distribution, the properties of sandstone follow Weibull distribution.

5 Conclusions

In this research, a 3D geological model of Changshou stratigraphy is demonstrated for geotechnical engineering. This study allows a beneficial record of statistical analysis of rock properties for building up database and the spatial variability of rock properties in Changshou rock. The principal conclusions and recommendations of this study are in this direction:

A 3D geological model of Changshou subsurface can be utilized to determine the outcomes from the geotechnical engineering properties.

The lognormal distribution function and Weibull distribution function is utilized to evaluate the rock variability. The values of mean, standard deviation and the coefficient of variation for the studied rock characteristics are supplied in this paper. The empirical correlations of strength can be also established. For the mudstone, the properties follow the lognormal distribution, while the properties of sandstone follow the Weibull distribution.

As for further study, it is suggested to collaborate the 3D geological modelling with variability of properties for widespread application in geotechnical engineering.

References

1. Zhang JZ, Liu ZQ, Zhang DM et al (2022) Improved coupled Markov chain method for simulating geological uncertainty. Eng Geol 298. ARTN 106539. https://doi.org/10.1016/j.enggeo.2022.106539
2. Elkateb T, Chalaturnyk R, Robertson PK (2003) An overview of soil heterogeneity: quantification and implications on geotechnical field problems. Can Geotech J 40(1):1–15. https://doi.org/10.1139/T02-090

3. Myers DE (2005) Reliability and statistics in geotechnical engineering. Technometrics 47(1):103–104. https://doi.org/10.1198/tech.2005.s838
4. Nguyen TS, Ngamcharoen K, Likitlersuang S (2023) Statistical characterisation of the geotechnical properties of Bangkok subsoil. Geotech Geol Eng 41(3):2043–2063. https://doi.org/10.1007/s10706-023-02390-z
5. Mohammed A, Mahmood W (2018) Statistical variations and new correlation models to predict the mechanical behavior and ultimate shear strength of gypsum rock. Open Eng 8(1):213–226. https://doi.org/10.1515/eng-2018-0026
6. Cai M (2011) Rock mass characterization and rock property variability considerations for tunnel and cavern design. Rock Mech Rock Eng 44(4):379–399. https://doi.org/10.1007/s00603-011-0138-5
7. Mahmoudi E, Stepien M, König M (2021) Optimisation of geotechnical surveys using a BIM-based geostatistical analysis. Smart Sustain Built Environ 10(3):420–437. https://doi.org/10.1108/Sasbe-03-2021-0045
8. Lemon AM, Jones NL (2003) Building solid models from boreholes and user-defined cross-sections. Comput Geosci 29(5):547–555. https://doi.org/10.1016/S0098-3004(03)00051-7
9. Aladejare AE (2020) Evaluation of empirical estimation of uniaxial compressive strength of rock using measurements from index and physical tests. J Rock Mechan Geotechn Eng 12(2):256–268. https://doi.org/10.1016/j.jrmge.2019.08.001
10. Shi Z, Zhang W, Wang Z (2021) Correlation of physical and mechanical properties of Jurassic sandstone in Jining, Shandong province. Arab J Geosci 14(13):1254. https://doi.org/10.1007/s12517-021-07655-6
11. Chang CD, Zoback MD, Khaksar A (2006) Empirical relations between rock strength and physical properties in sedimentary rocks. J Petrol Sci Eng 51(3–4):223–237. https://doi.org/10.1016/j.petrol.2006.01.003
12. Aladejare AE, Wang Y (2017) Evaluation of rock property variability. Georisk: assessment and management of risk for engineered systems and geohazards. 11(1):22–41. https://doi.org/10.1080/17499518.2016.1207784

Open Access This chapter is licensed under the terms of the Creative Commons Attribution 4.0 International License (http://creativecommons.org/licenses/by/4.0/), which permits use, sharing, adaptation, distribution and reproduction in any medium or format, as long as you give appropriate credit to the original author(s) and the source, provide a link to the Creative Commons license and indicate if changes were made.

The images or other third party material in this chapter are included in the chapter's Creative Commons license, unless indicated otherwise in a credit line to the material. If material is not included in the chapter's Creative Commons license and your intended use is not permitted by statutory regulation or exceeds the permitted use, you will need to obtain permission directly from the copyright holder.

A Development Method of Geotechnical Testing Instrument Based on 3D Printing Technology

Hungchou Lin, Mengyue Wang, Xiaodong Zhou, Yanbo Cao, and Jianbing Peng

Abstract In engineering investigation, geotechnical testing is the main method of obtaining the basic physical parameters of regional engineering geology bodies. Geotechnical testing is a main approach to understand the mechanical properties and the engineering characteristics of earth materials. Hence, many researchers devoted to exploit new techniques or develop new apparatus and devices of geotechnical testing to acquire the accurate data. Among these techniques, 3D printing is a promising technique for improving the apparatus and devices of geotechnical testing. In this article, we designed and developed three experimental devices (the laser water level gage, the vapor saturation ring for the pressure plate apparatus, and the lid of vacuum saturation chamber) by fused deposition modeling (FDM) 3D printing technology to discuss the application and prospect of 3D printing for geotechnical testing. The results indicate that the development time of device can be shorten significantly when the 3D printing technology is applied. Moreover, if FDM 3D printing technology is used to develop devices of geotechnical testing that is under loading conditions, the cracks would be generated between one and another fused filament surface due to the bending deformation, and then the watertightness and the airtightness of devices would not be guaranteed. The work demonstrates that if we understand the principle of 3D printing more clearly, 3D printing is not only an effective technology to develop laboratory or field devices in soil science, geo-engineering, and geotechnical engineering But it also permits a more efficient information and an irreplaceable opportunity to promote the development of experimental soil mechanics.

Keywords Geotechnical testing · 3D printing · Laser water level gage · Pressure plate apparatus · Vacuum saturation

---

H. Lin (✉) · M. Wang · X. Zhou · Y. Cao · J. Peng
School of Geological Engineering and Geomatics, Chang'an University, Xi'an, China
e-mail: linhz@chd.edu.cn

H. Lin · J. Peng
Institute of Yellow River Scientific Research, Chang'an University, Xi'an, China

Y. Cao
Shanxi Institute of Geology Survey, Xi'an, China

1 Introduction

Engineering investigation is a fundamental prerequisite for both engineering design and construction, facilitating the identification of regional engineering geological conditions and the comprehension of soil's mechanical and engineering properties. Geotechnical testing serves as a crucial means in engineering investigations to recognize these properties, forming the fundamental basis for revealing soil's mechanical laws and mechanisms [1]. Consequently, geotechnical testing often necessitates the development of suitable instrumentation based on the test's objectives, combined with the generalization and analysis of test data, to achieve the desired goals.

However, with the ever-growing comprehension of soil mechanical properties, complexity of encountered engineering challenges, and the need for timely responses, traditional geotechnical testing instruments, and equipment often fall short of meeting the demands of scientific research and engineering applications. For instance, balancing accuracy and convenience remains a challenge for current water level meters. In low-humidity laboratory environments, conventional pressure plate instruments fail to achieve sufficient vapor saturation. Additionally, the existing method for processing saturated containers for vacuum saturation of diverse porous media in laboratories is highly inefficient. Consequently, test equipment must be optimized and adapted to specific test conditions to enhance accuracy and ensure test results align with actual engineering scenarios. Therefore, the crucial developmental focus for geotechnical test instruments and equipment lies in streamlining the prototype production process and achieving high-precision test outcomes.

In recent decades, 3D printing technology has emerged as the most rapidly advancing additive manufacturing technique. Its widespread application across numerous domains is well documented [2]. Among various 3D printing technologies, fused deposition modeling (FDM) stands out due to its simplicity, versatility in material choice, rapid prototyping speed, and cost-effectiveness [3]. Therefore, this study introduces the integration of FDM 3D printing technology into the development of test equipment, aiming to significantly expedite the production process and yield high-precision equipment.

This paper reports the design and development of a high-precision laser water level meter, facilitated by FDM 3D printing technology, and a vapor saturation ring for pressure plate instruments, which significantly enhances vapor saturation within the chamber. Additionally, a vacuum saturation chamber top cover was rapidly fabricated using 3D printing. And in the development of equipment, the practical merits and constraints of 3D printing in geotechnical experimental research are examined in depth, aiming to foster advancements in experimental geotechnics.

2 Materials and Methods

2.1 Applications of 3D Printing

Performance Requirements for Equipment Development

In the design of instrumentation and components essential for geotechnical testing, the foremost consideration lies in evaluating the strength and deformation properties of candidate materials to guarantee their compatibility with the working environment's performance standards. For components mandated to satisfy particular functionalities, including water and airtightness specifications in geotechnical testing, the selection of 3D printing materials and their processing methodologies must be scrutinized.

This study leverages the distinct advantages of 3D printing technology, tailored to the performance requirements (watertightness, airtightness, and deformation strength) of geotechnical test instruments. Three focal areas are explored: ① examining 3D printing's applicability for watertight designs in laser water level gages, ② addressing challenges in saturated air pressure applications of conventional pressure plate gages by proposing improvements to the steam saturation ring and evaluating 3D printing materials for watertightness, and ③ analyzing the strength, deformation, and airtightness of 3D printed materials for the top cover of vacuum saturators commonly used in tensiometer saturation tests. Detailed design and planning for specific models are outlined in Sects. 2.2 and 2.3.

3D Printing Technology and Material Selection

Metal 3D printing technology is ideally suited to fulfill the strength and deformation requirements of stressed components subjected to the stress amplitude encountered in geotechnical tests. However, the significant cost of equipment and consumables associated with metal 3D printing currently hinders its widespread adoption. Conversely, FDM 3D printing, utilizing engineering plastics, exhibits more economical benefits in geotechnical test research involving low stress variation amplitudes, albeit with a reduced deformation strength [4].

Engineering plastics such as acrylonitrile–butadiene–styrene copolymer (ABS), polylactic acid (PLA), polycarbonate (PC), and nylon (PA) are robust and commonly utilized for 3D printing. However, ABS exhibits poor weather resistance and shrinkage upon molding, PLA lacks heat resistance and brittleness, PC is vulnerable to acids and alkalis, and PA readily absorbs water, softens, and is nonresistant to strong acids. Given the neutral acid–alkali nature of the testing environment, PC emerges as a preferred 3D printing material for instrument development. This study selects PC as the material for FDM 3D printing technology and investigates whether its strength, deformation, watertightness, and airtightness satisfy the requirements of geotechnical test instrument development under low stress amplitude conditions.

2.2 Development of a Laser Water Level Gage

Purpose of Laser Water Level Meter Development

Water level monitoring is extensively employed in hydraulics, hydrogeology, and geotechnical engineering [5, 6], prompting the development of numerous tools for this purpose. Despite its fundamental role as a water level monitoring instrument, manual reading is necessitated by the water level meter for data acquisition. Alternatively, pressure and sonic water level meters automate data collection, yet their accuracy is inadequate. Consequently, this study employs 3D printing technology to integrate a laser displacement meter with a float hydrometer, thereby enhancing measurement accuracy and enabling automatic, real-time tracking of water level variations.

Concept for the Development of a Laser Water Level Meter

Utilizing the principles of laser displacement measurement, a laser-based water level meter is developed and presented in Fig. 1, including its design schematic and physical photographs. The water level meter tube's wall incorporates drainage holes to facilitate rapid water inflow and outflow, enabling prompt and sensitive responses to water level variations. Moreover, the top of the tube is opened to accommodate a pin, preventing float excursions beyond the design limit and safeguarding the laser displacement meter from damage. Furthermore, in cases of turbid water or impurities, a nylon mesh can be installed to mitigate their influence on float movement.

To evaluate the measurement error and significance of the laser water level meter, it was positioned in a calibrated cylinder with a regulated water level rate of 0.3 cm/min.

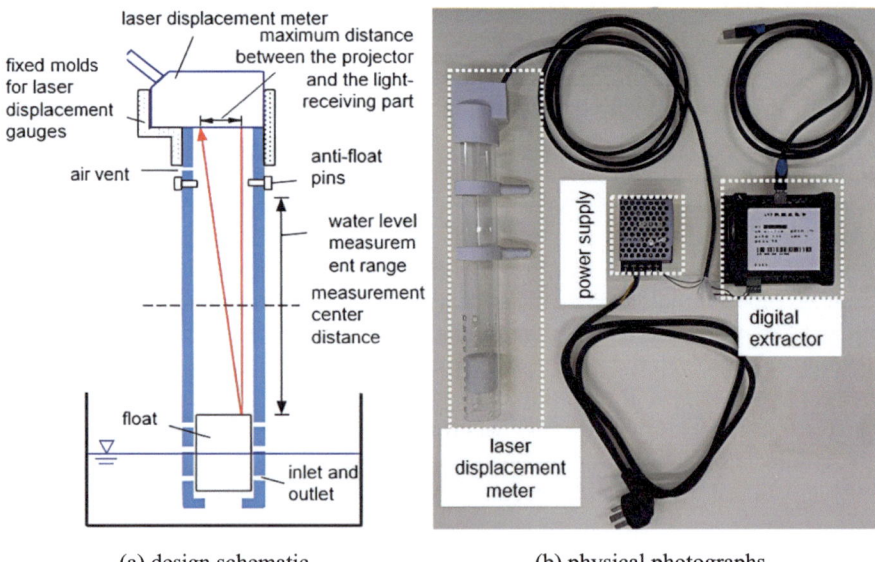

Fig. 1 Laser water level gage

Observed variations ranged from − 75 to 75 mm, and discrepancies with naked-eye observations were noted. Furthermore, the effect of float roughness on measurement was assessed by comparing a resin float with a smoother surface, fabricated via SLA 3D printing, to another float.

2.3 Improvement of Pressure Plate Instrument

Deficiencies of the Current Pressure Plate Instrument

The pressure plate apparatus, a pivotal testing device for investigating the soil water behavior in the saturated–unsaturated state, has been observed to facilitate water evaporation from specimens in low-humidity environments despite the presence of a vapor saturator (Fig. 2). This evaporation, stemming from insufficient air saturation, leads to significant inaccuracies. Consequently, it is imperative to devise optimization and enhancement strategies to mitigate the challenges posed by steam saturation in order to minimize the errors associated with moisture evaporation.

Improvement Program

An analysis of existing pressure plate apparatuses reveals that a water reservoir ring integrated with filter paper, forming a vapor-saturated ring system (Fig. 3), is the most effective method to maintain near-saturated conditions within the chamber. The ring can be precisely fabricated using 3D printing technology, tailored to the chamber's dimensions. Following fabrication, the ring undergoes a rigorous leak test at 800 kPa. If impermeability is confirmed, a hygrometer is utilized to evaluate the apparatus's improved performance.

Evaluation and Validation of Improvement Effects

To assess the effectiveness of the enhanced pressure plate apparatus, a prepared sandy soil specimen was placed in a sealed box, with only the top surface exposed to evaporation. Subsequently, the specimen was introduced into the pressure plate apparatus, both before and after the improvements, to quantify changes in humidity within the pressure chamber and soil sample evaporation. Additionally, the soil–water

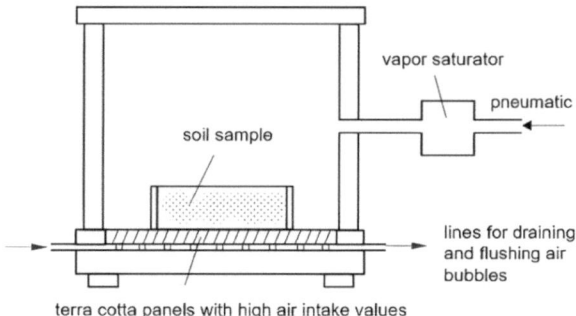

Fig. 2 Pressure plate gage before improvement

Fig. 3 Pressure plate gage after improvement

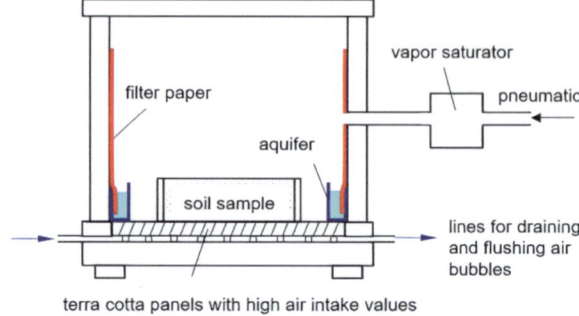

characteristic curve (SWCC) of the specimen was tested to visualize the impact of the apparatus' modifications on the resulting SWCC.

Design and Analysis of Vacuum Saturator Top Cover

Vacuum saturation methods are commonly employed in laboratories to saturate porous media including soil samples, clay plates, and clay heads for tensiometers [7]. To facilitate this saturation process, customized saturation vessels are often manufactured. When utilizing 3D printing for vessel fabrication, it is crucial to evaluate its strength, deformation, and airtightness under 1 atm pressure. This study leverages ANSYS finite element software to simulate the stress–strain properties of the vessel, and subsequently analyzes the processing of the top cover based on these results. Specific boundary conditions are depicted in Fig. 4, with applied loads (q) set at 100, 90, and 60 kPa to investigate the deformation characteristics of the top cover under varying vacuum conditions. Furthermore, the amplitude and distribution of von Mises stress on the top cover at 100 kPa are analyzed to determine if its strength satisfies design specifications. The material parameters utilized for the finite element analysis are outlined in Table 1.

The processed top cover, equipped with valves and vacuum gages, is mounted on a plexiglass drum with an outer diameter of 20 cm and a wall thickness of 10 mm for

Fig. 4 Model of FEM analysis for the lid of vacuum saturation chamber

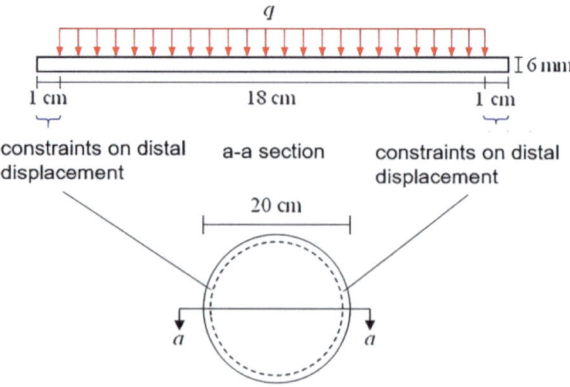

Table 1 Properties of polycarbonate (PC)

Density (g cm^{-3})	Elastic modulus (MPa)	Poisson's ratio	Yield strength (MPa)
1.2	2.38	0.36	62.1

evacuation purposes. This setup enables the evaluation of the strength, deformation, and airtightness of the top cover of the vacuum saturation chamber.

3 Results and Discussion

3.1 Error Analysis and Application of Laser Water Level Meter

Figure 5 illustrates the deviations between visual readings and those recorded by the laser water level meter, revealing that all differences fall within ± 1.0 mm, with a majority within ± 0.5 mm. A comparative analysis of micrographs reveals that the resin float exhibits the smoothest surface, whereas the PC float displays the roughest surface, among the PC float, polished PC float, and resin float. Consequently, the resin float exhibits superior resistance to the accumulation of microscopic air bubbles, thus achieving a higher level of measurement accuracy compared to the PC float.

The test results demonstrated the effectiveness of the laser water level meter in real-time measurement of minute water level fluctuations, exhibiting significant practical value and promising applications compared to traditional meters. Furthermore, the utilization of resin floats produced via SLA 3D printing technology yields slightly improved measurement outcomes compared to PC floats fabricated by FDM 3D printing, thus eliminating the need for grinding processes.

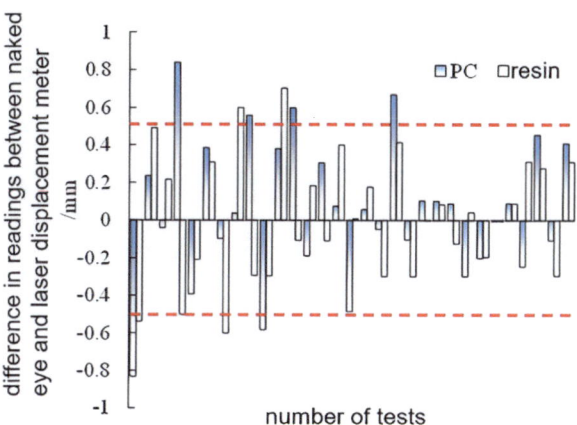

Fig. 5 Comparison between naked-eye observation and laser water level measurement

3.2 Effectiveness of Pressure Plate Instrument Improvements

Watertightness Assessment

To assess the efficacy of the enhanced pressure plate instrument, a 3D-printed water storage ring was positioned in a pressure chamber with a sealing lid, following water storage and filter paper placement. The chamber was subjected to 800 kPa of air pressure for 24 h to monitor for leakage. The results indicated no water leakage from the ring, thereby confirming the watertightness of the 3D-printed water ring and its potential application in enhancing the pressure chamber of the pressure plate instrument.

Improvement Effect

Figure 6 illustrates the pre- and postimprovement results. At a relative humidity of 100% in the pressure chamber, the time required was approximately 19 h prior to improvement and only 10 h postimprovement. For soil sample evaporation within the initial 40 h of the pressure chamber, the evaporation rate was 0.18 g before the improvement, reduced to 0.05 g after. The findings indicated that relying solely on the vapor saturator of the pressure plate instrument to saturate the air can lead to more pronounced initial evaporation of soil samples, potentially affecting the instrument's test results.

A comparative analysis of the soil–water characteristic curves of sand soil samples pre- and postimprovement (Fig. 7) revealed that for moisture contents exceeding 4%, the curves differ significantly due to moisture evaporation. Specifically, under identical matrix suction, the preimprovement moisture content is lower than the postimprovement value. Consequently, for accurate determination of the soil–water characteristic curve using a single soil sample, it is recommended to adopt the method presented in this study to enhance the pressure chamber, particularly in environments with low indoor humidity (typically below 60%). This approach mitigates water evaporation and enhances the accuracy of soil–water characteristic curve determination.

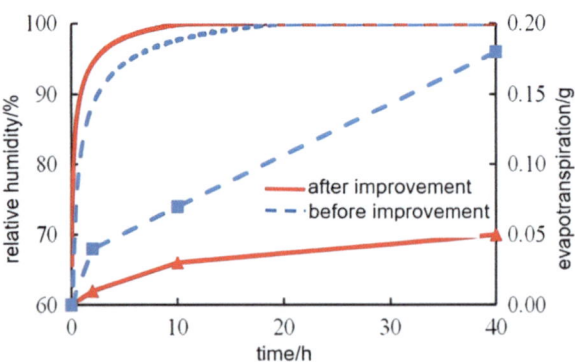

Fig. 6 Airtightness test results before and after pressure chamber modification

Fig. 7 Comparison of SWCC between before and after air pressure chamber improvement

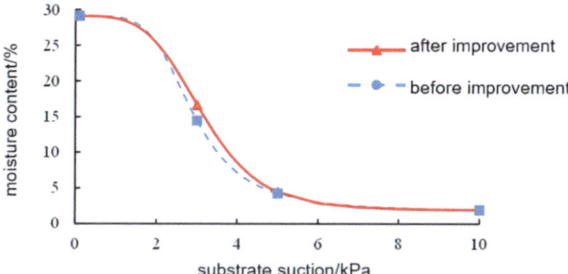

3.3 Analysis and Evaluation of Vacuum Saturation Chamber Headers

Calculation Results of Strength and Deformation

The ANSYS simulation of the vacuum saturation chamber roof cover data is summarized in Table 2. As evident from the table, under a load of 100 kPa, while the maximum deflection measures 10.44 mm, the safety coefficient of the top cover remains at 1.96, affirming that the strength of the PC top cover adheres to the design specifications, thus validating the feasibility of the method. Nevertheless, further investigation into the potential for airtightness issues following top cover deformation is warranted.

Actual Deformation and Airtightness Effect of the Top Cover

The direct utilization of the 3D-printed lid results in a vacuum drum pressure limitation of 60 kPa, which is insufficient for maintaining the requisite airtightness. The vacuum pressure is enhanced to 90 kPa by the application of a petroleum jelly coating and plastic wrap wrapping, fulfilling the airtightness criteria for soil sample saturation via pumping.

In summary, minimal discrepancy exists between the ANSYS-calculated deformation of the top cover and the observed results. When 3D printing technology is employed for fabricating airtight or watertight components, force-induced deformations lead to gaps at the fused filament deposition interfaces, compromising airtightness and watertightness. Additional sealing measures are thus necessary to satisfy design specifications.

Table 2 Numerical analysis results of the cover lid

Load (kPa)	Maximum von Mises equivalent force (MPa)	Factor of safety	Maximum deflection (mm)	
			Finite element calculated values	Measured value
100	31.7	1.96	10.44	–
90	28.5	2.18	9.40	9.0
60	19.0	3.27	6.26	7.0

4 Conclusion

This study examines the merits and drawbacks of FDM 3D printing in the fabrication of geotechnical test equipment through three case studies. The primary findings are as follows:

(1) FDM 3D printing is capable of fulfilling the essential performance criteria for geotechnical testing, including watertightness, airtightness, and deformation strength.
(2) Real-time digital laser water level meters are facilitated by the 3D printing technology, offering substantial advantages in monitoring scenarios involving minor water level fluctuations. However, for scenarios with significant water level variations, further exploration of more suitable water level meters remains necessary.
(3) The utilization of 3D printing technology in the production of water storage rings, composed of steam-saturated rings, enables the air within the pressure chamber of the pressure plate meter to be rapidly saturated, thereby being effectively enhanced for the accuracy of soil–water characteristic curve determinations.
(4) During the processing of watertight and airtight components using FDM 3D printing technology, significant deformations are caused by external forces, leading to the formation of gaps at the filament stack deposition, thereby compromising the watertightness and airtightness of the designed component. Consequently, in high-pressure applications, the exploration of high-strength yet economical printing materials is imperative.
(5) In conclusion, FDM 3D printing technology offers notable advantages in the development of geotechnical test instruments. However, its application must be tailored to its underlying principles and characteristics to yield more suitable components.

Acknowledgements National Key R&D Program of China (2022YFC3003400); National Natural Science Foundation of China (41272283)

References

1. Li GX, Lin HC (2017) Tutorials on advanced soil mechanics. Wuhan University of Technology Press. Wuhan. http://www.wutp.com.cn
2. He Y, Gao Q, Liu A et al (2019) Bio-3D printing—From shape to shadow. J Zhejiang Univ (Eng Ed) 53(3):407–419. https://kns.cnki.Net/kcms/detail/33.1245.T.20181116.1640.004.html
3. Ngo D, Kashani A, Imbalzano G et al (2018) Additive manufacturing (3D printing): a review of materials, methods, applications and challenges. Composites Part B 143. https://doi.org/10.1016/j.compositesb.2018.02.012
4. Arrieta-Escobar AJ, Derrien D, Ouvrard S et al (2020) 3D printing: an emerging opportunity for soil science. Geoderma 378:115588. https://doi.org/10.1016/j.geoderma.2020.114588

5. Kossieris S, Tsiakos V, Tsimiklis G et al (2024) Inland water level monitoring from satellite observations: a scoping review of current advances and future opportunities. Remote Sens 16(7). https://doi.org/10.3390/RS16071181
6. Purnell D, Gomez N, Minarik W et al (2024) Real-time water levels using GNSS-IR: a potential tool for flood monitoring. Geophys Res Lett 51(5). https://doi.org/10.1029/2023GL105039
7. Guo Y, Lin HC, Peng JB (2023) Response characteristics of an insertion tensiometer investigated via the vacuum saturation method. CATENA 224:106966. https://doi.org/10.1016/J.Catena.2023.106966

Open Access This chapter is licensed under the terms of the Creative Commons Attribution 4.0 International License (http://creativecommons.org/licenses/by/4.0/), which permits use, sharing, adaptation, distribution and reproduction in any medium or format, as long as you give appropriate credit to the original author(s) and the source, provide a link to the Creative Commons license and indicate if changes were made.

The images or other third party material in this chapter are included in the chapter's Creative Commons license, unless indicated otherwise in a credit line to the material. If material is not included in the chapter's Creative Commons license and your intended use is not permitted by statutory regulation or exceeds the permitted use, you will need to obtain permission directly from the copyright holder.

Centrifuge Model Test on the Seismic Response of the Pile-Supported Wharf Slope

Jiarui Zhang, Yu Shao, Long Lü, Xianlin Liu, and Xilin Lü

Abstract Pile-supported wharf is a quintessential wharf structure in soft soil areas, and it may suffer severe damage during earthquake. This paper studied the seismic response of the pile-supported wharf system by dynamic centrifuge model test. The variation of the acceleration, displacement, and excess pore water pressure with time at different positions in the wharf slope was studied. The results indicate that there is a risk of inducing soil liquefaction due to the continuously increase of pore pressure in the soil, and it may lead to slope instability.

Keywords Pile-supported wharf · Slope · Earthquake · Centrifuge model test

1 Introduction

Pile-supported wharf is a common form of wharf structure, and it is suitable for deep water and soft soil area. Pile-supported wharf may suffer severe damage during earthquakes, i.e., Takahama wharf of Kobe Port, Japan destoried by 1995 Great Hanshin earthquake [1] and the 7th Street Terminal wharf located in Port of Oakland, USA which was shaken by 1989 Loma Prieta earthquake [2]. Since earthquake may induce failure of pile-supported wharf, it is indispensable to study the seismic stability of wharf supported by slope piles.

J. Zhang · X. Lü (✉)
Department of Geotechnical Engineering, Tongji University, Shanghai, P. R. China
e-mail: xilinlu@tongji.edu.cn

Y. Shao · X. Liu
Guangxi Communications Design Group Co., Ltd, Nanning Guangxi, China

L. Lü
College of Environment and Civil Engineering, Chengdu University of Technology, Chengdu, China

X. Lü
Key Laboratory of Geotechnical and Underground Engineering of Ministry of Education, Tongji University, Shanghai, P. R. China

Centrifuge model test was a valuable method to explore the impact of earthquake on pile-supported wharfs. Takahashi [3] studied the dynamic response of a pile-supported wharf using centrifuge model tests, the failure characteristics of piles and the effect of liquefaction on the foundation were investigated. Mc Cullough et al. [4] launched earthquake centrifuge tests on five pile-supported wharf models. Yun and Han [5] investigated the mechanism of kinematic load induced by slope deformation during earthquake and its influence on pile-supported wharfs by using centrifuge model tests. Li et al. [6] carried out dynamic centrifuge tests of pile-supported wharf on sloping saturated sandy soil to investigate the pressure on piles. Besides, the seismic response and stability of pile-supported wharf was studied by numerical simulation in the recent years [7–10].

Previous studies mainly focus on the effects of earthquakes on piles of the pile-supported wharf. In this study, centrifuge model test was carried out to investigate the seismic response of the pile-supported wharf slope, including the time histories of the acceleration, displacement, and pore water pressure, the result is useful to assess the stability of slope.

2 Centrifuge Shaking Table Test

The centrifuge test was conducted by TLJ-150 geotechnical centrifuge apparatus in Tongji University, as shown in Fig. 1. The centrifuge capacity is 150g-t, the maximum centrifugal acceleration is 200 g, and the effective rotation radius is 3 m. The maximum acceleration of centrifuge shaking table is 20 g.

The prototype of the slope has a height of 6m and slope ratio of 1:2. The wharf consists of steel tube piles with an external diameter of 1.2 m and a thickness of 12 mm. The self weight load on the dock deck is 30 kPa. The pile-supported wharf structure has three rows of piles spaced 5 m apart. The acceleration rate in the test is

Fig. 1 Centrifuge model test apparatus (photographed during test)

50 g, so the model scale is 1/50, as shown in Fig. 2. The model piles are aluminum pipes with an external diameter of 24 mm and a thickness of 0.7 mm, and the deck is a 20 mm thick aluminum plate.

The slope is made of ISO standard sand; the specific gravity is 2.65, dry density is 1.72 g/cm^3, void ratio is 0.540, relative density is 45%. In order to maintain the consistency of the dynamic time and consolidation time, dimethicone with a kinematic viscosity of 50 cst was chosen for the saturation of sand. The filtered and amplitude-modulated Northridge earthquake wave is used as the earthquake input for the test. The input earthquake ground motion and recorded response of the shaking table are plotted in Fig. 3. It shows that the acceleration response of the shaking table matches the input ground motion acceleration. The peak acceleration occurred at the same time, but the peak acceleration of the shaking table is slightly lower than that of the input earthquake ground motion.

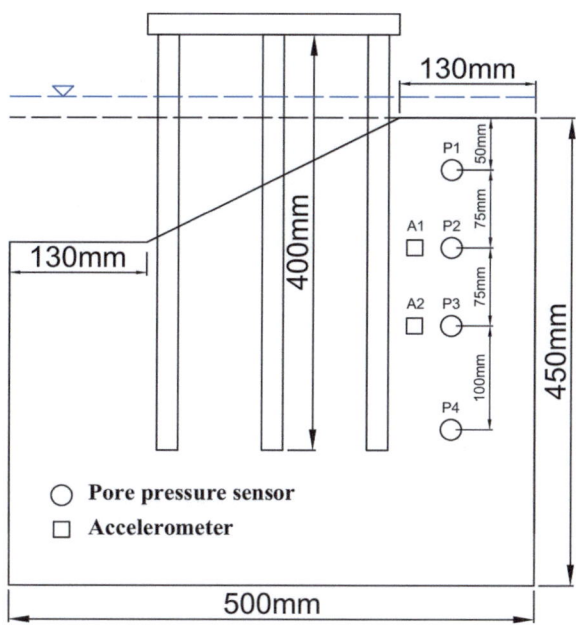

Fig. 2 Centrifuge test model of pile-supported wharf

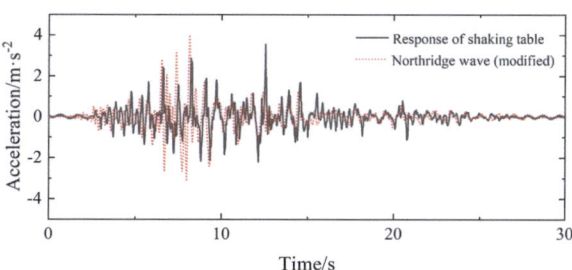

Fig. 3 Time histories of input ground motion

3 Test Results Analysis

The acceleration time history and displacement time history of accelerometer A1, which is 4 m from the top of the slope and 6.25 m deep in prototype, are shown in Fig. 4. The peak acceleration at A1 is 4.22 m/s², which is larger than the peak input acceleration. The acceleration waveform at A1 is also different from the input acceleration. The residual displacement is 0.123 m after the earthquake, indicating that the slope slips at this location due to earthquake.

The acceleration and displacement time histories of accelerometer A2, which is 4 m from the top of the slope and 10 m deep, are shown in Fig. 5. The peak acceleration at A2 is 0.93 m/s², which is lower than the input peak acceleration. The residual displacement at A2 is close to zero. It indicates that the earthquake does not cause the permanent displacement of the foundation at A2.

The time histories of excess pore water pressure at different positions (P1–P4) are shown in Fig. 6, in which $r_u = 1$ means the occurrence of liquefaction. The excess pore water pressure at P1 had several negative values in the early loading stage, and maintained a relatively small positive value after 15 s. The excess pore water pressure at P2–P4 increased before 10s, and then the excess pore water pressure at P2 and P3 remains stable until the end of earthquake. Liquefaction occurred as the excess pore water pressure reached the vertical effective stress of foundation at P4; however, there is no large-scale liquefaction, the slope still maintains stability. After the earthquake ground motion weakened, the excess pore water pressure slowly decreased.

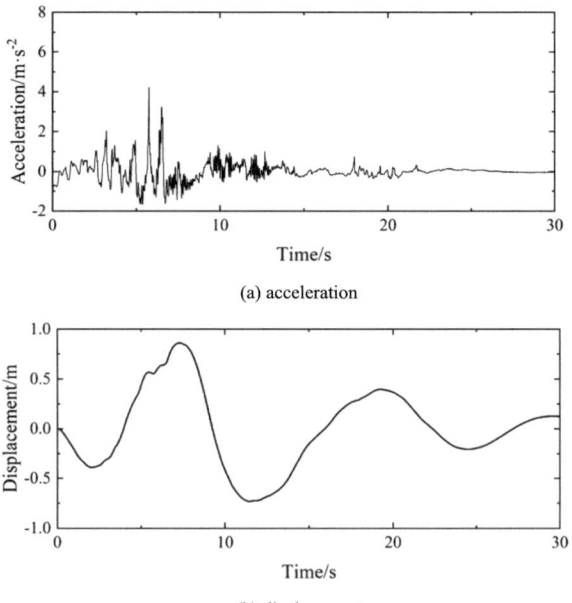

Fig. 4 Acceleration and displacement at A1

(a) acceleration

(b) displacement

Fig. 5 Acceleration and displacement at A2

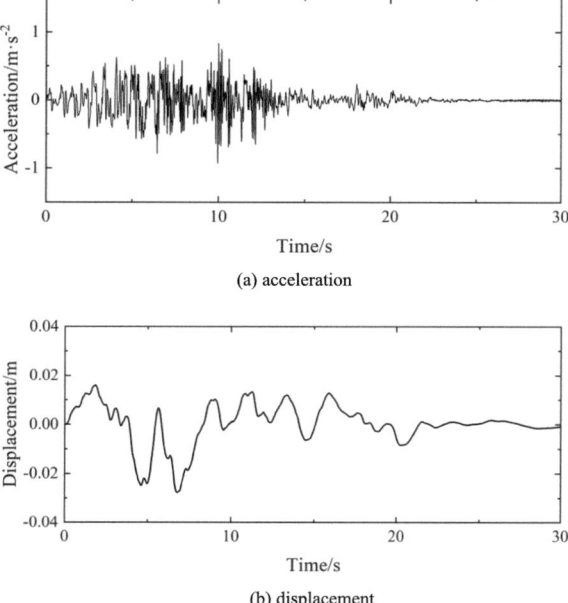

(a) acceleration

(b) displacement

4 Conclusions

The seismic response of the pile-supported wharf was investigated by centrifuge model test, and time histories of the acceleration, displacement, and excess pore water pressure of the slope were investigated. The conclusions are as follows.

The peak acceleration in shallow subsoil is larger than that in deep subsoil. Little residual displacement is found in deep subsoil, but there is significant residual displacement in shallow subsoil. Excess pore water pressure in foundation rises rapidly due to the strong earthquake ground motion, and it is relatively large in deeper soils. A risk of inducing soil liquefaction is caused by earthquake ground motion, and it may induce large displacements.

Fig. 6 Time histories of excess pore water pressure (P1-P4)

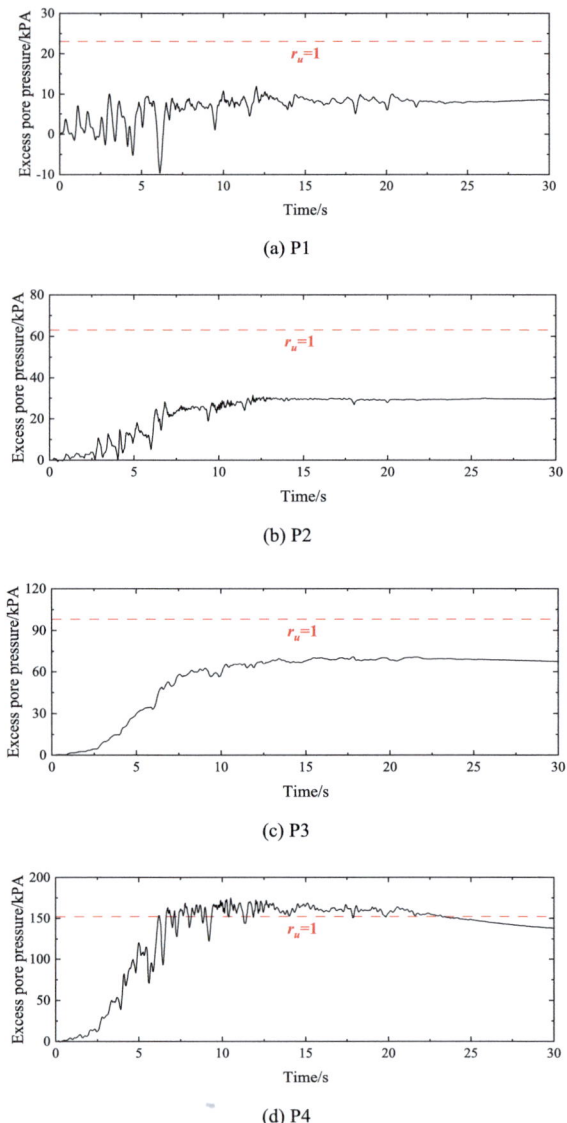

(a) P1

(b) P2

(c) P3

(d) P4

Acknowledgments The financial supports by National Key R&D Program Projects (through grant No. 2021YFB2600700) and Nanning Yongjiang Program (through grant No. RC20230108) are gratefully acknowledged.

References

1. Iai S (1998) Seismic analysis and performance of retaining structures. Geotechnical Earthquake Engineering and Soil Dynamics III, Geotechnical Special Publication No. 75, ASCE, pp 1020–1044
2. Egan JA, Hayden RF, Scheibel LL et al (1992) Seismic repair at seventh street marine terminal, grouting, soil improvement and geosynthetics. Geotechnical Special Publication No. 30, ASCE, pp 867–878
3. Takahashi A, Takemura J (2005) Liquefaction-induced large displacement of pile-supported wharf. Soil Dyn Earthq Eng 25(11):811–825
4. McCullough NJ, Dickenson SE, Schlechter SM et al (2007) Centrifuge seismic modeling of pile-supported wharves. Geotech Test J 30(5):349–359
5. Yun JW, Han JT (2021) Dynamic behavior of pile-supported wharves by slope failure during earthquake via centrifuge tests. Int J Geo-Eng 12(1):33
6. Li X, Tang L, Man X et al (2021) Liquefaction-induced lateral load on pile group of wharf system in a sloping stratum: a centrifuge shake-table investigation. Ocean Eng 242:110–119
7. Donahue MJ, Dickenson SE, Miller TH et al (2005) Implications of the observed seismic performance of a pile-supported wharf for numerical modeling. Earthq Spectra 21(3):617–634
8. Su L, Wan HP, Lu J et al (2021) Seismic performance evaluation of a pile-supported wharf system at two seismic hazard levels. Ocean Eng 219:108333
9. Nagao T, Lu P (2020) A simplified reliability estimation method for pile-supported wharf on the residual displacement by earthquake. Soil Dyn Earthq Eng 129:105904
10. Tran NX, Gu KY, Yoo M et al (2022) Numerical evaluation of slope effect on soil–pile interaction in seismic analysis. Appl Ocean Res 126:103291.s

Open Access This chapter is licensed under the terms of the Creative Commons Attribution 4.0 International License (http://creativecommons.org/licenses/by/4.0/), which permits use, sharing, adaptation, distribution and reproduction in any medium or format, as long as you give appropriate credit to the original author(s) and the source, provide a link to the Creative Commons license and indicate if changes were made.

The images or other third party material in this chapter are included in the chapter's Creative Commons license, unless indicated otherwise in a credit line to the material. If material is not included in the chapter's Creative Commons license and your intended use is not permitted by statutory regulation or exceeds the permitted use, you will need to obtain permission directly from the copyright holder.

Study on Mechanical Experimental Characteristics of Loess Under Different Water Content and Confining Pressure

Dehuan Sun

Abstract In order to solve the mechanical characteristics in the construction of pile-beam-arch method in loess stratum, this paper carried out laboratory tests, focusing on the mechanical behavior of loess under different water content. Through uniaxial and triaxial compression tests, the variation patterns of key parameters such as compressive strength, elastic modulus, cohesive force, and internal friction angle of loess samples were analyzed in detail. The results show that the compressive strength and elastic modulus of loess decrease significantly with the increase of water content, and the cohesion and internal friction angle also show a downward trend. These experimental results not only reveal the internal mechanism of the change of mechanical properties of loess with water content, but also provide an important theoretical and experimental basis for the construction of subway stations in the loess area.

Keywords Loess stratum · Water content · Pile-beam-arch method

1 Introduction

In the construction of subway stations, the PBA method, as a commonly used construction method, has the advantages of fast construction speed and small impact on the surrounding environment, and has been widely used in loess areas. During the construction of this method, the guide tunnel is excavated first, and then the strip (pile) foundation, bottom longitudinal beam, side pile, central column, crown beam, and top longitudinal beam are constructed inside the guide tunnel. Then, excavation and arch construction are carried out, and finally, the soil is excavated layer by layer and the internal structure is constructed in the support system formed by the Pile-Beam-Arch, also known as the PBA method.

However, as a special soil type, the mechanical properties of loess are affected by many factors such as water content and density, which brings many challenges

---

D. Sun (✉)
China Communications Construction First Highway Engineering Bureau Third Engineering Co., Ltd., Beijing, China
e-mail: WH18899319911@126.com

Table 1 Physical parameters of samples

Physical parameters	ρ/g.cm^{-3}	w/%	ρ_{sat}/g.cm^{-3}	w_{sat}/%
Verage value	1.72	23.3	1.88	35.07

to the construction of pile-beam-arch method. At present, research on the tunnel pile method has achieved certain results in the fine sand layer [1–3], water rich sand layer [4, 5], and water rich pebble layer. However, research on the tunnel pile method for subway construction in loess layers is rare. Therefore, this article aims to systematically study the changes in mechanical properties of loess under different moisture content conditions through indoor experiments, providing theoretical basis and experimental parameters for the design and construction of subway stations using the pile-beam-arch method in loess areas.

2 Physical Parameters of the Loess

Sample the undisturbed loess at the station location, the basic physical parameters of the sample were measured, as given in Table 1.

In order to study the effects of different moisture content on the mechanical properties of loess, soil samples were humidified and placed in a vacuum saturator for moisture. After many times weighing and adjusting, the moisture content of soil samples reached 8%, 16%, 23%, 30% and saturation state 35%, respectively.

3 Experimental Results

Conduct uniaxial and triaxial compression tests on samples with different moisture contents. Figure 1 shows the uniaxial stress–strain diagram of the samples under different rate of water content. Figure 2 shows the triaxial stress–strain curve of the sample under different rate of water content and confining pressure conditions.

3.1 Unconfined Compressive Strength Under Different Moisture Content

The following conclusion can be drawn from Fig. 3. The lower the water content of loess is, the greater its unconfined compressive strength will be. When the water content is lower, the faster its unconfined compressive strength will increase, as shown in.

Study on Mechanical Experimental Characteristics of Loess Under … 315

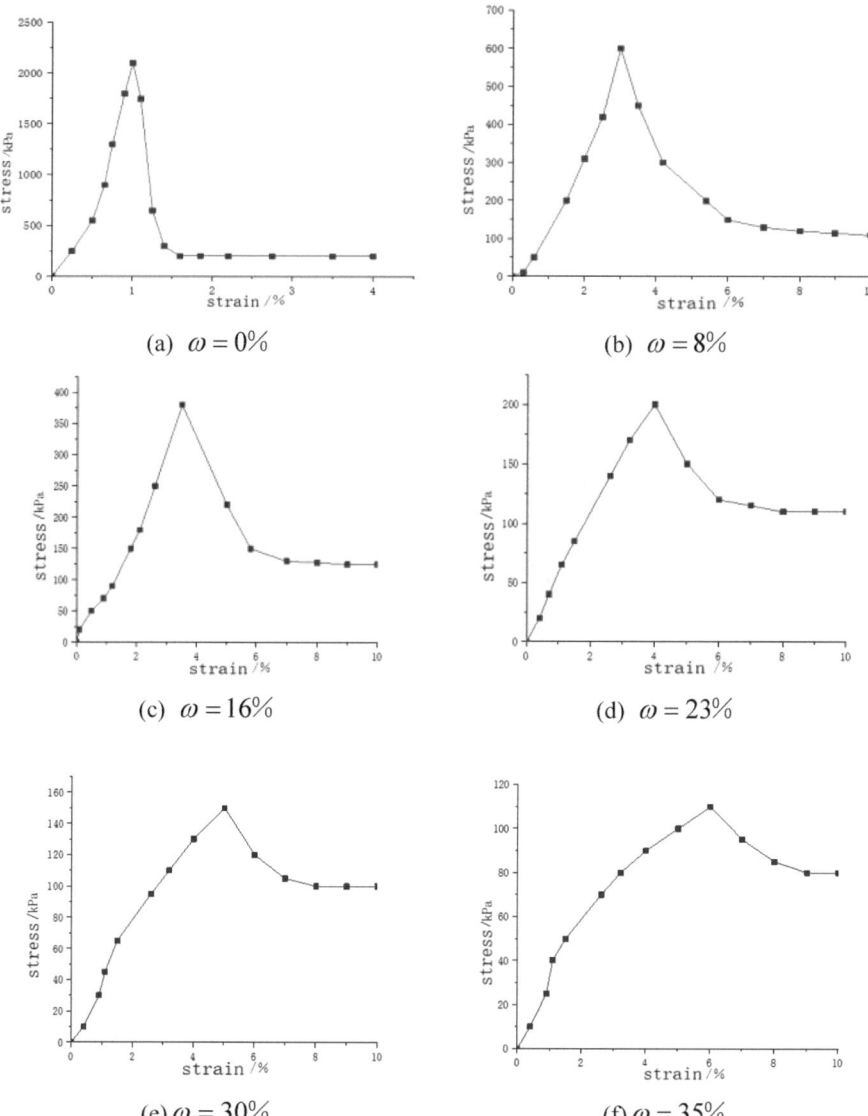

Fig. 1 Uniaxial stress–strain diagram of samples under different rate of water content

3.2 Variation of Elastic Modulus of Loess Under Different Water Content

According to the lower stress–strain diagram of loess under different rate of water content, the elastic modulus of loess under various water content conditions is

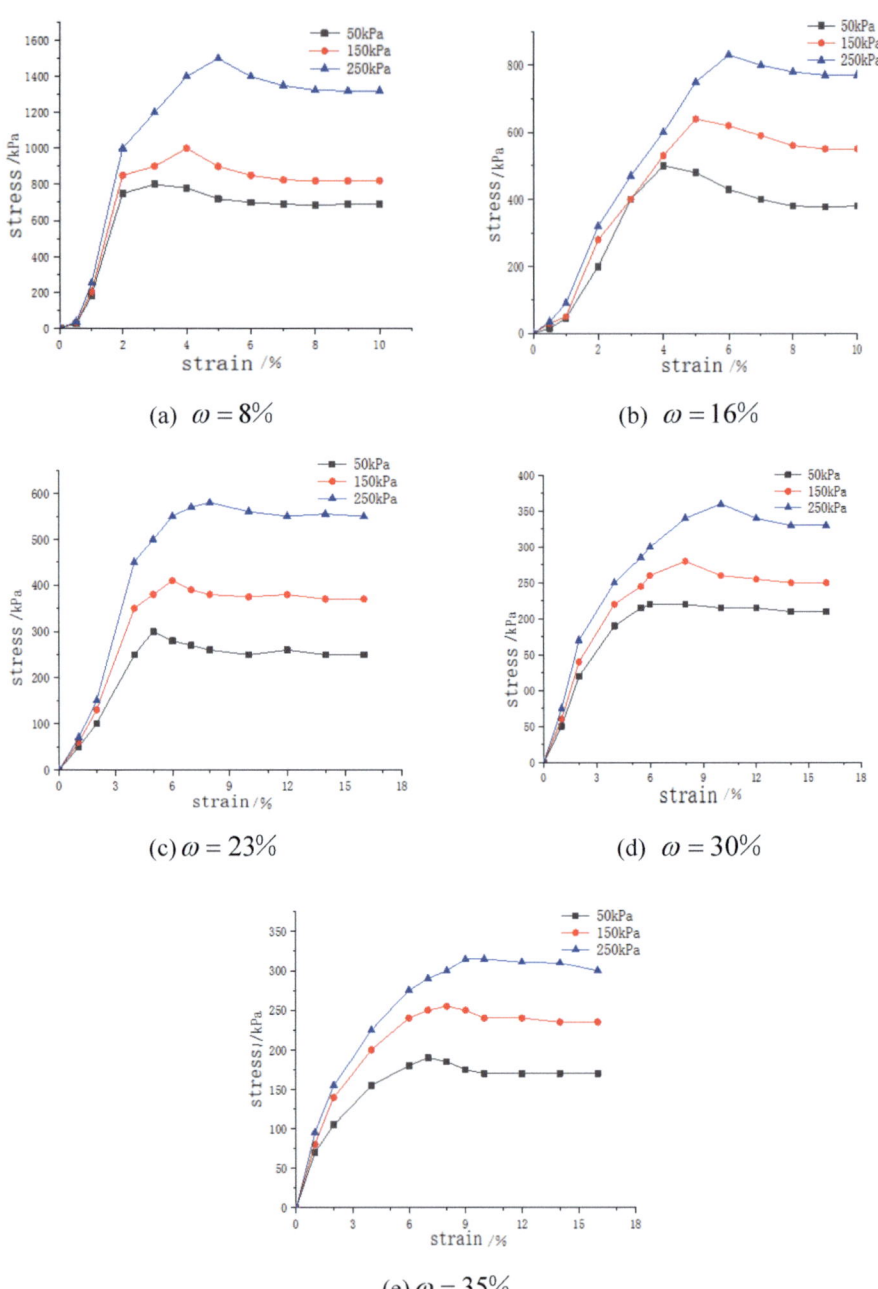

Fig. 2 Triaxial stress–strain diagram under different rate of water content

Fig. 3 Relation between compressive strength and moisture content

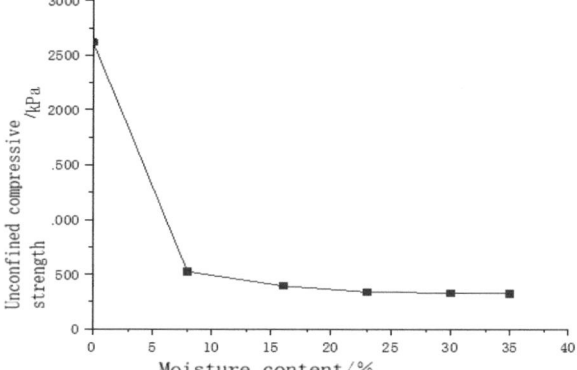

obtained, the curve relationship is shown in Fig. 4. It can be seen that when the moisture content of the sample increases, the elastic modulus of loess will decrease. When the moisture content rises from 0 to 8%, the elastic modulus decreases by 90.82%. When the moisture content rises from 30 to 35%, it only decreases by 20.94%, indicating that the greater the moisture content, the slower the change rate of elastic modulus and the smaller the slope of the curve.

Fig. 4 Relation between moisture content and elastic modulus

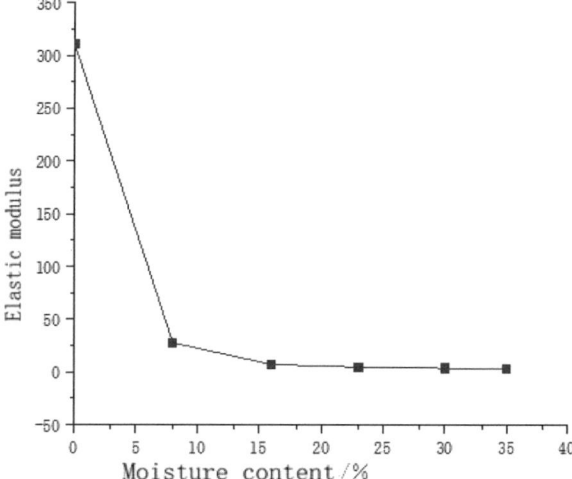

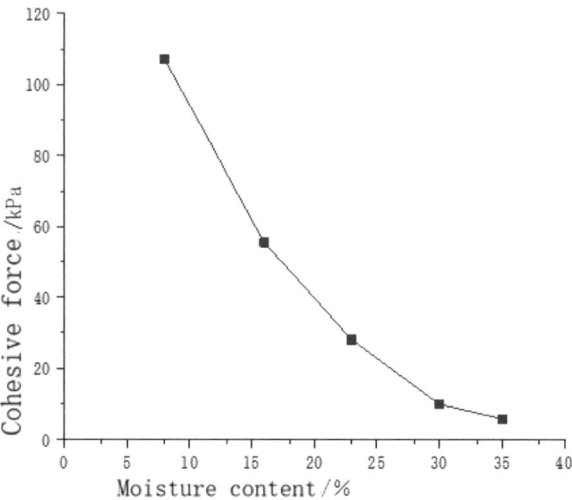

Fig. 5 Connection between water content and cohesion

3.3 Changes of Cohesive Force and Internal Friction Angle of Loess Under Different Water Content

The cohesive force and internal friction angle of loess under triaxial pressure were summarized, and the variation law of mechanical properties of loess under triaxial pressure was obtained.

As shown in Figs. 5 and 6, with the continuous increase of water content, the cohesion and internal friction angle of the specimen decreased, and the decreasing trend of cohesion gradually slowed down with the increase of water content. However, the decreasing trend of internal friction angle remains unchanged as water content increases, and the slope can be approximated as a constant value of −0.65.

3.4 Variation of Triaxial Strength of Loess Under Different Water Content

As can be seen from Fig. 7, under the same confining pressure, the triaxial compressive strength of loess gradually decreases with the increase of water content, and the lower the water content, the increasing range of triaxial compressive strength increases. With the water content increasing from 8 to 35%, the triaxial compressive strength decreases by 80.4% under 50 kPa confining pressure. Under 150 kPa confining pressure, the triaxial compressive strength decreases by 76.3%. Under 250 kPa confining pressure, the triaxial compressive strength decreased by 75.6%.

Study on Mechanical Experimental Characteristics of Loess Under … 319

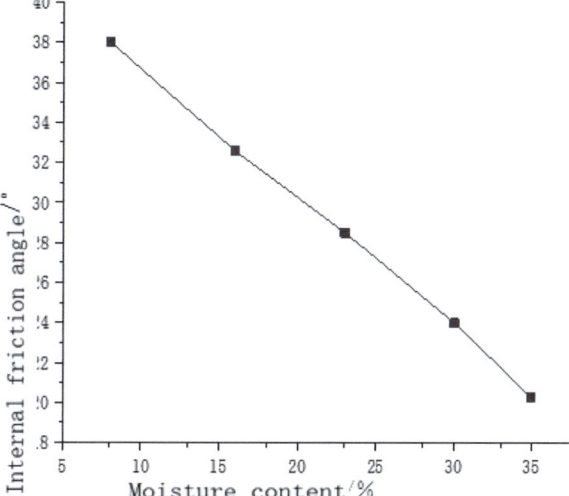

Fig. 6 Connection between water content and internal friction angle

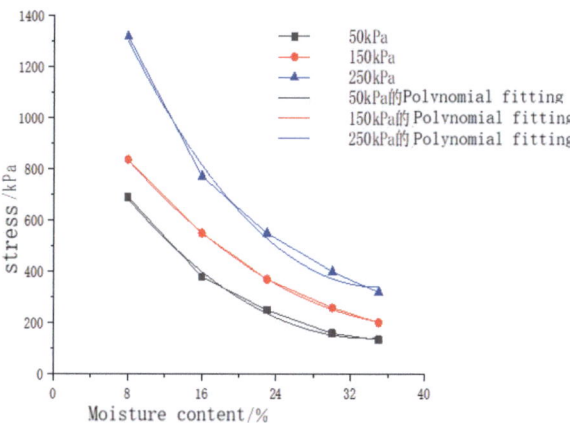

Fig. 7 Triaxial compressive strength of samples with different moisture content

As can be seen from Fig. 7, the effect of water content on the triaxial compressive strength of soil is very significant. When the loess with low water content is humidified, the triaxial compressive strength of the loess will drop sharply, thus bringing great adverse effects on the safety of the project.

4 Conclusion

In this paper, after undergoing uniaxial and triaxial compression tests, the changes of mechanical properties of loess under different water content conditions were deeply explored, and it was found that:

As the confining pressure increases, the compressive strength of loess increases and the strain during failure increases. However, the increase of rate of water content leads to the weakening of compressive strength, the decrease of failure strain and the decrease of elastic modulus, and the decreasing trend of elastic modulus becomes slower with the increase of water content.

The increase of water content reduces the cohesion and internal friction Angle of loess, especially the decrease of strength of low water content loess after humidification, which easily leads to soil instability and structural damage. During the construction, the moisture content of loess should be strictly monitored and controlled to ensure the construction safety.

References

1. Ağbay ET, Topal T (2014) Evaluation of twin tunnel-induced surface ground deformation by empirical and numerical analyses (NATM part of Eurasia tunnel, Turkey).Comp Geo 119. https://doi.org/10.1016/j.compgeo.2019.103367
2. Meng F, Li C, Wu JS et al (2014) Ground movement analysis based on stochastic medium theory. Sci World J 4:702561. https://doi.org/10.1155/2014/702561
3. Yuan YK, Jiao Q, Muchi L et al (2020) Load-structure model of deformation of hollow pile method. J Lanzhou Univ Technol 046(001):123–128
4. Li T, Gao Y, Shao W et al (2019) Study on the influence of inclined load on the deformation rule of hollow pile with normal edge. J Undergr Space Eng 15(S2):666–672+686
5. Deng XH, Cao WP, Yang DS (2019) Effect of rainfall infiltration on the stability of shallow buried loess tunnel. J Civ Environ Eng 42(02):45–55

Open Access This chapter is licensed under the terms of the Creative Commons Attribution 4.0 International License (http://creativecommons.org/licenses/by/4.0/), which permits use, sharing, adaptation, distribution and reproduction in any medium or format, as long as you give appropriate credit to the original author(s) and the source, provide a link to the Creative Commons license and indicate if changes were made.

The images or other third party material in this chapter are included in the chapter's Creative Commons license, unless indicated otherwise in a credit line to the material. If material is not included in the chapter's Creative Commons license and your intended use is not permitted by statutory regulation or exceeds the permitted use, you will need to obtain permission directly from the copyright holder.

Deformation Analysis of Subway Pit Excavation Under Diaphragm Wall and Steel Support

Qi Chen, Xuezhu Li, and Lingfeng Wan

Abstract In this paper, ABAQUS numerical software is used to simulate the excavation of deep foundation pit of subway in a city. The enclosing structure of the foundation pit is diaphragm wall, and the internal support is concrete support and steel support. Based on the numerical simulation, the effect of steel support on the deformation of diaphragm wall and surface settlement control was analyzed under the action of 0, 50, 60, 70, 80% preloaded axial force, and the optimal preloaded axial force value of 60% was determined. Under the optimal preloaded axial force, the steel support was analyzed for the change of support axial force and horizontal displacement, deformation of the top of diaphragm wall, maximum settlement of the ground surface, and the amount of pit bottom uplift in the process of foundation pit excavation. The results show that the support axial force and deformation, diaphragm wall deformation and surface settlement during pit excavation are within the specification requirements. The amount of the pit bottom uplift slightly exceeded the limit value, and this part of the soil should be reinforced to ensure the construction safety.

Keywords Steel support preloaded axial force · Enclosure system deformation · Surface settlement · Pit floor upliftment

1 Introduction

With the significant improvement of urban economic development in China, the problem of traffic congestion on the ground is inevitable. In order to alleviate the operation pressure of urban ground traffic, underground rail transit has developed rapidly. By May 2020, there were 47 cities in mainland China that had opened urban rail transit, of which 41 were in mainland China and 6 were in Hong Kong, Macao and Taiwan, China [1]. It is a common practice to build new buildings near existing buildings. New buildings require deep foundation pit excavation, resulting in

Q. Chen (✉) · X. Li · L. Wan
School of Civil Engineering and Architecture, Wuhan Polytechnic University, Wuhan, China
e-mail: qchenqqmail123@qq.com

uneven ground displacement, which has a negative impact on the safety and technical conditions of adjacent buildings [2, 3].

The excavation of foundation pit causes surface subsidence and deformation of adjacent buildings, which will lead to the overturning of buildings in severe cases. In particular, subway foundation pits are generally excavated to a greater depth. Due to the diversity of geological conditions, soft soil layers such as silty loam may be encountered [4]. A new subway station is proposed to be built in a city to meet the traveling needs of the public. This paper is based on ABAQUS software to simulate the excavation process of the subway pit and analyze the surface settlement of the pit and the deformation of its supporting structure.

2 Engineering Overview

The excavation depth of the foundation pit is 18.9 m, the width is 27.1 m, and the surrounding site is open. Based on the geological report, the surrounding environment and relevant standards, the safety level of the foundation pit is determined as Grade II. The top soil layer of the site is artificial fill, and the middle soil layer is powdery clay and silty clay, and the parameters of the soil body are given in Table 1. Selection of suitable enclosure structure is an important guarantee for safe excavation and construction of the foundation pit. Compared with the basic strength design, the control of pit deformation design is more demanding at this stage.

The types of pit enclosure structure mainly include column type, sheet pile type, soil nail wall, hydraulic soil retaining wall, diaphragm wall and combined type, etc. Different enclosure structures have their own advantages [5]. In the excavation

Table 1 Soil parameters

Serial number	Ground level	Cohesive force (kPa)	Friction angle (°)	Modulus of elasticity (MPa)	Poisson ratio	Intensity (kg/m^3)	Soil layer elevation (m)
1	Miscellaneous fillings	9	14	3	0.40	1745	22.2
2	Plain filling soil	10	8	4	0.41	1745	15.1
3	Silty clay	21	10	17	0.41	1855	12.7
4	Silty clay	12	4	9	0.35	1740	11.6
5	Silty clay	25	13	22	0.33	1900	3.22
6	Silty clay	42	14	38	0.32	1955	1.01
7	Highly weathered argillaceous siltstone	39	15	128	0.24	2200	−3.19

of deep foundation pit, the deformation requirements of the retaining structure are high. Due to the large filling thickness and loose soil structure of the project, to avoid excessive deformation of the foundation pit excavation support structure, it is proposed to adopt the foundation pit support program of diaphragm wall combined with internal support. A column is set between the two walls, and the support is concrete support and steel support.

3 Numerical Simulation

3.1 Modeling

ABAQUS is used to establish a two-dimensional finite element model as shown in Fig. 1. The total thickness of the model soil is 44 m and the width direction is 80 m. The Mohr–Coulomb model is selected as the constitutive relation of the material. The thickness of the underground continuous wall is 1000 mm, the concrete strength grade is C40, and a total of 4 supports are set. The first one is the concrete crown beam, and the section height is 1000 mm. The other three are steel supports, and the section height is 800 mm. The left and right sides of the model are horizontally constrained, and the grid is delimited by CPE4 cells.

According to the design standards and excavation conditions, the design values of the axial forces of the four supports are determined to be 5300 KN, 4339 KN, 2621 KN and 3325 KN, respectively. The pre-axial force of steel support can be taken as 0.5 ~ 0.8 times the design value of axial force [6]. In this paper, the design values of 0.5,0.6,0.7 and 0.8 times axial force are applied in turn in the simulation, and the deformation of the supporting structure under different pre-axial forces is analyzed to determine the most suitable pre-axial force value. The ultimate bearing capacity of the concrete support is 12053 KN, the steel support is 6738 KN, and the axial force warning value is 0.7 times the ultimate bearing capacity.

The whole excavation is carried out in five parts. In the first part, the ground stress is balanced and the initial displacement of the whole model is zeroed; in the second part, the vertical support structure diaphragm walls and columns are added into the

Fig. 1 Pit support model

model; in the third part, the first layer of soil is excavated and a concrete support is set; in the fourth part, the layers of soil are excavated one by one and three steel supports are applied; finally, the third steel support is excavated to the soil at the bottom of the pit.

3.2 Optimum Preloaded Axial Force

The axial force design values of 50, 60, 70 and 80% were applied to the first, second and third steel supports, and the deformation control effect of the support structure under different pre-axial forces was analyzed [7]. The deformation requirements of the secondary foundation pit for the retaining system are that the horizontal and vertical displacements at the top of the diaphragm wall are less than 30 mm, the deep horizontal displacement of the diaphragm wall is less than 40 mm, and the maximum surface settlement is less than 35 mm [8].

Deformation of Diaphragm Wall Under Different Preloaded Axial Forces

A path is created between the top and bottom of the two walls on both sides [9]. The horizontal deformation of both sides of the diaphragm wall under four preloaded axial forces applied by the steel support is obtained in comparison with the deformation when no axial force is applied, as shown in Fig. 2.

It can be seen that the deformation of both sides of the diaphragm wall is significantly improved after the steel support is applied with the preloaded axial force applied by the steel support. There is a maximum deformation at the top of the ground diaphragm wall, and with the increase of the distance from the top of the wall, there is a second maximum deformation that decreases to zero with the increase of the depth of the wall. The deformation of diaphragm wall decreases with the increase of

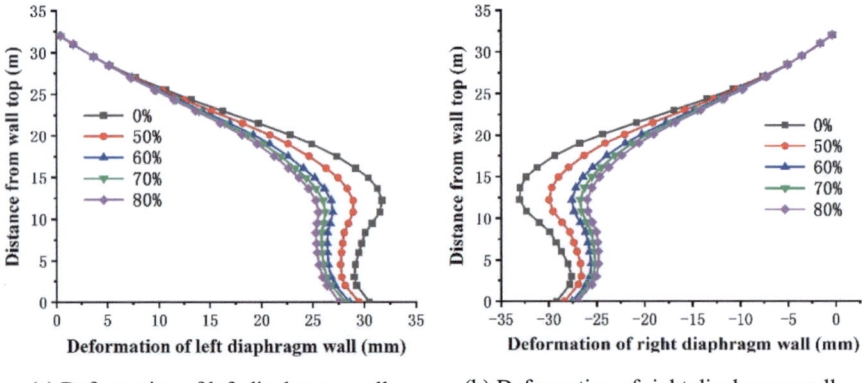

(a) Deformation of left diaphragm wall   (b) Deformation of right diaphragm wall

Fig. 2 Deformation of diaphragm wall under different preloaded axial forces

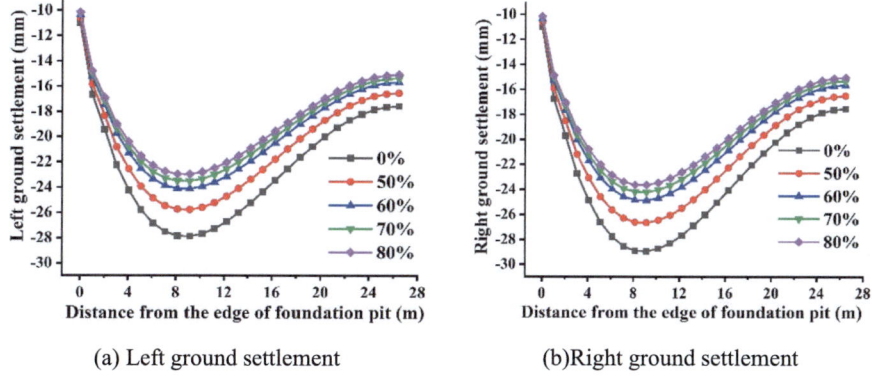

(a) Left ground settlement   (b) Right ground settlement

Fig. 3 Surface settlement under different preloaded axial forces

preloaded axial force, and the deformation control effect is better when 60, 70 and 80% preloaded axial force is applied, and the two maximum deformations are less than 30 mm, but the distribution of the three curves is closer, and 60% preloaded axial force can be preferred to consider the construction cost.

Surface Settlement Under Different Preloaded Axial Forces

The trend of surface settlement on the left and right sides after the preloaded axial force of the steel support is shown in Fig. 3. The surface settlement on both sides is notched with the increase of horizontal distance, and the maximum position is at 9 m from the edge of the foundation pit. When the steel support is not applied axial force, the maximum settlement on the right side is -28.94 mm, which is reduced by 2.3 mm after applying 50% preloaded axial force, and the maximum settlement is reduced by 5.3 mm after applying 80% preloaded axial force; when applying 60, 70 and 80% preloaded axial force, the best effect of controlling the surface settlement is achieved on the two sides, and the difference between the three curves in settlement is small, so 60% axial force is the optimal value to be chosen. The optimal preloading value is 60% axial force.

3.3 Support Axial Force and Horizontal Displacement During Excavation

The monitoring items for the secondary pit support structure mainly include the axial force and deformation of the support, the displacement of the enclosure structure, the amount of surface settlement, and the amount of the bottom of the pit rises. According to the research of Yu Qinqin et al., the surface settlement of the pit with groove-shaped distribution, with the increase of excavation depth and the number of supports, the surface settlement around the pit and the maximum surface settlement value show an increasing trend [10]. From the previous analysis, the maximum settlement value of the surface and the deformation of the diaphragm wall meet the design value, and the change of the remaining index will be analyzed later.

A total of 12 analysis steps were set up in the simulation of pit excavation, in which the first part of the soil was excavated to 2 m in the 4th–6th analysis step, followed by the setting of the first support. In the 6th–8th analysis steps, the second part of the soil was excavated to 7.8 m and the second support was set. At analysis steps 8–10, the third portion of the earth is removed to 12.69 m and the third support is set. In the 10th–11th analysis steps, the fourth part of the soil is removed to 16.09 m, the fourth support is set in the 11th–12th analysis steps, and finally, the excavation is carried out to the base.

The variation of the four support axial forces with analysis steps is shown in Fig. 4, with positive values for tension and negative values for pressure. The first support axial force gradually increases in the 5, 6 and 7th analysis steps and reaches the maximum value of −953.95 kN, and rapidly decreases in the 8th analysis step. The reason is that after the soil excavation in the 4th and 6th analysis steps, the two sides of the wall extruded on the support, which made the axial force grow gradually. Meanwhile the second support was set in the 7th to improve the stress state of the first support.

The excavation of the neighboring soil layer had the strongest influence on the axial force of the steel support, and the subsequent excavation of the soil body had a slightly weaker influence on the axial force. The maximum axial force of the second support slightly exceeds the axial force warning value. However, the support is squeezed by the ground connecting walls on both sides, and the maximum horizontal displacements are at the end of the support. From the analysis of optimal preloaded axial force, the displacement of the wall at the support under 60% preloaded axial force is not more than 30 mm, so the deformation of the end of the support is also small.

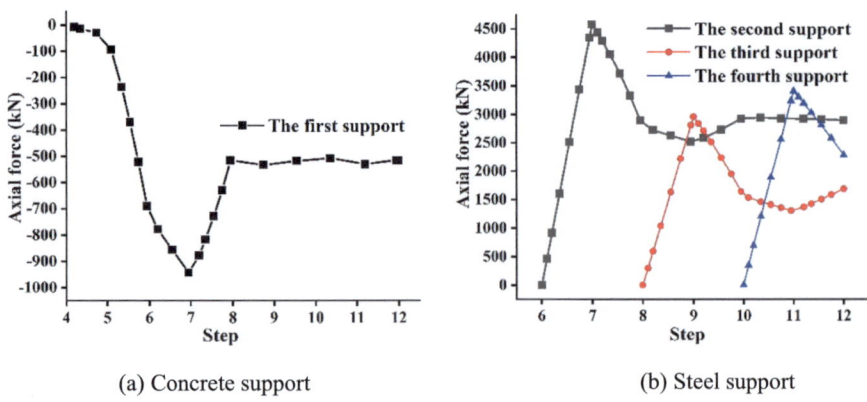

Fig. 4 Variation curve of support axial force

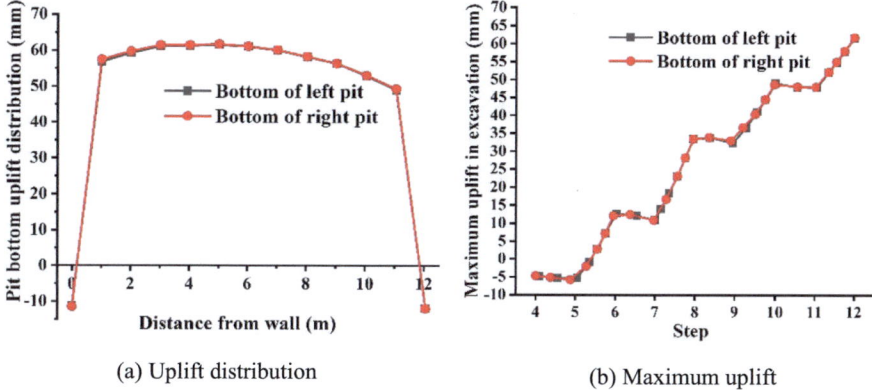

(a) Uplift distribution (b) Maximum uplift

Fig. 5 Curve of change of elevation at the bottom of the pit

3.4 Analysis of Soil Bulging at the Bottom of the Pit During Excavation

The distribution of pit bottom bulge values along the horizontal direction on both sides after excavation, and the change of maximum bulge values on both sides during excavation are shown in Fig. 5. As can be seen in Fig. 5, the distribution curves of the two sides of the pit bottom bulge basically overlap, and with the increase of the distance from the ground connecting wall, the pit bottom bulge grows gradually and decreases after reaching the maximum value.

The maximum value of the right side of the pit bottom is 61.72 mm, which is slightly larger than that of the left side of 61.57 mm, and the location of the maximum bulge is 5.02 m away from the ground wall on each side. During excavation, the maximum bulge of the soil body, with the increase of excavation depth, shows a stepwise growth. The value of pit bottom bulge decreased slightly after soil excavation, and the soil body reached the initial equilibrium state, whereas the soil body bulge at the pit bottom increased significantly after the support effect, indicating that the addition of the support greatly changed the stress state of the soil body.

4 Conclusion

(1) This paper analyzes the deformation of the enclosure structure and surface settlement under different preloaded axial force of steel support in the excavation process of foundation pit to determine the optimal preloaded axial force value. Under the action of optimal preloaded axial force, the support axial force, support deformation, deformation of the top of the ground connecting wall, the

maximum settlement change of the ground surface, and the rise of the pit bottom were analyzed in the excavation process.
(2) The preloaded axial force of the steel support has obvious limiting effect on the deformation of the ground wall and surface settlement. 60, 70, 80% preloaded axial force has the best effect on the deformation control of the support system, and the deformation of the ground wall and surface settlement are not more than 30 mm, and the optimal preloaded axial force of 60% is finally determined.
(3) The axial force of the four supports during excavation is within the bearing range, and the axial force of the second and fourth supports is the largest, which contributes more to the pit support. The maximum deformation of the four supports and the maximum horizontal displacement of the top of the diaphragm wall were less than 30 mm.
(4) The surface settlement on both sides is maximum at 9 m from the edge of the pit. The settlement decreases from this to both sides, and the maximum settlement is -25.46 mm. During excavation, the maximum bulge value of the soil body at the bottom of the foundation pit shows a stepwise growth with the increase of excavation depth. The location of the maximum bulge on both sides is 5.02 m from the ground connecting wall on each side, and the maximum bulge value is 61.72 mm.
(5) In the excavation process, the deformation of diaphragm wall and surface settlement meet the specification requirements, and the axial force and deformation of the support are also within the normal range. In summary, the pit can effectively control the deformation of the maintenance system and surface settlement under the action of 60% preloaded axial force of the diaphragm wall and steel support, and the excavation of the pit can be carried out smoothly. However, the value of soil bulge at the bottom of the pit exceeds 60 mm, and the soil at the bottom of the pit should be reinforced during actual excavation.

In this paper, the ground settlement near the foundation pit and the deformation of the supporting structure are analyzed, and the feasibility of the excavation scheme is verified. However, there is a lack of analysis of the settlement of surrounding buildings. The settlement of adjacent buildings should be predicted and corresponding monitoring schemes should be formulated during construction.

References

1. Chen YW (2021) Research on the construction of urban rail transportation standard system. China Stand S2:34–37. https://doi.org/10.3969/j.issn.1674-5698.2021.11.007
2. Vamsi NV (2023) Health and safety Concerns in excavation and the measures to mitigate risk. Int J Res Appl Sci Eng Technol 11:188–192. https://doi.org/10.22214/ijraset.2023.49400
3. Jasiński R, Harabinova S, Kotrasova K et al (2023) Assessment of safety of masonry buildings near deep excavations: ultimate limit states. Buildings 13(11):2803. https://doi.org/10.3390/buildings13112803
4. Deng JS, Pan EM, Xiang WM (2021) Research on deformation characteristics of subway pit excavation. Civ Eng 10(10):1013–1025. https://doi.org/10.12677/HJCE.2021.1010112

5. Cheng BR (2022) Analysis of deformation characteristics of combined support system of diaphragm wall and ring support in deep foundation pit. Munic Technol 40(1):123–129. https://doi.org/10.19922/j.1009-7767.2022.01.123
6. JGJ 120–2012, Technical specification for foundation pit support of construction. blob: https://s.wanfangdata.com.cn/b849dd7e-0392-454a-b944-96f1f1dd1615
7. Wang S (2021) Research on the performance of subway diaphragm wall and steel support structure for subway foundation pit. Eng Technol Res 6(04):101–103. https://doi.org/10.19537/j.cnki.2096-2789.2021.04.044
8. GB50497-2019, Technical standards for monitoring of construction pit works. blob: https://s.wanfangdata.com.cn/e632b50b-f1bd-4940-99cf-468e4ac176f5.
9. Fei K, Peng J (2017) ABAQUS geotechnical engineering examples in detail. People's Posts and Telecommunications Press, Beijing
10. Yu QQ, Wang LF, Pang J et al (2019) Research on surface settlement law around deep foundation pit in subway station. Sci Technol Bull 35(2):94–100, 110. https://doi.org/10.13774/j.cnki.kjtb.2019.02.020

Open Access This chapter is licensed under the terms of the Creative Commons Attribution 4.0 International License (http://creativecommons.org/licenses/by/4.0/), which permits use, sharing, adaptation, distribution and reproduction in any medium or format, as long as you give appropriate credit to the original author(s) and the source, provide a link to the Creative Commons license and indicate if changes were made.

The images or other third party material in this chapter are included in the chapter's Creative Commons license, unless indicated otherwise in a credit line to the material. If material is not included in the chapter's Creative Commons license and your intended use is not permitted by statutory regulation or exceeds the permitted use, you will need to obtain permission directly from the copyright holder.

Experimental Study on Water Resistance of Microbial-Magnesium Oxide Improved Red Clay

Haodong Qin, Shangbin Wu, Yicong Wang, Yueguang Yang, Wenrong Li, Xiaoqing Wang, and Yuqin Liao

Abstract Red clay has the undesirable properties such as softening and disintegration when encountering water, and its slopes are damaged by long-term erosion, often resulting in soil erosion. In the recent years, the microbial soil improvement technology has emerged, and the application prospect is promising. In this study, the disintegration test of microbial-magnesium oxide cured red clay and the slope erosion model test was conducted to compare the disintegration rate of vegetative soil and cured soil under dry and wet cycles, and the slope morphology and erosion rate of vegetative soil and cured soil slopes under a single rainfall condition and under dry and wet cycles. The results showed that the disintegration rate of the vegetative soil was accelerated under dry and wet cycles, but had little effect on the disintegration of the cured soil; the erosion rate of cured slopes was only about 2.8% of the erosion rate of the vegetative soil slopes under single rainfall conditions; under dry and wet cycles, the erosion rate of the vegetative soil slopes decreased and then increased with the increase of the number of dry and wet cycles, while the erosion rate of the cured soil was extremely small. The curing effect of microorganisms-magnesium oxide on red clay is good, and the research results are of guiding significance for the practical application of cured red clay in open-pit mine slopes and other projects.

H. Qin · Y. Yang · W. Li · X. Wang · Y. Liao
Electric Power Research Institute, CSG EHV Power Transmission Company, Guangzhou, China
e-mail: 1289480349@qq.com

Y. Yang
e-mail: 393217467@qq.com

W. Li
e-mail: 635659129@qq.com

X. Wang
e-mail: wangxiaoqing@im.ehv.csg

Y. Liao
e-mail: 291673198@qq.com

S. Wu · Y. Wang (✉)
School of Engineering and Technology, China University of Geosciences (Beijing), Beijing, China
e-mail: wyc2372@163.com

Keyword Microbial-induced precipitation of magnesium carbonate · Solidified red clay · Slope erosion · Modeling tests

1 Introduction

Red clay is widely distributed in hot and humid areas in the south, and under the effect of frequent rainfall and evaporation, the fissure development of red clay slopes can easily lead to disasters such as collapses and landslides. In the recent years, more and more researchers have focused on slope protection problems of side slopes and explored various slope protection techniques such as curing agent slope protection [1–5]. Adding ameliorators to the soil to make the fine-grained soil cemented to form a larger agglomerate structure can effectively enhance the resistance of the slope soil to scouring [6–9].

Microbial-induced calcium carbonate precipitation technology has been utilized in clay soil improvement, but some studies have shown that this technology is not effective in clay soil treatment, and the concentration of carbon dioxide is difficult to control in indoor test [10–12]. Since magnesium carbonate also has gelling properties, and at the same time, the strength of magnesite in nature is much greater than that of calcium ore, the study combines the microbial curing technology with active magnesium oxide curing technology, and proposes a microbial-active magnesium oxide curing technology in order to achieve the purpose of ameliorating the red clay soil and enhancing the erosion resistance of red clay slopes. Through the test, the deterioration law of physical and mechanical properties of plain red clay and cured soil under the action of dry and wet cycle; the development law of erosion pattern of plain red clay slope and cured soil slope under rainfall condition and dry and wet cycle, and the effect of microorganism-induced magnesium carbonate precipitation in improving the anti-erosion effect of red clay slopes were evaluated. This is a useful supplement to the existing ecological protection technology in practical projects such as open-pit mine slopes, roadbed slopes and power grid tower slopes, and has a good application prospect in the field of slope protection.

2 Experimental Principle and Design

2.1 Strain Activation and Culture

The strain used in this study was Sporosarcina pasteurii, No. ATCC 11,859, which originated from China Strain Collection Center. Under the aseptic environment of the ultra-clean table (see Fig. 1), solid medium was used for activation and liquid medium was used for expansion culture, and the solid and liquid medium compositions are given in Table 1.

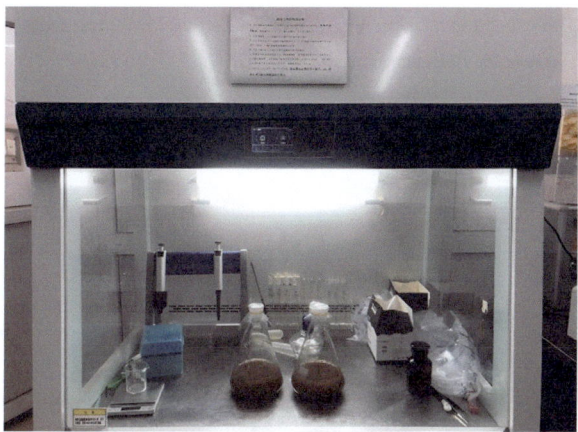

Fig. 1 Ultra-clean worktable (Shangbin Wu, 2022)

Table 1 Medium composition table

Solid media composition table		Liquid culture media composition list	
Reagent name	Dosages/g	Reagent name	Dosages/g
Urea	20	Yeast extracting powder	20
Casein peptone	15	NH4Cl	10
Soybean peptone	5	Deionized water	1000
NaCl	5	pH	9
Deionized water	1000		

Sporosarcina pasteurii is an aerobic bacterium, which is widely distributed in nature and has strong tolerance and environmental adaptability. At the same time, the bacterium is a common urease bacteria, can be secreted through the metabolic process of high activity urease, with the role of hydrolysis of urea.

2.2 Measurement of Urease Activity

The urea used in the test was purchased from Sinopharm Group Chemical Reagent Co. Urease activity is an index for evaluating the efficiency of urease-catalyzed hydrolysis of urea, and Whiffin [13] first proposed to measure the urease activity by measuring the change in the conductivity of the solution. The urease activity can be characterized by the amount of change in conductivity of the measured solution:

$$U = 1.11 f \tag{1}$$

Table 2 Relationship between urease activity of bacterial sap and incubation time

Time/h	Urease activity/(ms/cm/min)	Time/h	Urease activity/(ms/cm/min)
0	0	28	11.3
4	1.5	32	10.6
8	3.1	36	9.8
12	4.3	40	9.4
16	5.9	44	9.1
20	7.8	48	8.8
24	11.8		

where U denotes the enzymatic activity of the bacteria and f denotes the rate of change of conductivity in the bacterial fluid. The present study was based on this method to measure urease activity. 3 mL of the bacterial solution was thoroughly mixed with 27 mL of 1.5 mol/L urea solution in a 5 mL beaker using a sterile pipette gun at 25 °C, and placed in a constant temperature laboratory at 25 °C until the temperature of the mixed solution was the same as room temperature. Then the conductivity value was measured every minute with a conductivity meter for 5 times and the average conductivity change value was calculated. The change in solution conductivity was measured for 5 min and the average change in conductivity value was multiplied by the dilution of the bacterial solution to obtain the urease activity of the bacterial solution as given in Table 2.

The growth curve of Sporosarcina pasteurii was plotted using incubation time as the horizontal coordinate and OD600 as the vertical coordinate as shown in Fig. 2.

2.3 Selection of Soil Material and Magnesium Oxide

(1) The red clay soil used in the test was taken from the slope of a mine in Nanping, Fujian Province, with an average annual temperature of 18–26 °C. The depth range of the soil was 10–100 cm, which was red clay. Before the test, the soil samples were mixed and air-dried, sieved through 2 mm, placed in a drying oven at a temperature of 130 °C for 24 h, taken out and cooled, and then divided into sets with sealing bags. According to the Standard for Geotechnical Test Methods (GB/T50123-2019) to determine the basic physical property indicators of red clay as shown in Table 3.

(2) The test was carried out using Japanese Concord 150 activated magnesium oxide with an iodine adsorption value of 120 mg/g, which was placed in a dry environment and sealed for storage, and its main chemical composition and main parameters are given in Table 4.

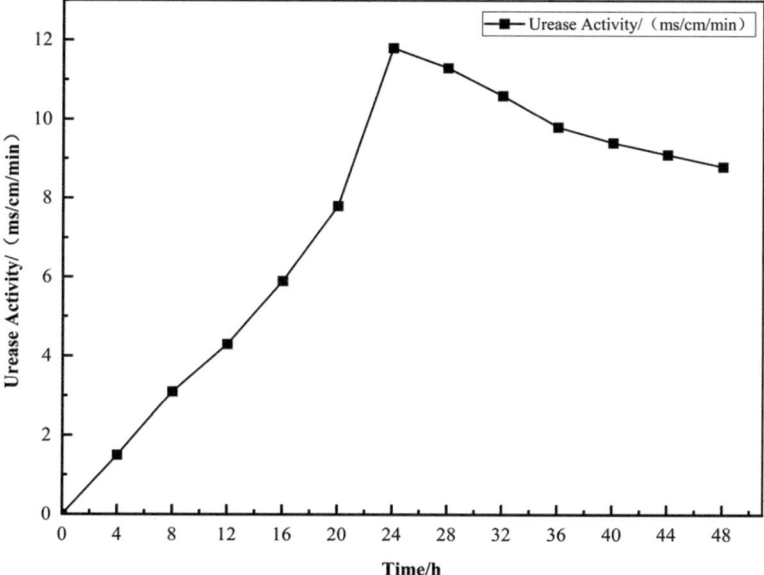

Fig. 2 Growth curves of *Sporosarcina pasteurii*

Table 3 Indicators of basic physical properties of red clay

Natural density / (g·cm^{-3})	Natural water content/%	Quid limit/%	Plastic limit/%	Soil grain	Void ratio
1.72	30.98	49.16	35.73	2.67	1.075

Table 4 Main chemical composition of activated MgO and its main parameters

MgO/%	Fe/%	Mn/%	CaCl$_2$/%	Bulk density /(g·cm^{-3})
98	0.05	0.003	1	0.12

2.4 Test Principle

The following reaction occurs when MgO, urea, bacterial solution and red clay are thoroughly mixed together [14]:

$$MgO + H_2O \rightarrow Mg^{2+} + OH^- \qquad (2)$$

$$CO(NH_2)_2 + H_2O \xrightarrow{urease} HCO_3^- + NH_4^+ + OH^- \qquad (3)$$

$$HCO_3^- \leftrightarrow H^+ + CO_3^{2-} \qquad (4)$$

$$Mg^{2+} + CO_3^{2-} + H_2O \rightarrow MgCO_3 3H_2O \tag{5}$$

$$Mg^{2+} + CO_3^{2-} + H_2O \rightarrow Mg_5(CO_3)_4(OH)_2 5H_2O \tag{6}$$

$$Mg^{2+} + CO_3^{2-} + H_2O \rightarrow Mg_5(CO_3)_4(OH)_2 4H_2O \tag{7}$$

$$Mg^{2+} + CO_3^{2-} + H_2O \rightarrow Mg_2(CO_3)(OH)_2 3H_2O \tag{8}$$

The $Mg(OH)^2$ generated in the first step of the reaction has poor cementing properties and is not ideal for curing red clay [15]. As the reaction proceeds, the generated hydrated magnesium carbonate has good cementing properties, which can cement the red clay particles and connect them into a cured whole through the bonding cooperation between the soil particles–cement–soil particles, and the hydrated magnesium carbonate can seal the pores of the red clay so that the soil structure becomes denser, and the integrity of the soil is enhanced [16, 17].

2.5 Test Methods

Disintegration Test Design

In order to ensure the comparability between the specimens, the compaction of the specimens was controlled to be 96%, and the dry density was controlled to be 1.30 g/cm^3. The specimens were ring knife samples with a diameter of 61.8 mm and a height of 40.0 mm. The test formulation was 20% magnesium oxide, bacterial liquid concentration OD600 = 2.0, water content 33%, and urea content 6 mol·L-1. Based on the disintegration test requirements [18], a homemade disintegration test device was made. As shown in Fig. 3, the device includes a porous plate with 1 cm^2 eyelets of standard size, a glass bucket and so on. The bucket was filled with water and the specimen was immersed in the water. The specimen was placed in a curing cylinder with a porous plate for 24 h.

Two sets of specimens were prepared, one for the blank control sample (plain red clay) and one for the test group (cured soil). The wet and dry cycles of the specimens were processed as shown in Fig. 4, reducing the water content from 33 to 10%, and then increasing the water content of the specimens to 33%, and maintaining them for 24 h so that the water content of each part of the specimens was the same. The above is a dry and wet cycle process, a total of five dry and wet cycles were performed, and the cohesion and internal friction angle of the specimens in each dry and wet cycle were measured, respectively. The disintegration test was conducted after the two groups of soils were processed for 5 wet and dry cycles, respectively.

Fig. 3 Schematic diagram of disintegration test device (Shangbin Wu, 2022)

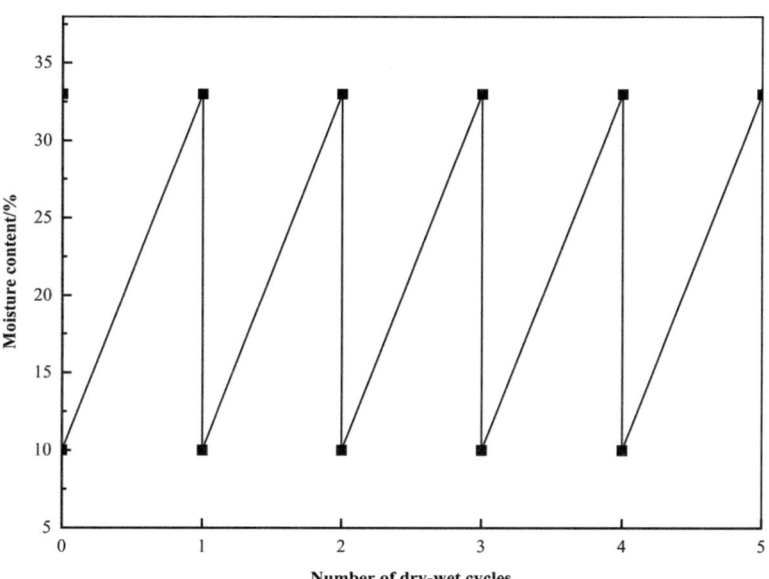

Fig. 4 Control chart of moisture content during wet and dry cycles

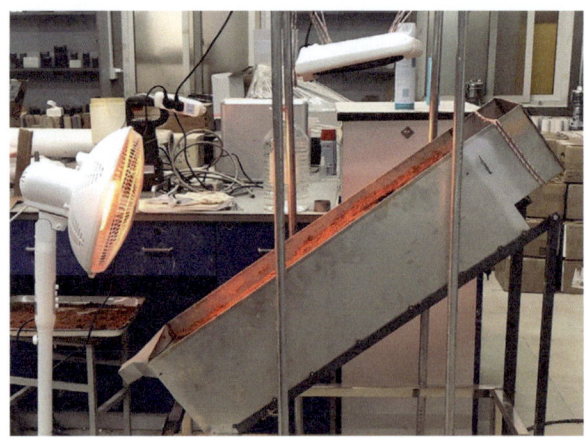

Fig. 5 Photographs of erosion resistance model test (Shangbin Wu, 2022)

Erosion Test Design

Model Case Design

The test needs to be carried out under rainfall conditions as well as dry and wet cycle conditions of the slope erosion test, the use of homemade model device, as shown in Fig. 5. The main body of the model box is a 100 × 30 × 30 cm soil tank, which is supported by a bracket and can be adjusted in angle; a water tank is set up above the soil tank, and the water flow is controlled by overflow to simulate rainfall; the water flow overflows from the water tank to the soil tank, and the slope surface is damaged by erosion, and finally the water flow flows into the collection bucket through the outlet; a heating fan is placed in front of the model to simulate the light system, and to accelerate the wet and dry cycles. The wet and dry cycles are accelerated.

Simulated Rainfall Washout Test Design

In this study, the optimal ratio of microorganism-MgO cured soil in this test was determined through indoor tests, using MgO content of 20%, bacterial liquid concentration OD600 of 2.0, moisture content of 35%, and urea concentration of 6.0 mol/L. Slope erosion tests were set up for two working conditions, namely rainfall erosion and dry and wet cycling, and a simple simulated rainfall and scouring test set was developed independently. A simple simulated rainfall erosion test device was developed independently.

(1) Slope erosion tests under rainfall conditions: The slope of the plain soil trench was 30°, the water content of the soil sample was 33%, and the compaction degree was 0.85. The cured soil was fully mixed and uniformly applied on the red clay slope with the thicknesses of 0 mm (plain soil), 5 mm, 10 mm, and 15 mm, respectively. The flow rate of the test was set at 3L/min, and the water flow eroded for 1 h, and then naturally air-dried. During the test, the characteristics of the erosion pattern of the slope were observed.

(2) Slope erosion tests under wet and dry cycling conditions: Vegetative soil and cured soil trough setting and vegetative soil flow rate are the same as (1). However, the flow rate of cured soil was set to 3L/min, 5L/min and 7L/min, respectively, and the first water flow overflow (simulated rainfall) was 1 h, followed by static maintenance for 24 h, and finally, the heating fan was turned on to warm up and dry the soil for 24 h, which was a dry and wet cycle process. At the end of each drying, measure the length, width, and depth of the slope fissure, a total of 4 dry and wet cycle test.

3 Analysis of Test Results

3.1 Disintegration Rates of Vegetated and Consolidated Soils Under Dry and Wet Cycling Conditions

According to the wet and dry cycle conditions of the disintegration of vegetal soil photographic (Fig. 6) can be observed, the initial surface of the soil samples appeared many bubbles, soil particles slowly collapse; after the water surface quickly become turbid, and floating more bubbles; rapid disintegration process, the disintegration rate of the soil samples gradually slowed down; disintegration stabilization, the moisture layer of the lower part of the beaker, the basket is still a small amount of soil samples remain, the soil samples in the state of the application of a small external force can be quickly disintegrated. The soil sample can be rapidly disintegrated and precipitated by applying a small external force in this state. Therefore, the disintegration rate of the red clay is 100%.

As can be seen in Fig. 7, bubbles could be observed emerging from the surface of the soil samples during the contact of water with the cured soil after the wet and dry cycles, but the soil samples always remained intact and did not disintegrate. The water at the bottom of the beaker remained clear at the end of the test. After removing the soil samples, it can be measured that the soil samples still have high strength.

From the disintegration amount versus time graph (Fig. 8), it can be seen that the dry–wet cycle accelerated the disintegration rate of the vegetal soil, and in the 120th s, the two kinds of vegetal soil disintegrated equally; after 210 s, the disintegration amount stabilized, and the amount of the vegetal soil disintegrated after the dry–wet

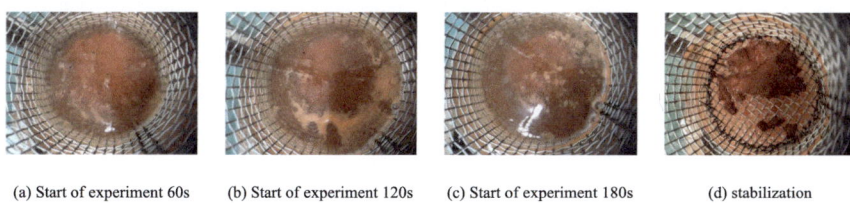

(a) Start of experiment 60s (b) Start of experiment 120s (c) Start of experiment 180s (d) stabilization

Fig. 6 Photographs of vegetal soil disintegration under dry and wet cycling conditions

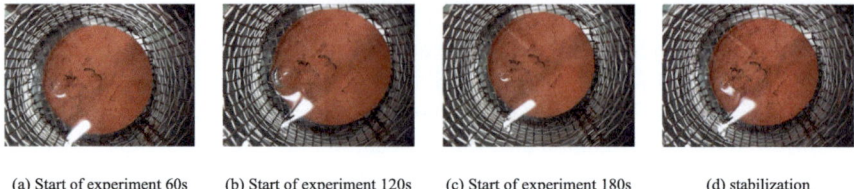

(a) Start of experiment 60s (b) Start of experiment 120s (c) Start of experiment 180s (d) stabilization

Fig. 7 Photographs of solidified soil disintegration under dry and wet cycling conditions

cycle was 32.0 g, and that of the normal vegetal soil disintegrated 33.0 g. However, the dry–wet cycle did not affect the disintegration rate of the curing soil, and in the 120th s, the curing soil after the dry–wet cycle remained in a non-disintegration state, the normal cured soil began to disintegrate, the disintegration amount was 1.0 g and continued to stability has been 1.0 g. This shows that the dry and wet cycle cannot destroy the cementing effect of magnesium carbonate hydrate on the soil particles, the magnesium carbonate hydrate has a very good stability and durability.

A scanning electron microscope was used to scan and test fresh sections of disintegrated soil samples. As can be seen from Fig. 9, in the plain soil sample, the clay minerals were disorganized and many tiny pores could be observed; in contrast, the content of agglomerates in the cured soil was significantly increased, and the microbial-induced generation of hydrated magnesium carbonate cemented the soil particles into agglomerates with a larger particle size, which improved the stability

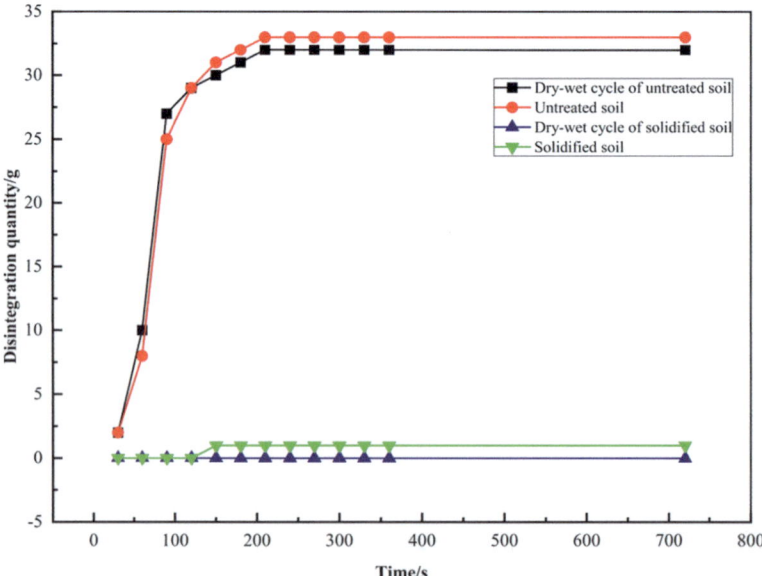

Fig. 8 Volume of disintegration plotted against time

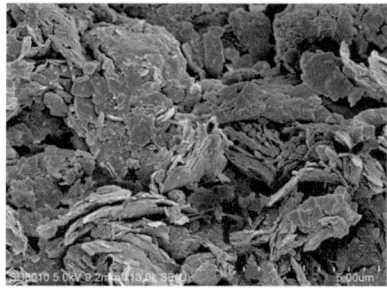
(a) vegetable eed clay sample

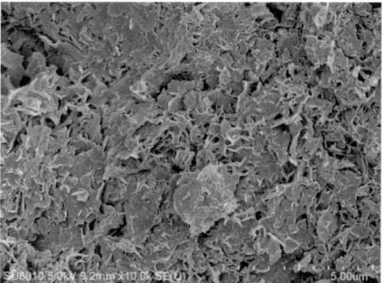
(b) solidified soil

Fig. 9 Scanning electron micrograph of the sample

of the soil body. In addition, the hydrated magnesium carbonate is insoluble in water, so the specimen can still maintain its integrity after curing in the presence of water.

3.2 Characteristics of Slope Erosion Patterns Under Different Curing Thicknesses

In order to study the effect of curing thickness on the erosion morphology characteristics of red clay slopes, the maximum length, maximum width, and maximum depth of slope erosion gullies under different curing thicknesses were measured, and the results of the measurements are given in Table 5.

According to the actual photographs of the slope erosion tests with different curing thicknesses (Fig. 10), it can be observed that the red clay slopes generally show the appearance of surface erosion first, followed by gully erosion. The curing thickness affects the onset time of surface and gully erosion on slopes, and the greater the curing thickness, the later the surface and gully erosion appear.

Table 5 Characteristic parameters of erosion gully morphology under different curing thicknesses

	Length of erosion gully/cm	Width of erosion gully/cm	Depth of erosion gully/cm
Curing thickness 0 mm	115.6	6.5	2.3
Curing thickness 5 mm	0	0	0
Curing thickness 10 mm	0	0	0
Curing thickness 15 mm	0	0	0

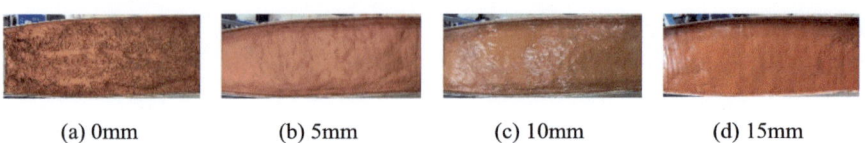

(a) 0mm (b) 5mm (c) 10mm (d) 15mm

Fig. 10 Photographs of slope erosion with different curing thicknesses

When the test soil is unconsolidated, surface erosion occurs first on the slope, and then gradually develops to form a drop can, and a large number of drop cans continuously undercut erosion, and finally connected to form a gully erosion. After the formation of the erosion ditch, the runoff converged to the ditch, due to the significant disintegration of red clay, the erosion ditch deepened, lengthened, steepened, and the ditch wall produced the phenomenon of falling and collapsing. Until the runoff at the outlet is clear, the erosion gully development basically stops. After the completion of the test, it can be observed that the erosion gully at the top of the slope is wide and deep, while the erosion gully in the middle and lower part of the slope is thin and narrow, and the erosion gully is approximately in the shape of a U shape.

After curing, under the scouring effect of the water flow, only a few elongated fissures with a width of less than 0.2 mm could be observed on the slope surface, and the erosion was suppressed. In addition, the elevated flow rate during the test did not result in the appearance of erosion gullies on the slope, and no sedimentation was observed at the catchment bucket.

3.3 Slope Erosion Rates at Different Curing Thicknesses

Figure 11 shows the variation curve of slope erosion rate with curing thickness, and it can be seen that the erosion rate of cured slopes is significantly lower than that of plain soils, which is only about 2.8% of the erosion rate of plain red clay slopes. The hydrated magnesium carbonate generated by microbial induction enhances the soil cohesion and can better resist erosion and disintegration by water flow.

3.4 Characteristics of Slope Erosion Patterns Under Different Numbers of Wet and Dry Cycles

The actual photographs of the vegetated soil and cured soil slopes under different number of wet and dry cycles are shown in Fig. 12.

Figure 12 shows that after the first dry–wet cycle, the fissures at the top of the slope are generally wider than those in the middle and bottom of the slope, and the erosion gully is approximately U-shaped; after the second dry–wet cycle, the number of erosion gully grows, and its width and depth continue to increase; after the third

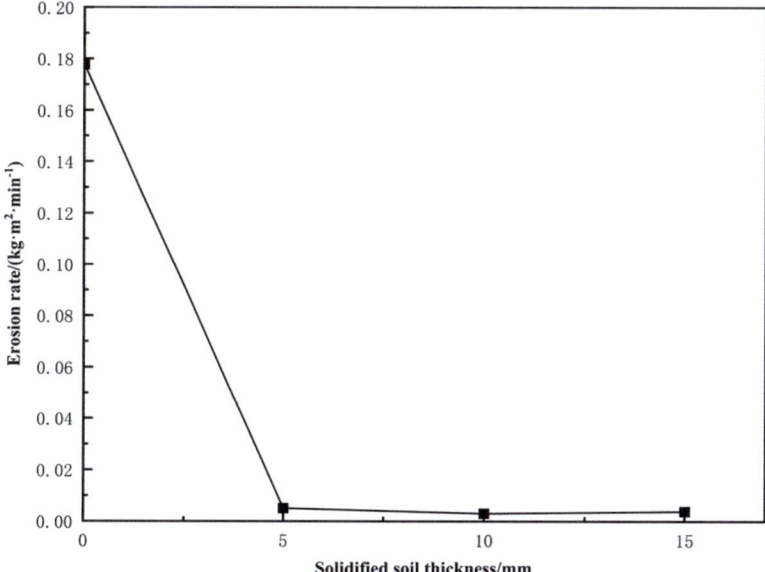

Fig. 11 Curve of slope erosion rate with curing thickness

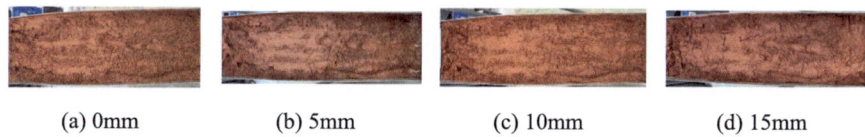

(a) 0mm (b) 5mm (c) 10mm (d) 15mm

Fig. 12 Curing thickness 0 mm slope erosion actual photo

dry–wet cycle, the number of erosion gully no longer increases, and the depth and width of the erosion gully continue to increase; after the fourth dry–wet cycle, the soil mass at the top of the slope completely disintegrates, and the maximum depth and the maximum depth and width of the erosion gully increased.

Curing thickness of 5 mm slope erosion of the actual photographs shown in Fig. 13, after different wet and dry cycle treatment of the slope surface did not appear erosion gully. After the first dry–wet cycle, several slender cracks were observed on the slope, but the dry–wet cycle treatment did not widen or extend the cracks, and the water flowed down the slope into the water collection bucket.

The actual photographs of erosion on the slopes with curing thicknesses of 10 mm and 15 mm are shown in Figs. 14 and 15, and none of the cured soil slopes showed erosion gullies.

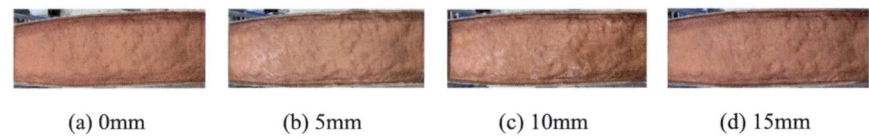

(a) 0mm   (b) 5mm   (c) 10mm   (d) 15mm

Fig. 13 Curing thickness 5 mm slope erosion actual photo

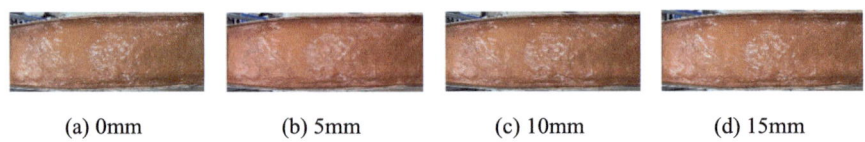

(a) 0mm   (b) 5mm   (c) 10mm   (d) 15mm

Fig. 14 Curing thickness 10 mm slope erosion actual photo

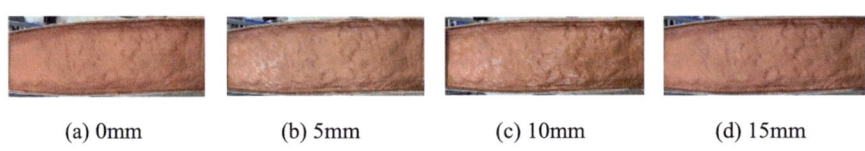

(a) 0mm   (b) 5mm   (c) 10mm   (d) 15mm

Fig. 15 Curing thickness 15 mm slope erosion actual photo

3.5 Slope Erosion Rates at Different Numbers of Wet and Dry Cycles

The curve of slope erosion rate in relation to the number of wet and dry cycles is shown in Fig. 16.

Vegetative soil (curing thickness of 0 mm) slopes in the first erosion process, the water flow to take away a large number of soil particles to form erosion gullies, so the initial dry and wet cycle corresponds to the highest erosion rate, erosion rate of 0.178 kg/(m^2·min); the second erosion process, the water flow flushing deepened the erosion gullies, takes away soil particles, but the erosion rate is relatively reduced, erosion rate of 0.095 kg/(m^2·min); during the third erosion process, the width of the erosion gully was still slightly expanded, but the erosion rate decreased greatly compared with the corresponding erosion rate of the first two wet and dry cycles, which was 0.036 kg/(m^2·min); during the fourth erosion process, accompanied by the disintegration of the soil blocks at the top of the slope, the erosion rate increased slightly, which was 0.075 kg/(m^2·min).

The erosion rate corresponding to the first wet and dry cycles was 0.005 kg/(m^2·min) for a cured thickness of 5 mm, and about 0.003 kg/(m^2·min) for cured thicknesses of 10 mm and 15 mm. The degree of erosion of the cured soil before and after the wet and dry cycles was very small and did not affect the erosion rate of the slope.

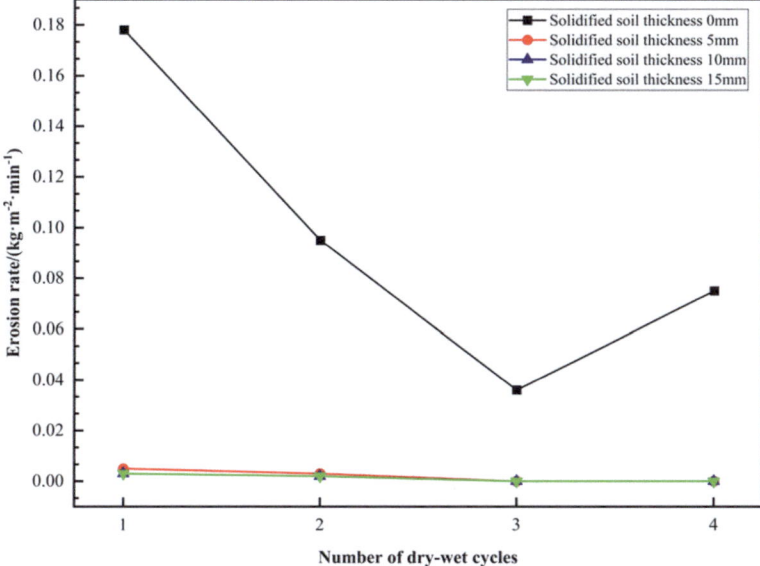

Fig. 16 Curve of slope erosion rate in relation to the number of wet and dry cycles

Therefore, when microbial-induced magnesium carbonate precipitation technology is used to treat red clay slopes, a curing thickness of 5 mm can enhance the erosion resistance of the slopes without the need to use a higher curing thickness, which can save economic costs.

4 Conclusions

The experiment explored the water resistance test of microbial-magnesium oxide cured red clay, set up disintegration test as well as the erosion test of vegetal soil and cured soil slopes under single rainfall condition and dry–wet cycle condition, compared the effects of different working conditions, curing thickness, flow rate and other factors on the erosion morphology characteristics and erosion rate of red clay slopes, and came up with the following conclusions:

(1) The disintegration resistance of cured soils treated with microbial-induced precipitation of magnesium carbonate was almost unaffected by wet and dry cycles. Macroscopic and microscopic levels showed that dry and wet cycles could not destroy the cementing effect of hydrated magnesium carbonate on soil particles, and hydrated magnesium carbonate had good stability and durability.

(2) Using microorganisms cured soil, under the scouring effect of water flow, only slender fissures with a width of less than 0.2 mm can be observed on the slope, and the width of the fissures did not change significantly during the scouring

process, and the erosion rate of microorganisms–magnesium oxide cured slopes was significantly reduced to only about 2.8% of the erosion rate of the vegetal red clay slopes.
(3) Under wet and dry cycling conditions, no erosion gullies appeared on the cured slopes, and only a few elongated fissures were observed, which did not extend with the increase in the number of wet and dry cycles. The erosion rate of the vegetal red clay slopes decreased and then increased with the number of wet and dry cycles, while the erosion rate of the cured soil was very small.
(4) The hydrated magnesium carbonate induced by microorganisms enhances the cohesive force of the soil, so that the soil particles can still be tightly cemented together under the scouring effect of the water flow, restricting the spatial transport of soil particles, and the soil in the surface layer of the slope forms an integral whole, which can resist the erosion and scouring and disintegration and softening effect of the water flow, and this provides a guide and a reference for the bio-control technology for the projects of the slopes of the open-pit mines, the slopes of the road beds, and the slopes of the power grid poles and towers, and so on.

The suitable curing thickness given in this paper is 5 mm, which is too large and there is still room for further refinement, and the minimum curing thickness that can ensure the erosion resistance of slopes can be explored through model tests. This paper is the preliminary exploratory results of microbial-magnesium oxide curing technology in enhancing the erosion resistance of red clay slopes, which verifies that the technology can indeed enhance the physical and mechanical properties of red clay, but there is still a considerable distance from the field application of this technology, and the next step can be considered to adjust the three major indicators of magnesium oxide content, bacterial liquid concentration, and the initial water content, and to explore the proportion of formulations suitable for the growth of grass species.

References

1. Bao X, Liao W, Dong Z et al (2017) Development of vegetation-pervious concrete in grid beam system for soil slope protection. J Mater 10(2):96. https://doi.org/10.3390/ma10020096
2. Busari AA, Akinwumi II, Awoyera PO et al (2018) Stabilization effect of aluminum dross on tropical lateritic soil. J Int J Eng Res Afr 39:86–96. https://doi.org/10.4028/www.scientific.net/JERA.39.86
3. Zhao HY, Lei XW, Chen YJ (2020) Experimental study of red clay modified by activated MgO. J Highway 65(11):6. https://kns.cnki.net/KCMS/detail/detail.aspx?dbname=cjfd2020&filename=glgl202011060&dbcode=cjfq
4. Ouyang M, Wang GY, Deng RR et al (2023) A kind of ecological protection structure and method for expanding soil slope. P China Patent: CN116335166A. http://www.zhizhen.com/detail_38502727e7500f26886b7e48576501360c9c6640d4dc968a1921b0a3ea2551016bd6091b92ceac00c0c5f0f4712cb9e1f5b0902d47f4eabf1c305377cf413c3003546f79cf6b85a71922fae01738ca2e

5. Jiao LX, Zeng WK, Wang QY (2019) Microbial induced calcium carbonate precipitation technology for reinforcing red bed clay intercalation. J Ind Constr 49(05):93–97. https://doi.org/10.13204/j.gyjz201905017
6. Jie YX, Li GX, Chen L (1998) Study of centrifugal model tests on texsol and cohesive soil slopes. Chin J Geotech Eng (04):15–18. https://kns.cnki.net/KCMS/detail/detail.aspx?dbname=cjfd1998&filename=ytgc804.002&dbcode=cjfq
7. Han ZG (2024) Multi scale experimental research on anti-liquefaction performance of liquefiable sands strengthened by microorganisms. D. Tsinghua University. CNKI: CDMD: 1.1018.875866
8. Xie YH, Tang CS, Yin LY et al (2019) Mechanical behavior of microbial-induced calcite precipitation (MICP)-treated soil with fiber reinforcement. J Chin J Geotech Eng 4:8. https://doi.org/10.11779/CJGE201904010
9. Hu QZ, Huo WY, Ma Q et al (2023) Mechanical properties and water stability of MICP combined fiber reinforced loesses. J Yangtze River 54(8):227–232. https://doi.org/10.16232/j.cnki.1001-4179.2023.08.032
10. Xie YH, Tang CS, Yin LY et al (2019) Mechanical behavior of microbial-induced calcite precipitation (MICP)-treated soil with fiber reinforcement. J Chin J Geotech Eng 41(04):675–682. https://doi.org/10.11779/CJGE201904010
11. Xiao Y, He X, Evans TM et al (2019) Unconfined compressive and splitting tensile strength of basalt fiber-reinforced biocemented sand. J Geotech Geoenvironmental Eng 145(9):1–11. https://doi.org/10.1061/(ASCE)GT.1943-5606.0002108
12. Martinez A, Huang L, Gomez MG (2018) Thermal conductivity of MICP-treated sands at varying degrees of saturation. J Géotechnique Lett 8(4):1–23. https://doi.org/10.1680/jgele.18.00126
13. Whiffin VS (2004) Microbial $CaCO_3$ precipitation for the production of biocement. D. Perth: Murdoch University
14. Huang T, Fang XW, Zhang W et al (2020) Experimental study on solidified loess by microbes and reactive magnesium oxide. J Rock Soil Mech 41(10):3300–3306+3316. https://doi.org/10.16285/.rsm.2020.0151
15. Cai GH, Du YJ, Liu SY et al (2015) Physical properties, electrical resistivity, and strength characteristics of carbonated silty soil admixed with reactive magnesia. J Can Geotech J 52:150408143425002. https://doi.org/10.1139/cgj-2015-0053
16. He J, Qu SY, Hang L et al (2024) Experimental study on enzyme enhanced magnesia carbonation process for soil stabilization. J Civ Environ Eng, pp 1–8 [2024–06–14]. http://kns.cnki.net/kcms/detail/50.1218.TU.20231127.1749.008.html
17. Liu SY, Cao JJ, Cai GH (2018) Microstructural mechanism of reactive magnesia carbonated and stabilized silty clays. J Rock Soil Mech 39(05):1543–1552+1563. https://doi.org/10.16285/.rsm.2016.1308
18. China academy of building research (2012) Technical code for building in expansive soil regions. http://www.zhizhen.com/detail_38502727e7500f26686118d0dd70b1bee5793b98c3f0c2131921b0a3ea255101ec06fb7872c54b461459ac134165093210ac0de254643fdc4ff190cf1899537f0604b08bc5f3a07c7ed0e38f6ca0a798

Open Access This chapter is licensed under the terms of the Creative Commons Attribution 4.0 International License (http://creativecommons.org/licenses/by/4.0/), which permits use, sharing, adaptation, distribution and reproduction in any medium or format, as long as you give appropriate credit to the original author(s) and the source, provide a link to the Creative Commons license and indicate if changes were made.

The images or other third party material in this chapter are included in the chapter's Creative Commons license, unless indicated otherwise in a credit line to the material. If material is not included in the chapter's Creative Commons license and your intended use is not permitted by statutory regulation or exceeds the permitted use, you will need to obtain permission directly from the copyright holder.

Research on Anti-deformation Technology of High-rise Buildings in Coal Mining Subsidence Areas

Keyi Guo, Xing Li, and Yue Li

Abstract Due to continuous coal mining, extensive coal mining subsidence areas have formed. Constructing buildings, especially high-rise buildings, on these areas poses significant safety risks. Implementing necessary anti-deformation technical measures for high-rise buildings situated above coal mining subsidence areas can effectively address this issue. Using COMSOL finite element analysis software, a 3D framework model of a 13-story building was established. Targeted structural measures for the foundation and the upper structure were set, and the stress and strain conditions of the 13-story building were calculated. By analyzing the impact of surface deformation on the building, the permissible values of tilt deformation for anti-deformation high-rise buildings were discussed. The study examined a series of anti-deformation technical measures for constructing high-rise buildings above coal mining subsidence areas, including site selection, foundation selection, and building-structure types. The results show that high-rise buildings, due to their good integrity and large stiffness, are mainly affected by tilt deformation. When constructing above coal mining subsidence areas, foundation types with good integrity should be chosen. Strengthening the structure of high-rise buildings in subsidence areas should focus on the foundation and the connection between the foundation and the upper structure.

Keywords High-rise buildings · Anti-deformation · Coal mining subsidence area · Numerical simulation

1 Introduction

With population growth and the continuous expansion of urban areas, the demand for construction land is becoming increasingly tight. To meet the residential and living requirements of urban residents, more and more buildings are being constructed above coal mining subsidence areas. Coal mining can cause uneven surface subsidence, which poses certain safety risks for buildings above these areas [1–7].

K. Guo (✉) · X. Li · Y. Li
CCTEG Ecological Environment Technology Co., Ltd., Tianjin, China
e-mail: tsjxgky@163.com

To address the problems of constructing buildings above mined-out areas, many domestic scholars have conducted extensive research. Guangyun et al. [8, 9] summarized the surface deformation under dynamic and static load conditions in mined-out areas and proposed existing problems and future research directions. Qingfeng [10], through numerical simulation analysis of rock and soil in coal seam mined-out areas and comparative analysis with traditional calculation methods, combined with field measurement results, proved the feasibility of numerical simulation analysis of rock and soil. Shaojie et al. [11] studied the surface movement and deformation patterns of shallow longwall old mined-out areas under building loads and the mechanism of surface movement and deformation under building loads using field surveys, physical simulations, and numerical simulations. Zhanxin et al. [12] analyzed the activation factors of secondary rock mass in mined-out areas, identifying the main activation factors under specific engineering conditions as water filling and building loads. They obtained the surface movement and deformation patterns under different working conditions through numerical simulation schemes. Dehua et al. [13] studied the stress distribution and settlement patterns of building foundations in mined-out areas under additional loads using FLAC3D numerical software. They concluded that the greater the depth and thickness ratio of coal seams, the smaller the impact of mined-out areas on the surface. When the influence depth of additional building loads reaches the height of the collapse fracture zone in mined-out areas, reinforcement of the mined-out area is necessary. Wei et al. [14, 15] explored the structural types of buildings in mined-out areas. Yonghai et al. [16] proposed a complete solution for construction issues in mined-out areas.

High-rise buildings are usually determined by height and the number of floors, with a height above 28 m or 10 floors or more being classified as high-rise buildings [17]. This paper uses COMSOL finite element analysis software to establish a 3D frame structure model. Based on observed data summarizing the damage and failure forms of buildings in mined-out areas, the interaction mechanism between building and foundation deformation is analyzed [18]. The anti-deformation structural model of high-rise buildings is validated, the effects of various anti-deformation technical measures are studied, the permissible tilt deformation values for high-rise buildings above mined-out areas are discussed, and specific anti-deformation measures for constructing high-rise buildings above coal mining subsidence areas are proposed. This research supplements and improves the anti-deformation theory of constructing high-rise buildings above coal mining subsidence areas and has significant implications for urban construction in mining areas in China.1 Study on the Theoretical Study of Anti-deformation of High-rise Buildings.

2 Research on Anti-deformation Theory for High-rise Buildings

Using COMSOL finite element analysis software, a 3D frame structure model of a 13-story building was established. The dimensions of the upper structure are 60 m by 15 m, with a standard floor height of 3 m, including one basement level. To account for mining-induced surface deformation, a base layer tilt deformation of 6 mm/m and a base layer horizontal deformation of 4 mm/m were set.

2.1 Simulation Research on Foundation Type of High-rise Buildings

Independent Basis

The model adopts an independent foundation, and the foundation pile depth is 1 m. The results obtained by this model are shown in Figs. 1, 2, 3 and 4.

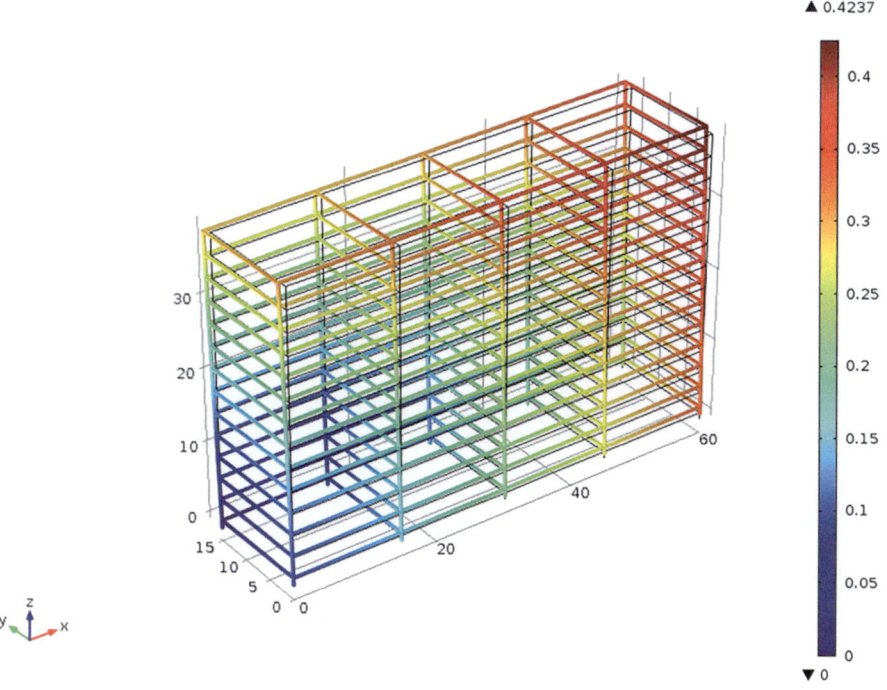

Fig. 1 Independent basis displacement diagram

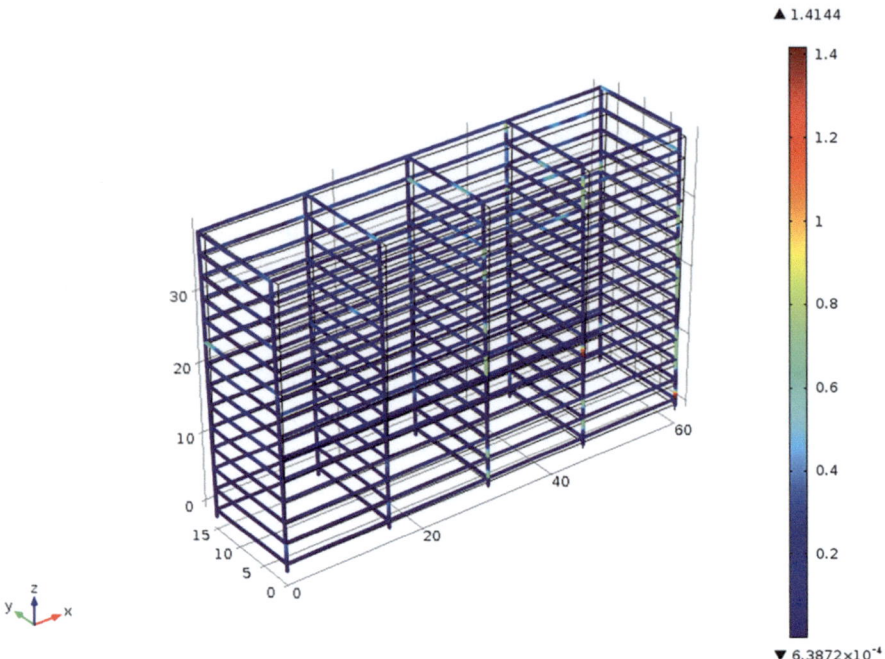

Fig. 2 Independent basis tress diagram

From the results in the figure, the maximum displacement occurs above the maximum deformation position at the bottom, and the maximum Mises Stress, strain and strain energy occur at the two spans on the right.

Overall foundation (Large Slab Foundation)

The model adopts a large slab foundation, and the foundation pile depth is 1 m, cross-connect the piles to form a large slab foundation. The result is shown in the Figs. 5, 6, 7 and 8.

From the results in the figure, it can be seen that the maximum displacement occurs above the maximum deformation position at the bottom, and the maximum Mises Stress, strain. The strain energy and strain energy occur at the two spans on the right. By comparing the independent foundation analysis, it can be seen that the large slab foundation can effectively reduce the lowest maximum deformation.

Addition of Elastic Supports to the Foundation

The model adopts elastic support foundation, and the foundation pile depth is 1 m, which connect the piles and set spring supports at the bottom pile foundation. It can prevent stress concentration and promote overall deformation coordination. The results obtained by this model are shown in Figs. 9, 10, 11 and 12.

From the results in the figure, it can be seen that the maximum displacement occurs above the maximum deformation position at the bottom, and the maximum

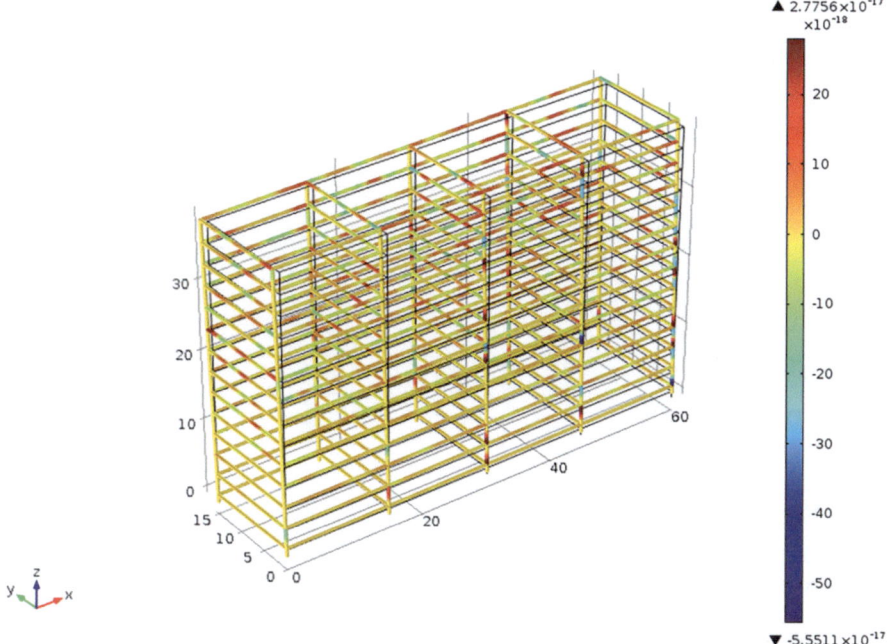

Fig. 3 Independent basis strain diagram

Mises stress, strain. The strain energy and strain energy occur at the two spans on the right. By comparing and analyzing, it can be seen that the elastic support foundation can effectively improve the low average levels and distributions of deformations and stresses, etc.

2.2 Simulation of High-rise Building Superstructure

Installation of Ground Floor Ring Beam

In the model with independent foundations, ground floor inter-column cross braces are installed to increase the stiffness of the ground floor and prevent excessive deformation. This consideration is crucial for accounting for mining-induced surface deformations. The results obtained by this model are shown in Figs. 13, 14, 15 and 16.

From the results in the figure, the maximum displacement occurs at the upper part of the area experiencing the greatest deformation at the base. The maximum Mises stress, strain, and strain energy are all observed in the position of the two spans at the right end. Comparative analysis shows that the installation of a ground floor ring

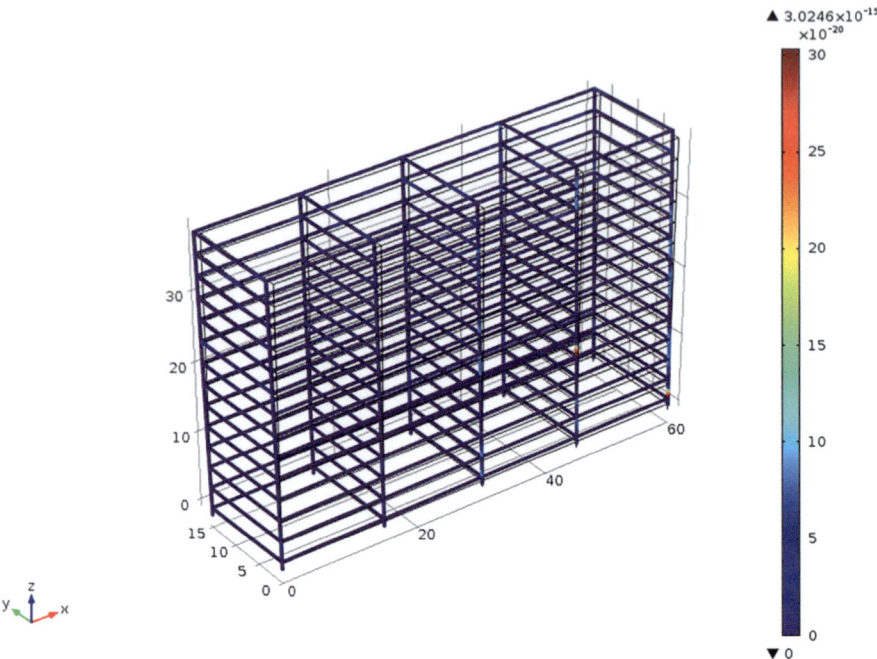

Fig. 4 Independent basis strain energy diagram

beam can effectively reduce and improve the average levels and distribution forms of deformation and stress.

Installation of Side Supports

In the areas of the model experiencing the greatest deformation, side supports are installed from the ground floor to the top floor. These inter-column supports increase the local stiffness of the structure and reduce the deformation caused by surface mining activities. The results obtained by this model are shown in Figs. 17, 18, 19 and 20.

From the results in the figure, the maximum displacement occurs at the upper part of the area experiencing the greatest deformation at the base. The maximum Mises stress, strain, and strain energy are all observed in the position of the two spans at the right end. Comparative analysis shows that side supports can also effectively reduce the maximum deformation.

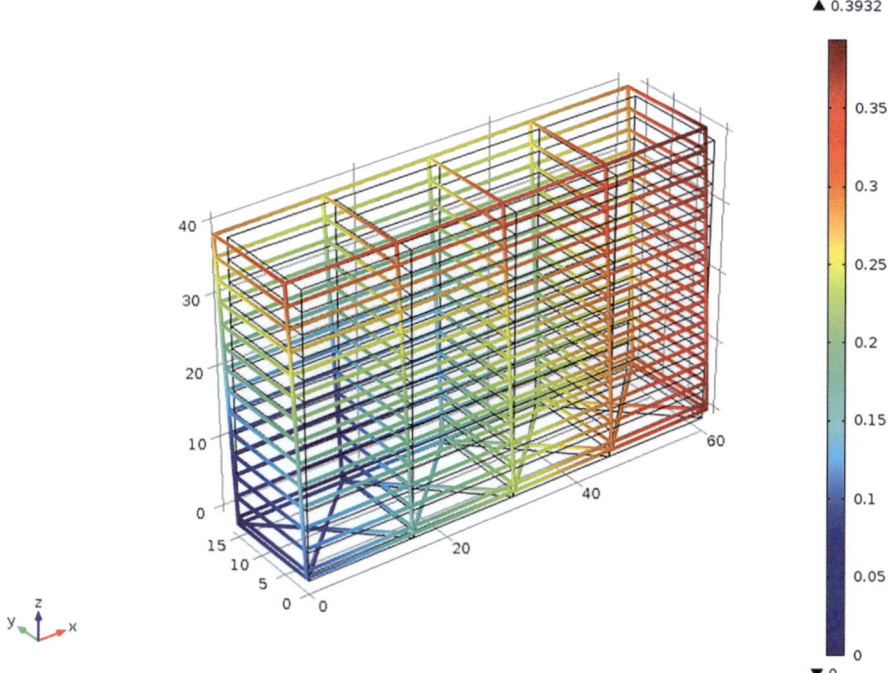

Fig. 5 Large slab foundation displacement diagram

3 Discussion on Deformation Requirements for Anti-deformation Buildings

3.1 Impact of Surface Deformation on Buildings

Years of research and practice have established that surface movement and deformation induced by underground mining activities affect buildings within the influenced area. This impact is generally transmitted from the surface to the building's upper structure through its foundation. Different magnitudes and types of surface deformation result in varying impacts on buildings.

Uneven surface subsidence can cause buildings to experience various forms of deformation, such as tension, compression, tilting, and horizontal movement. Among these, horizontal surface deformation (tensile and compressive deformation) is a significant factor contributing to building damage. This is particularly true for brickwood structures, which have minimal resistance to tensile deformation. Such buildings tend to develop cracks in their weaker sections under tensile deformation, sometimes even before noticeable cracks appear on the surface, and severe damage can lead to the collapse of the structure. Compressive deformation, on the other hand, can crush walls, buckle floors, and cause shear or compression cracks, leading to door and

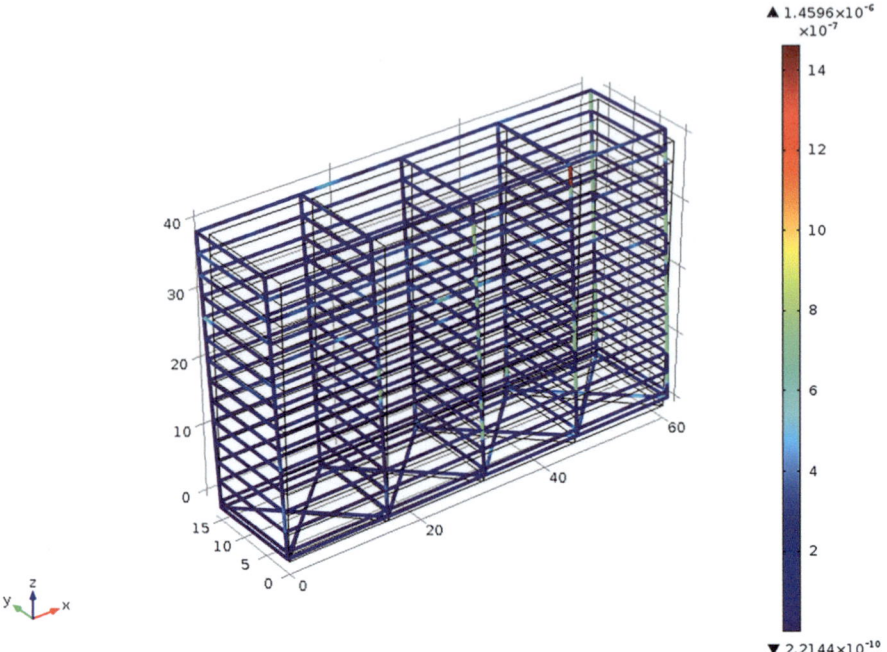

Fig. 6 Large slab foundation tress diagram

window misalignments and operational difficulties. Furthermore, when coal mining causes surface subsidence, even buildings located at the center of the subsidence area can experience substantial overall settlement. If the subsidence value is significant, especially in areas with high groundwater levels, the resultant basin formation can lead to water accumulation, submerging the building and rendering it unusable despite the absence of structural damage.

For high-rise buildings, which typically adopt structural forms such as frame structures, frame-shear wall structures, shear wall structures, tubular structures, and frame-core tube structures, the impact of surface curvature deformation and horizontal deformation is relatively minor due to their high stiffness. However, given the considerable height of these buildings, surface tilt deformation poses a significant threat to their stability.

3.2 Deformation Requirements for Buildings

While surface curvature deformation and horizontal deformation can be mitigated through anti-deformation methods and technical measures, surface tilt deformation is predominantly transmitted to the building. Excessive tilt deformation not only

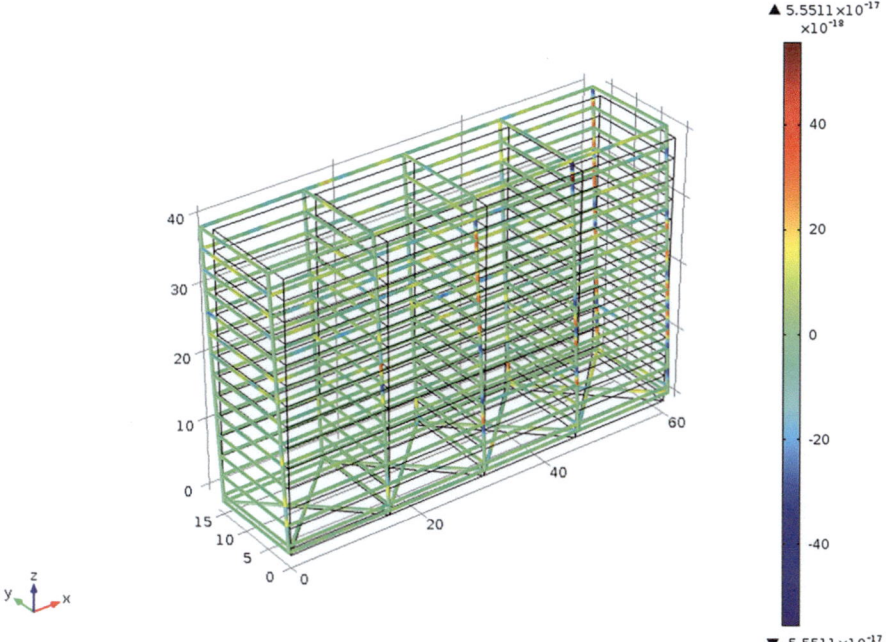

Fig. 7 Large slab foundation strain diagram

affects the aesthetic appearance of the building but also threatens the stability of high-rise structures.

Table 1 presents the correlation between surface tilt deformation values and building deformation values for an anti-deformation experimental house, brick-concrete structures, and typical rural dwellings in a coal mining area. It is evident that, regardless of the building type—be it a standard rural house, a brick-concrete structure, or an anti-deformation house—the tilt deformation of the building closely mirrors that of the surface. Anti-deformation technical measures cannot significantly reduce the tilt deformation experienced by buildings.

The "Code for Design of Building Foundations" and "Code for Design of Concrete Structures". Specific regulations have been issued [19, 20], including the overall tilt allowance of high-rise buildings and the tilt allowance of the foundation of tall structures. The same applies to the technical requirements for building buildings in coal mining subsidence areas. Of course, most new buildings in coal mining subsidence areas are the anti-deformation structural technology measures are adopted, and its ability to resist overall tilt is strengthened. Based on practical experience and the requirements in the code, the allowable values of ground inclination deformation and the anti-deformation have been detemined. The allowable value of ground tilt deformation for high-rise structures can be calculated according to Table 2. Regulations shall be implemented.

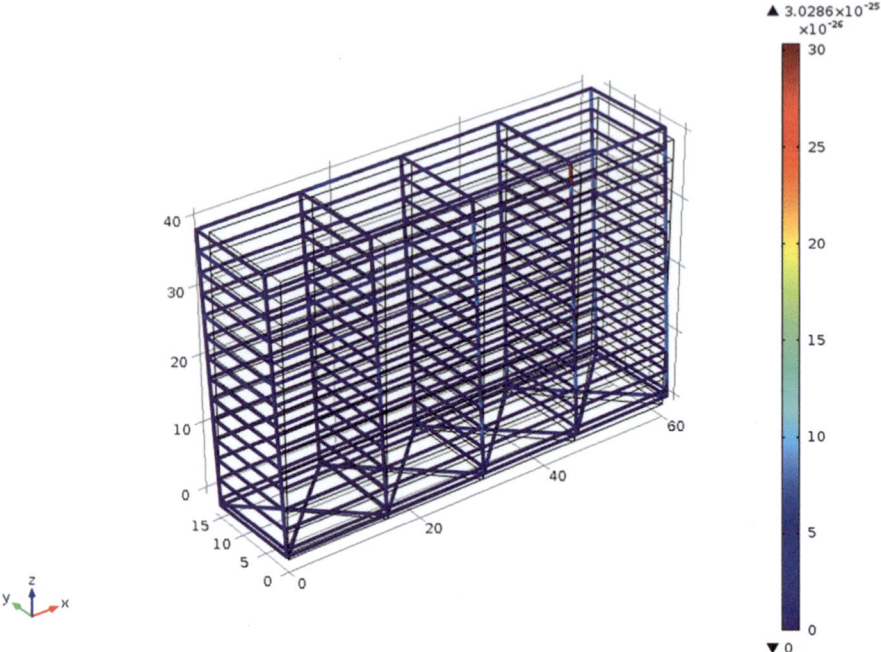

Fig. 8 Large slab foundation strain energy diagram

4 Technical Measures for Anti-deformation of High-rise Buildings

Under the influence of various surface deformations caused by mining, surface inclination deformation is the main factor affecting the safety of high-rise buildings. When the ground tilts and deforms, the building tilts accordingly, causing its center of gravity to shift, thus changing the original structures. The stress state and redistribute the foundation reaction force can cause insufficient cross-sectional strength, unstable foundation, and insufficient bearing capacity of high-rise building components, leading to damage or collapse [21–23].

4.1 Site Selection for High-rise Buildings

High-rise buildings must first comply with general building site selection principles. Due to their height, high-rise buildings are susceptible to significant lateral displacement caused by surface tilt deformation. Excessive lateral movement can induce additional internal forces in the structure, leading to cracks or damage in infill walls and main structures, thus affecting normal use and potentially causing structural

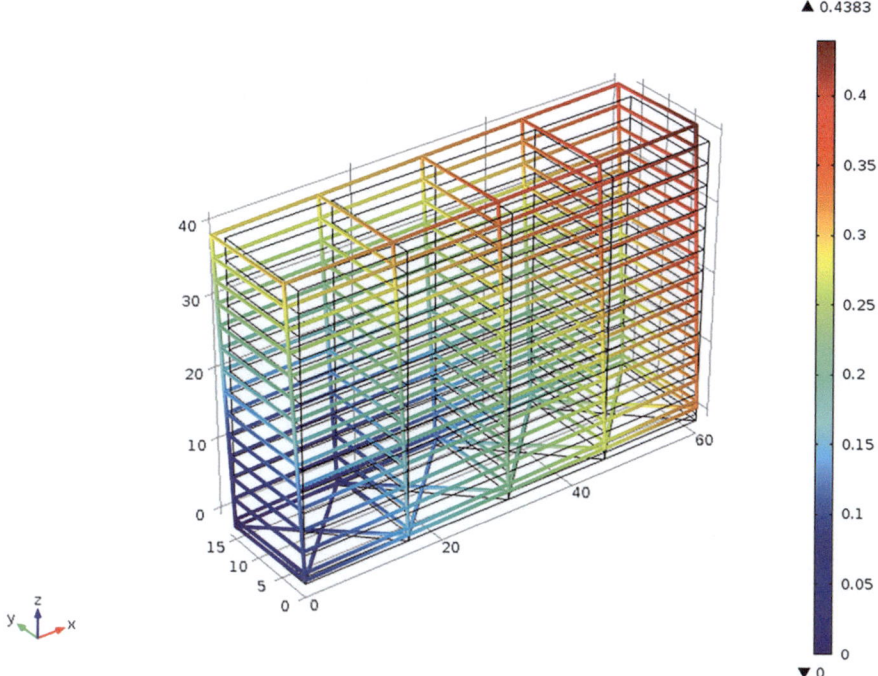

Fig. 9 Elastic support displacement diagram

failure. Generally, the taller the building, the smaller the permissible tilt deformation value. When selecting sites for high-rise buildings, areas with greater mining depth and minimal surface tilt deformation should be preferred.

4.2 Anti-deformation Measures for High-rise Buildings

The Structural Layout of High-rise Buildings

It should be conducive to resisting both horizontal and vertical loads, ensuring clear force transmission with direct load paths, and aiming for uniform symmetry to minimize torsional effects.

Within each independent structural unit of a high-rise building, the structural plane shape should be simple and regular, with uniform distribution of stiffness and load-bearing capacity, minimizing eccentricities. Severe irregularities in plane layout should be avoided, and the length of the plane should not be excessively large.

High-rise buildings are advised to adopt plane shapes that minimize the effects of wind action. Shapes such as circular, regular polygons, elliptical, or barrel shapes are advantageous in resisting wind.

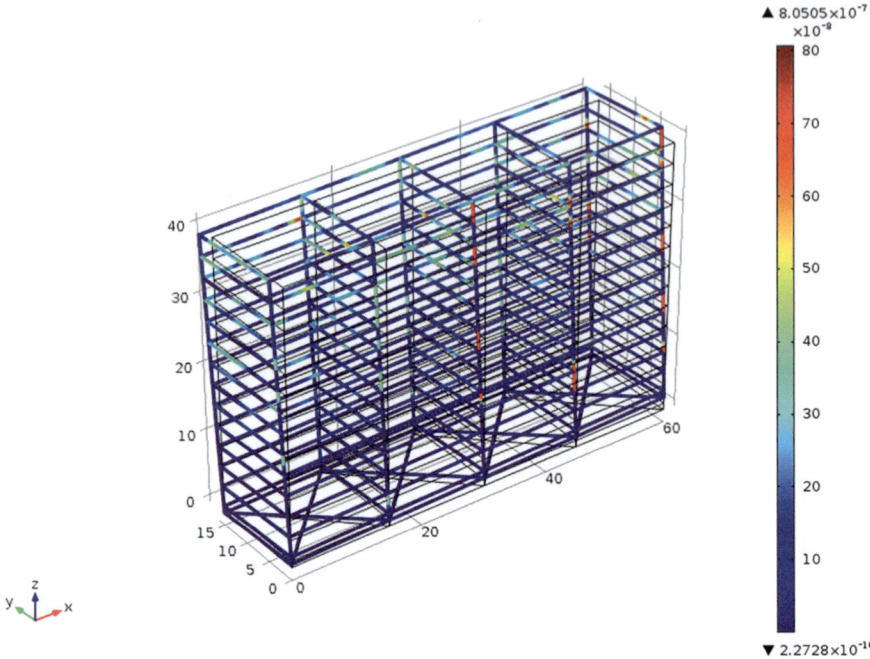

Fig. 10 Elastic support tress diagram

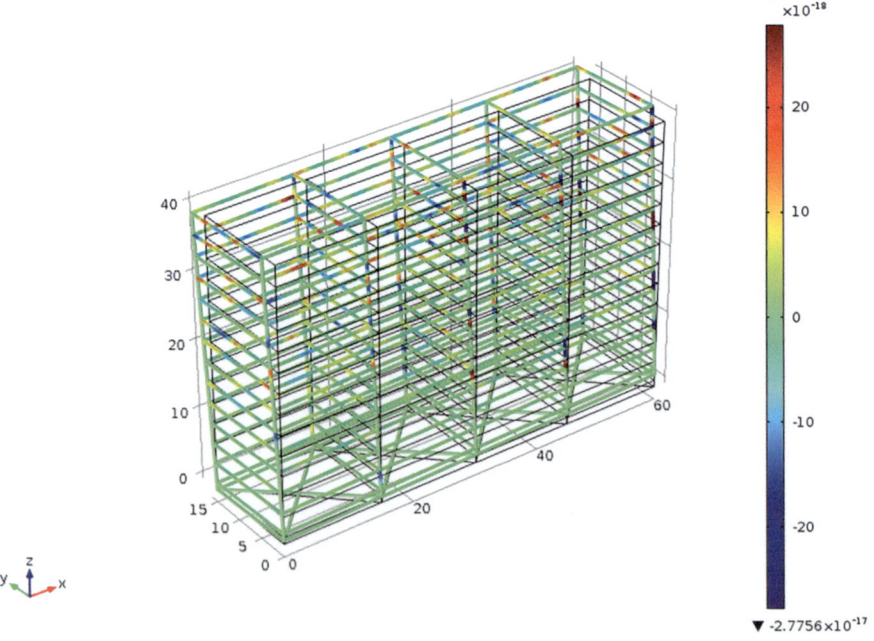

Fig. 11 Elastic support strain diagram

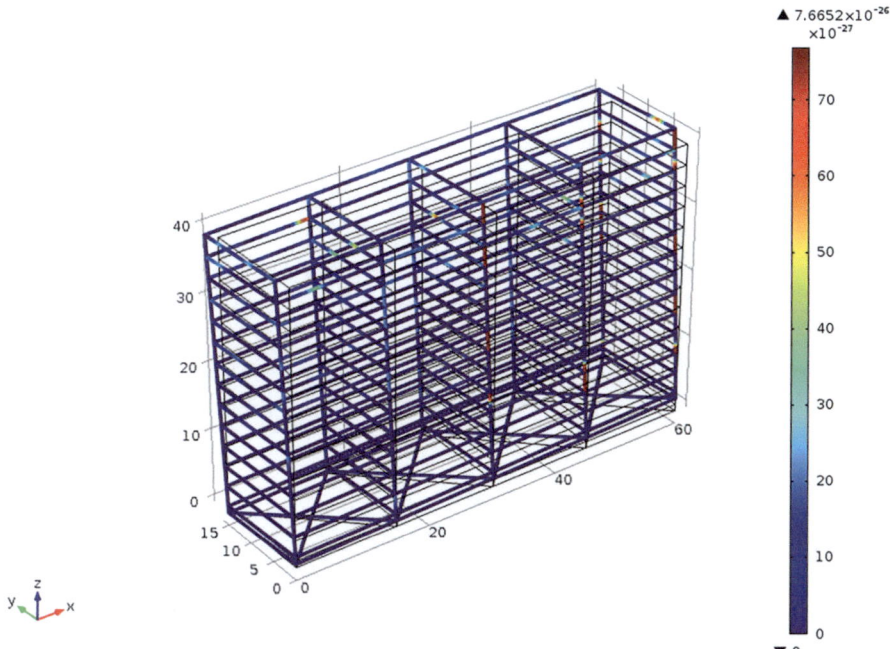

Fig. 12 Elastic support strain energy diagram

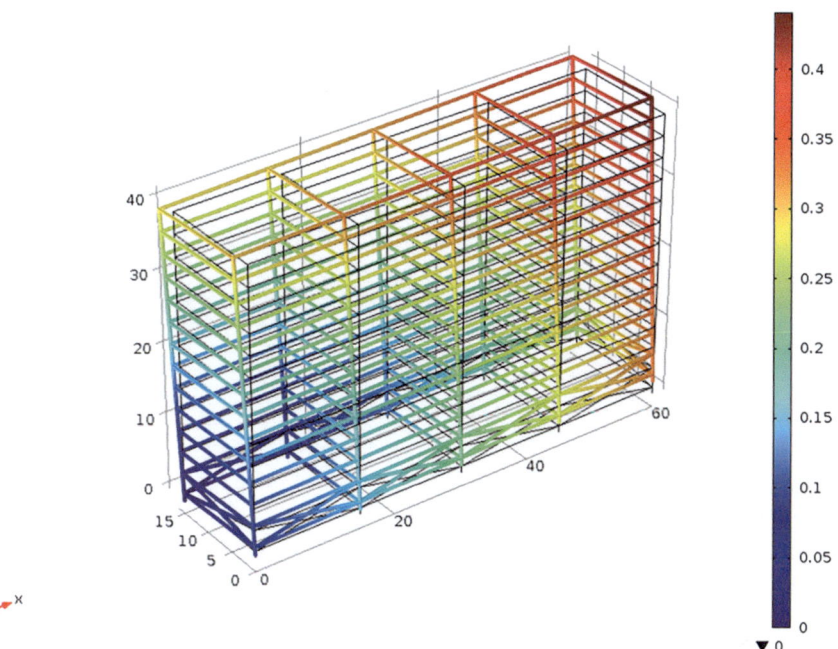

Fig. 13 Ground floor ring beam displacement diagram

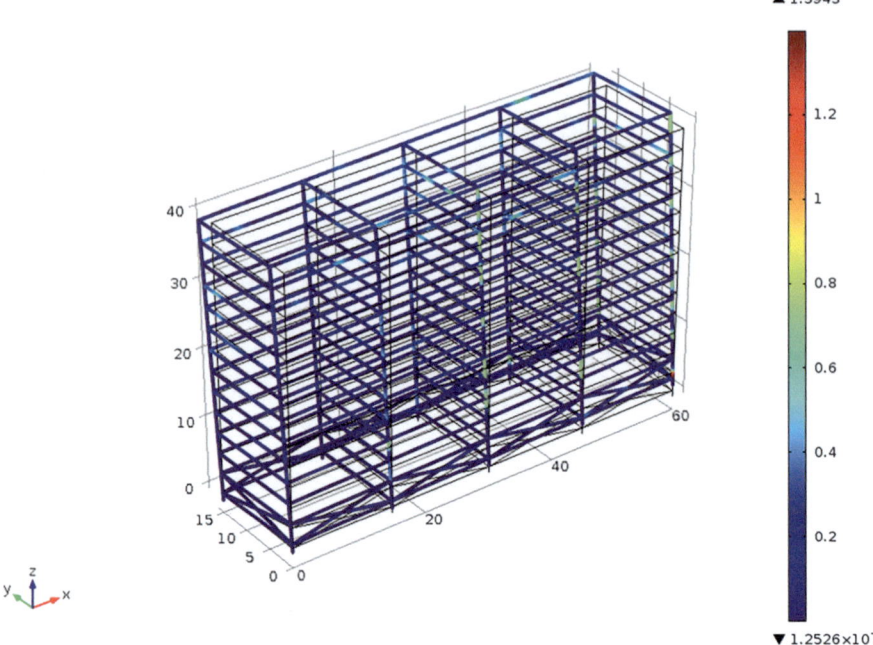

Fig. 14 Ground floor ring beam tress diagram

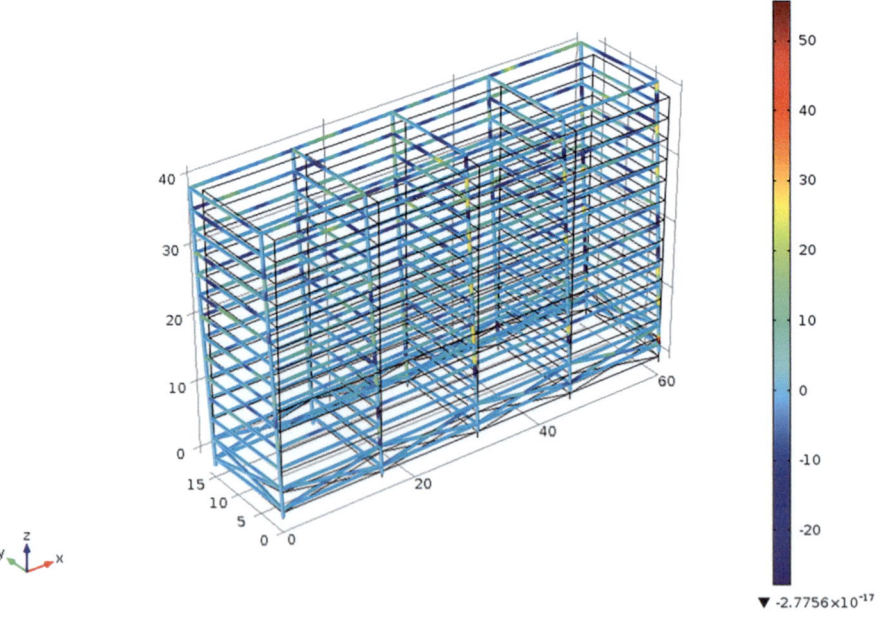

Fig. 15 Ground floor ring beam strain diagram

Research on Anti-deformation Technology of High-rise Buildings … 363

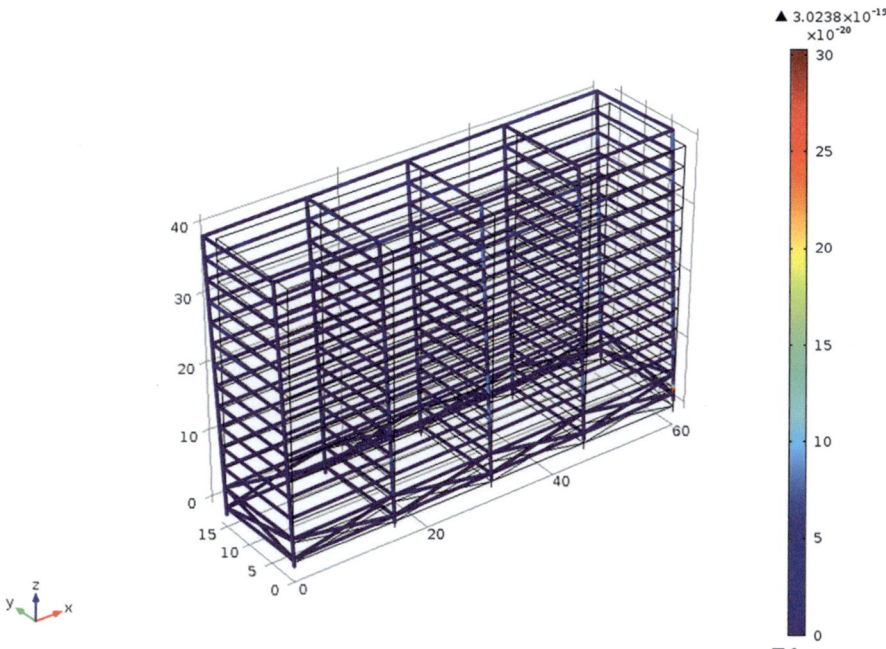

Fig. 16 Ground floor ring beam strain energy diagram

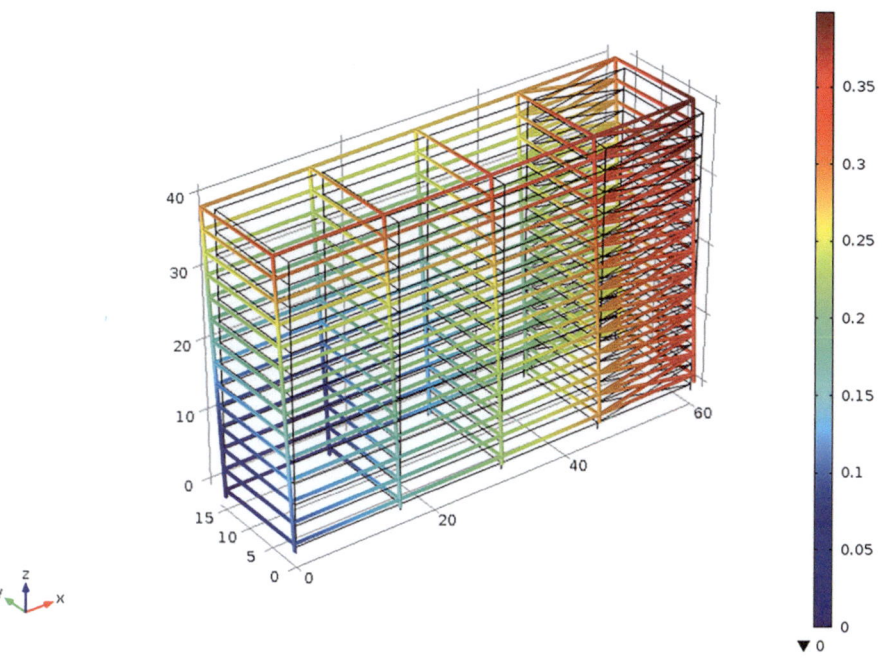

Fig. 17 Side supports displacement diagram

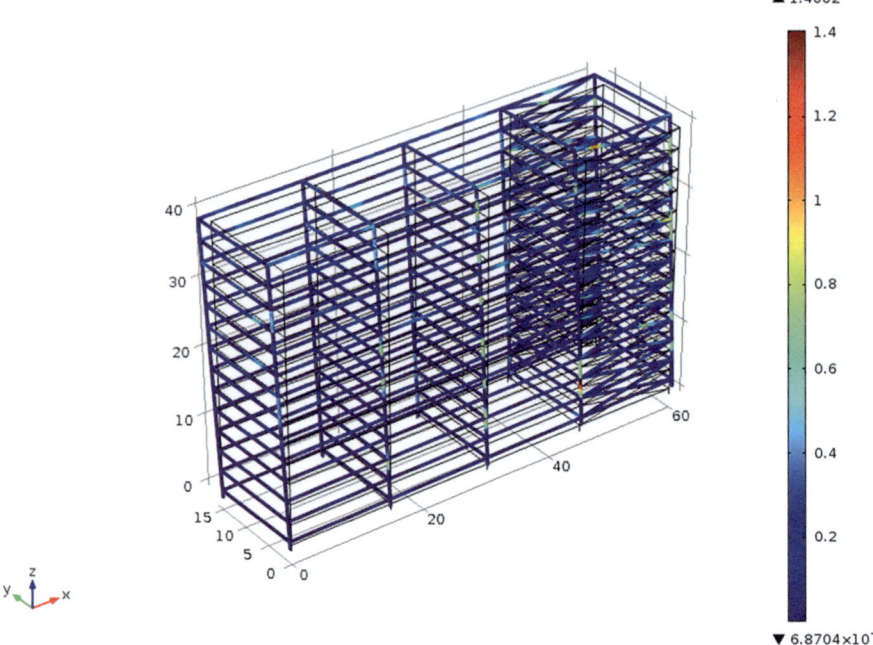

Fig. 18 Side supports tress diagram

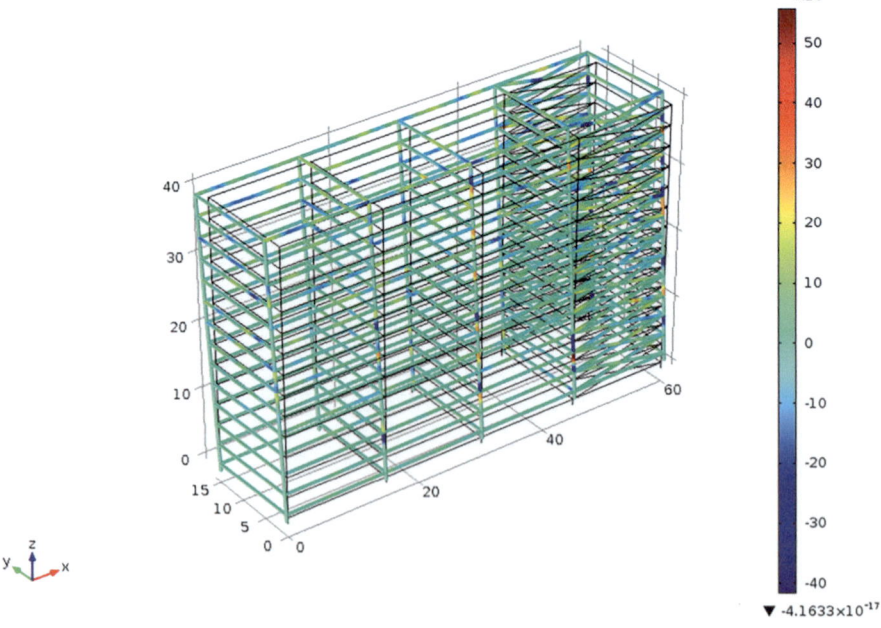

Fig. 19 Side supports strain diagram

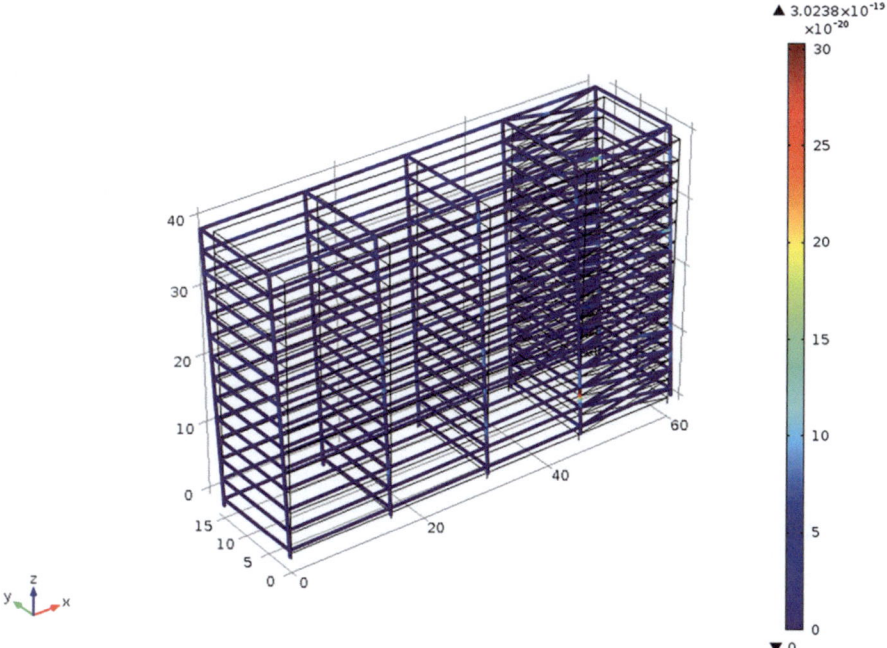

Fig. 20 Side supports strain energy diagram

Table 1 Relationship between ground surface and building tilt deformation

building	Tilt deformation relationship	Correlation coefficient	Number of data sets
Anti-deformation test room	$i\,g = 0.953\,i\,d + 1.16$	0.954	72
Brick-concrete structure house	$i\,g = 0.916\,i\,d - 0.72$	0.997	9
Rural ordinary houses	$i\,g = 0.826\,i\,d + 0.85$	0.934	8

Note g represents building and d represents ground surface

Given their high stiffness, high-rise buildings possess greater capacity to resist horizontal ground deformations. When incorporating expansion joints, the length of individual sections can be appropriately increased.

Vertical Structural Arrangement

The load-bearing capacity and stiffness of high-rise structures should gradually decrease from bottom to top in a uniform and continuous manner, avoiding abrupt changes. The vertical arrangement should be regular and uniform, minimizing outward projections and inward recesses. The lateral stiffness should decrease upwards in a gradual and uniform manner, avoiding severely irregular structural

Table 2 Allowable value of ground tilt deformation for high-rise anti-deformation buildings

New deformation-resistant buildings		The ground surface allows tilt deformation
Multi-story and high-rise buildings the overall tilt	$Hg \leq 12$	0.008
	$12 < Hg \leq 24$	0.006
	$24 < Hg \leq 40$	0.004
	$40 < Hg \leq 60$	0.003
	$60 < Hg \leq 100$	0.002
Tilt of the foundation of a tall structure	$Hg \leq 20$	0.010
	$20 < Hg \leq 50$	0.008
	$50 < Hg \leq 100$	0.006
	$100 < Hg \leq 150$	0.004

arrangements. Vertical resistance components should be continuously connected from top to bottom.

Installation of Sand Cushion Layers

When the bearing capacity allows, the sand cushion layer or gravel layer can reduce uneven foundation reaction force, effectively absorbing the effects of horizontal deformation, curvature deformation, and local tilt deformation.

Installation of Sliding Layers

High-rise buildings typically utilize reinforced concrete integral foundations with sliding layers positioned between the reinforced concrete base and the plain concrete cushion layer. Considering seismic and deformation resistance, materials with a friction coefficient of approximately 0.2–0.4 are suitable when the building height-to-width ratio is less than 3. This choice prevents excessive swaying during earthquakes and mitigates mining-induced deformations. However, sliding layers are not recommended when the height-to-width ratio exceeds 3 (i.e., building heights exceeding 60 m).

Foundation Design

High-rise building structures should employ raft foundations, beam-raft foundations, or box foundations [24, 25]. Besides conventional calculations, foundation dimensions and reinforcement should be calculated based on the magnitude of surface deformations. Shallow foundations are not recommended. For tall structures, foundation plan dimensions and strength should be increased appropriately to resist surface tilt deformations.

Strengthening Structural Integrity

Surface deformations induce additional internal forces in upper structures. Structural calculations should consider enhancing the building's strength accordingly. Horizontal deformations primarily affect the foundation, while surface tilt deformations primarily affect the upper structure. Structural component strength and overturning stability should be verified during calculations.

5 Conclusion

(1) When constructing high-rise buildings above mining subsidence areas, integral foundations are preferred over isolated foundations, and shallow foundations are not recommended.
(2) Numerical simulations indicate maximum displacements occur at the base and upper sections of high-rise buildings. Therefore, strengthening structural integrity should focus on the foundation and connections between the foundation and upper structure.
(3) High-rise buildings are primarily affected by tilt deformations. In accordance with existing regulations, permissible values for surface tilt deformations affecting multi-story anti-deformation buildings are proposed.
(4) Constructing high-rise buildings above mining subsidence areas necessitates comprehensive considerations in site selection, building-structural typology, foundation selection, and ground treatment.

Acknowledgements This research was supported by "Science and Technology Innovation and Entrepreneurship Funds Special Project" of CCTEG. Research and Development of Key Technologies and Equipment for Coal-based Solid Waste Co-disposal Utilization (2022-2-ZD004).

References

1. Jia X, Qi L (2023) Research on geotechnical engineering survey of high-rise residential area in Goaf. Eng Technol Res 8(142):210–212
2. Jianjun D, Ying Z, Xin LI et al (2024) InSAR deformation monitoring and safety and stability evaluation on surface of coal mine goaf. China Saf Sci J 34(1):140–149
3. Dandan X, Daifeng S (2020) FLAC3D-based analysis of surface building stability above minedout areas. Build Safety 2020(9):4–7
4. Zhe Y, Yuanqiang L, Guokai S (2020) Influence of multi-seam Goafs on the stability of proposed buildings above. J Xi'an Univ Sci Technol 40(5):862–868
5. Zhe Y, Jingang Z (2021) Numerical simulation analysis of surface deformation evaluation in coal mine Goafs. Surv Sci Technol 1:6–10
6. Keyi G, Yangyang Z (2023) Foundation grouting treatment and safety technology practice for newly built large buildings in Goaf of small coal mines. Shang Dong Sci Technol 41(9):200–202
7. Weiqiang Y (2022) Evaluation of foundation stability for new buildings in the subsidence area of Lugou coal mine. Henan Polytechnic University
8. Guangyun G, Chunxiao N, Chao S et al (2014) Review of surface deformation in mined-out areas under static loading. Chin J Undergr Space Eng 10:1970–1975
9. Guangyun G, Chunxiao N, Tonghe Z (2014) Review of surface deformation in mined-out areas under dynamic loading. Chin J Undergr Space Eng 10:1963–1969

10. Qingfeng L (2021) Application of numerical simulation in forensic appraisal of building damage induced by overlying strata movement in goaf areas. West Prospect Eng 2021(7):157–161
11. Shaojie C, Weihao Z, Feng W et al (2022) Laws and mechanisms of surface movement and deformation above shallow longwall abandoned gob under building load. J China Coal Soc 47(12):4403–4416
12. Zhanxin L, Changxiang W, Lulin Y (2019) Simulation analysis of key factors of activation instability in Goafs. Safety in Coal Mines 50(6):240–244
13. Dehua L (2019) Study on interaction laws between new building loads and underlying coal mine Goafs. J Zhongyuan Univ Technol 30(5):57–64
14. Wei W, Tingwei C, Jilong L (2018) Research on anti-deformation mechanism of new building raft foundations in coal mine Goafs. J Henan Univ Urban Constr 27(6):38–44
15. Keyi G (2020) Study on goaf treatment and anti-deformation building technology in thick coal seam warehouse mining. Mine Surv 48(6):16–19
16. Yonghai T, Zhixin T, Keyi G et al (2021) Construction technology for building utilization in coal mining subsidence areas. Emergency Management Press, Beijing
17. JGJ3-2010, Technical code for concrete structures of high-rise buildings
18. Sihai Y, Yonghai T, Zhixin T et al (2020) Study on interaction mechanism between large buildings and ground in coal mining subsidence areas. Coal Sci Technol 48(10):166–172
19. GB50007-2011, Code for design of building foundations
20. GB50010-2010, Code for design of concrete structures
21. State Administration of Coal Industry (2000) Regulations on leaving coal pillars and pressure coal mining for buildings, water bodies, railways, and main shafts. Coal Industry Press, Beijing
22. GB51180-2016, Technical code for construction and foundation treatment in coal mining Goafs
23. Ministry of energy of the People's Republic of China. Regulations on coal mine surveying. Beijing: Coal Industry Press, 1989
24. Tianjin University, Tongji University, Southeast University, editors. Concrete structures. Beijing: China Architecture and Building Press, 1998
25. Weiguang Y (1998) Foundations and bases, 3rd edn. China Architecture and Building Press, Beijing

Open Access This chapter is licensed under the terms of the Creative Commons Attribution 4.0 International License (http://creativecommons.org/licenses/by/4.0/), which permits use, sharing, adaptation, distribution and reproduction in any medium or format, as long as you give appropriate credit to the original author(s) and the source, provide a link to the Creative Commons license and indicate if changes were made.

The images or other third party material in this chapter are included in the chapter's Creative Commons license, unless indicated otherwise in a credit line to the material. If material is not included in the chapter's Creative Commons license and your intended use is not permitted by statutory regulation or exceeds the permitted use, you will need to obtain permission directly from the copyright holder.

Multi-criteria Decision Analysis of TGS360Pro Advance Geological Forecasting Results Based on Deviation Maximization

Mingcai Zhang, Guanghong Ju, Dazhou Zhang, Zonggang Chen, Dong Li, and Song Han

Abstract The TGS360Pro system utilizes seismic wave technology to perform advanced geological forecasting for tunnels, generating multidimensional data parameters including relative stress changes, water content, elastic modulus, and danger levels. However, the complex interrelations among these parameters often come with significant redundant information and sometimes exhibit mutually exclusive relationships, noticeably increasing the difficulty of data processing and geological interpretation. To address this issue, this study effectively integrated multiple related features, significantly improving the accuracy of data analysis and the generalization ability of the predictive model through techniques such as deviation maximization and multi-criteria decision analysis. It revealed potential hidden patterns within the data, thereby enhancing the model's prediction accuracy and improving the accuracy of geological hazard identification. Through practical application in an underground plant project of a hydropower station, this study has demonstrated the efficiency and reliability of this method in identifying geological phenomena and predicting geological conditions.

Keywords TGS360Pro · Advance geological forecasting · Deviation maximization · Multi-criteria decision analysis

1 Introduction

In the excavation of underground chambers, the geological and hydrological conditions in front of the tunnel face are complex and changeable, often leading to sudden geological disasters such as collapses, inrush of water and mud (sand), and blowouts.

M. Zhang (✉) · G. Ju · Z. Chen · D. Li · S. Han
PowerChina, Northwest Engineering Corporation Limited, Xi'an, Shaanxi, China
e-mail: csuzmc@163.com

D. Zhang
School of Geosciences and Info-Physics, Central South University, Changsha, Hunan, China

Such incidents can cause minor delays in construction schedules or, in severe cases, significant loss of life and property. Due to various constraints, it is difficult for preliminary surveys to accurately identify potential geological disasters in the sections the underground chambers will pass through, making it challenging to fully meet the safety requirements for excavation. This is especially true for underground chambers with long tunnel lines and significant burial depths. Therefore, advanced geological forecasting is a critical component in the excavation process of underground chambers [1].

Due to its high detection accuracy, wide range, and low sensitivity to electromagnetic interference such as metal pipelines, the advanced prediction technology of seismic reflection waves has become the first choice for high-precision geological prediction in tunnel construction at this stage. With the development of the research on the propagation characteristics of seismic waves, it has become an important branch of advanced prediction to use the propagation and reflection laws of seismic waves in the rock strata to judge the surrounding rock conditions in front of the face. With the development of research on seismic wave propagation characteristics, many complete sets of technical equipment for advanced prediction of seismic wave methods have emerged at home and abroad, such as TSP (tunnel seismic prediction) [2–4] in Switzerland, TRT (tunnel reflection tomography) [5, 6] in the United States, TGS (tunnel geology system) [7, 8] in Russia, TST (tunnel seismic tomography) developed by Qingdao Guoke Marine Environmental Engineering Technology Co., Ltd., and TGP (tunnel geology prediction) produced by Beijing Hydropower Geophysical Prospecting Institute have been widely used.

TGS360Pro is a new generation of geological prediction system jointly developed by GEOTECH Company of Russia and Ural State University of Geology of Russia (founded in 1914), and it is also a leading and advanced geological prediction system in the world. The main technical concept of the prediction system has been continuously improved by Mr. Vladimir Pisetski, director of the Department of Geographic Information of Ural State University of Geology, doctor of mineralogy and geology, and professor. The prediction method is mainly protected by two international patents [7, 8]: (US005796678A: A method that can determine the location of underground fluids (oil, gas, water-rich areas); US6498989B1: A method that can predict the dynamic parameters of underground fluids (oil, gas, water-rich areas)), by artificially stimulating seismic waves on both sides of the side wall or face, the seismic waves propagate in the surrounding rock of the tunnel in the form of spherical waves. When the wave impedance of the surrounding rock changes (for example, when encountering karst, fault, or interface of rock strata), some seismic waves will be reflected, and another part of the seismic waves will continue to propagate forward. The reflected wave is received by a high-precision receiver and transmitted to the host to form a seismic wave record.

2 TGS360Pro Multi-parameter Data Features

The TGS360Pro advance forecasting system produces results with several data parameters, typically focusing on eight main parameters: "relative stress change," "water content probability," "dynamic Young's modulus," "Poisson's ratio," "P-wave velocity," "S-wave velocity," and "hazard level." Among them, the parameter "relative stress change" is measured in relative units, and through this parameter, it is possible to define the stress state of the surrounding rock and to analyze unfavorable geological entities such as karst, fractures, and caves. The "water content probability" is calculated based on the seismic wave reflectivity and measured in percentage. The "dynamic Young's modulus" is computed using P-wave, S-wave, and density values to characterize the material's tensile or compressive resistance during elastic deformation, measured in Pascals (Pa). Other parameters, such as P-wave velocity, S-wave velocity, and their ratios, on one hand, can be used to calculate main parameter values inversely, and on the other hand, the results obtained from the parameters themselves can be compared with the results from the main parameters for analysis. This comparison is then used to infer the information of the surrounding rock ahead of the tunnel face and the possible locations of poor geological distributions. Particularly, the "hazard level" parameter is a core parameter derived from a comprehensive assessment that integrates the other seven parameters. The rock mass classification corresponding to the Rock Mass Rating (RMR) is derived by classifying rock quality during the process of supporting a 10-m rock tunnel according to the variation of each parameter and the size of RMR, with the specific classification as shown in Table 1.

The results data for TGS360Pro contain many variables or features, which may lead to the curse of dimensionality during processing and analysis; the features have different dimensions and data types, such as some are continuous (parameters like wave velocity, elastic modulus, etc.), and some are categorical (hazard levels). Each parameter data have a complex internal structure, such as nonlinear relationships, hidden patterns, and dependencies. Additionally, there are redundant or highly correlated features among various parameters, which do not provide additional information, such as the ratio of wave velocities. Moreover, the actual measured data typically include noise and outliers, which may affect the accuracy of data analysis and model prediction.

To address these issues, this paper employs a multi-criteria decision-making approach to analyze TGS360Pro results data. On the one hand, this method can combine related features into fewer representative features, integrating essential information into fewer characteristics, and reducing the impact of irrelevant or noisy features. This reduces the dimensionality of the data and simplifies the model, enhancing the accuracy and predictive capability of the data analysis model. On the other hand, clustering features together can reveal potential patterns or structures within the data, reduce redundant features, and strengthen the representativeness of the features. This helps to improve the model's generalizability, allowing it to perform better when processing new datasets.

Table 1 Hazard class classification table

Hazard class	RMR	Excavation method	Rock Bolt 20 mm (Grouting Bolt)	Shotcrete	Steel bracket
I-Very good	81 ~ 100	Full-face excavation method. Excavate 3 m in advance	Except for special circumstances that require anchor bolt support, no other support is necessary		
II-Good	61 ~ 80	Full-face excavation method. Advance support of 20 m, excavate 1.0 ~ 1.5 m ahead	The anchor bolt length is 3 m, sometimes a 2.5-m wide steel reinforcement mesh is needed	The thickness of the second lining is 50 mm	/
III-General	41 ~ 60	Bench excavation method. Advance support of 10 m, excavate 1.5 ~ 3.0 m on the upper bench, and begin support after blasting	The anchor bolt length is 4 m, requiring a steel reinforcement mesh that is 1.5 ~ 2 m wide	The thickness of the secondary support is 50 ~ 100 mm, with an initial support thickness of 30 mm	/
IV-Poor	21 ~ 40	Bench excavation method, advance support of 10 m, excavate 1.0 ~ 1.5 m on the upper bench, conduct excavation and support simultaneously	The anchor bolt length is 4 ~ 5 m, requiring a steel reinforcement mesh that is 1 ~ 1.5 m wide	The thickness of the secondary support is 100 ~ 150 mm, with an initial support thickness of 100 mm	Requires lightweight or medium-weight threaded steel supports spaced at 1.5 m
V-Very poor	< 20	Multi-face excavation method. Excavate the Sect. 0.5 ~ 1.5 m in advance, carry out excavation and support at the same time, and immediately spray shotcrete after blasting	The anchor bolt length is 5 ~ 6 m, requiring a steel reinforcement mesh that is 1.0 ~ 1.5 m wide	The thickness of the secondary support is 150 ~ 200 mm, with an initial support thickness of 150 mm, and the excavation surface shotcrete layer thickness is 50 mm	The inverted arch requires a spacing of 0.75 m for medium-weight or heavy-weight steel backing plates

3 Multiple Criteria Decision Analysis

Multi-criteria decision analysis (MCDA) is an integral part of modern decision science, primarily exploring how to explicitly evaluate multiple interconnected criteria in decision-making processes. This method focuses on making decisions that involve trade-offs among multiple objectives (or criteria) [9]. Its theories and methods have broad applications across various fields such as engineering design, economics, management, and military, including investment decisions, project evaluation, factory siting, bidding, maintenance services, assessment of weapons system performance, ranking of industrial sectors, and the comprehensive evaluation of economic benefits [10]. MCDA provides a range of methods and procedures that help decision-makers choose the best feasible solution after considering all relevant criteria [11].

MCDA mainly includes two main steps: the acquisition and aggregation of decision information, and the acquisition of decision information mainly involves two aspects: attribute value and attribute weighting.

3.1 Attribute Normalization

Attribute types generally include benefit, cost, fixed, deviation, range, and out-of-range, among others. Benefit-type attributes are those where a larger attribute value is better. Cost-type attributes are those where a smaller attribute value is better. Fixed-type attributes are those where a value closer to a certain fixed value α is better. Deviation-type attributes are those where a value more deviant from a certain fixed value β is better. Range-type attributes are those where a value closer to a certain fixed range $[\alpha, \beta]$ (including falling within that range) is better. Out-of-range attributes are those where a value more deviant from a certain fixed range $[\alpha, \beta]$ is better. Based on the TGS360Pro advanced geological forecast results data, a decision matrix A_{ij} is constructed, as shown in Table 2, where parameters such as "dynamic Young's modulus," "longitudinal wave speed," and "transverse wave speed" are benefit-type parameters; "relative stress change," "water content probability," "Poisson's ratio," "longitudinal to transverse wave speed ratio," and "hazard level" are cost-type parameters.

Let a_{ij} the i attribute of the benefit type be normalized according to formula (1):

$$R_{ij} = \frac{A_{ij} - \min_i(A_{ij})}{\max_i(A_{ij}) - \min_i(A_{ij})}, i \in [1, n], j \in [1, m] \qquad (1)$$

For cost-type attributes, normalize according to formula (2):

$$R_{ij} = \frac{\max_i(A_{ij}) - A_{ij}}{\max_i(A_{ij}) - \min_i(A_{ij})}, i \in [1, n], j \in [1, m] \qquad (2)$$

Table 2 Data decision matrix A_{ij} of TGS360Pro advanced geological forecast results

Space coordinate point	Attributes							
	Benefit type			Cost type				Hazard class
	Dynamic young's modulus	Longitudinal wave velocity	Shear wave velocity	Relative stress change	Water content probability	Poisson's ratio	The ratio of longitudinal to transverse wave velocity	
P1	A11	A12	A13	A14	A15	A16	A17	A18
P2	A21	A22	A23	A24	A25	A26	A27	A28
...	...	...	...	...	...	...	...	...
Pn	An1	An2	An3	An4	An5	An6	An7	An8

Here, n represents the number of points, m represents the number of attributes, and in this study, take $m = 8$.

3.2 Attribute Weighting

The current methods for determining the distribution of weights across multiple attributes mainly fall into two categories [10]. The first category is the subjective weighting method, such as the eigenvector method, the least squares method, and the Delphi method. Although this method uses long-term accumulated experiential knowledge to reflect many years of comprehensive understanding of the system under inspection, it overlooks the current state of the objective system. It has deficiencies such as insufficient theoretical arguments and expert personal preferences affecting the objectivity of judgment results. The second category is the objective weighting method, such as principal component analysis, entropy method, and the method of maximizing deviation [9, 11]. This method determines the weight of each attribute based on the information currently collected, grounded in clear facts and mature theory, thus possessing strong persuasiveness. This study adopts the method of maximizing deviation to determine the weight of each attribute.

Let $V_{ij}(w)$ denotes the dispersion of a point P_i from individual attributes at other points [12, 13], i.e., as follows:

$$V_{ij}(w) = \sum_{k=1}^{n} |w_j R_{ij} - w_j R_{kj}| = \sum_{k=1}^{n} |R_{ij} - R_{kj}| w_j, i \in [1, n], j \in [1, m] \quad (3)$$

The selection of the weighting vector w should make the total deviation of all evaluation indicators to all decision schemes the largest. Based on this, the objective function is constructed. Combined with the characteristic that the sum of the weighting vector w is 1, the solution of the weighting vector w is transformed into the solution of the following optimization problem:

$$\begin{cases} \max F(w) = \sum_{j=1}^{m} \sum_{i=1}^{n} \sum_{k=1}^{n} |R_{ij} - R_{kj}| w_j \\ \sum_{j=1}^{m} w_j^2 = 1, w_j \geq 0 \end{cases} \quad (4)$$

Construct the Lagrange function by Eq. (4), and let its partial derivative be 0, which solves:

$$w_j = \frac{\sum_{i=1}^{n} \sum_{k=1}^{n} |R_{ij} - R_{kj}|}{\sqrt{\sum_{j=1}^{m} [\sum_{i=1}^{n} \sum_{k=1}^{n} |R_{ij} - R_{kj}|]^2}} \quad (5)$$

3.3 Decision Information Aggregation

The comprehensive attribute values at each point are obtained by solving the following formula:

$$D_i(w) = \sum_{j=1}^{m} R_{ij} w_j \tag{6}$$

After multi-attribute decision-making, the comprehensive attribute value can be normalized to the range of each attribute, and each attribute can be updated by weighted average. The first j attribute A_j can be updated as follows:

$$A'_j = A_j\left(1 - w_j^2\right) + \frac{[D_j - \min(D_j)][\max(A_j) - \min(A_j)] + \min(A_j)}{\max(D_j) - \min(D_j)} w_j^2 \tag{7}$$

4 Practical Engineering Case Analysis

4.1 Engineering Geological Overview

The underground powerhouse caverns of a hydropower station are located on the left bank of the mountain body, and the structures of the underground powerhouse caverns are relatively developed. Most of them are interlayer faults, and most of them are located below the groundwater level. To ensure the safe construction of the underground powerhouse caverns, grasp the deformation law and development trend of the surrounding rock, and dynamically adjust the excavation and support of the surrounding rock during the construction period, at the 0 + 053 m right of the main transformer room, TGS360Pro complete set of equipment is used to predict the rock mass structure and surrounding rock classification 100 m in front of the face (0 + 053 m right to 0 + 047 m left of the underground powerhouse cavern). The field photographs of the face and the catalog of rock occurrence are shown in Fig. 1a and b, respectively. Weakly weathered medium-thick stratified metamorphic fine sandstone with a small amount of carbonaceous phyllite is exposed on the face, mainly medium-hard rock; the dip of the rock layer is NW330SW∠88°, for the ① set NE45°NW∠35°, for the ② set NW340SW∠16°, and for the ③ set NW320SW∠55°. The structural surfaces are mostly straight and rough, filled with rock debris and rock powder, unconsolidated. The main category of the surrounding rock is IV1.

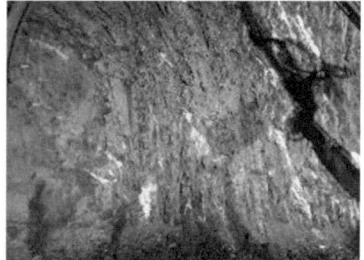 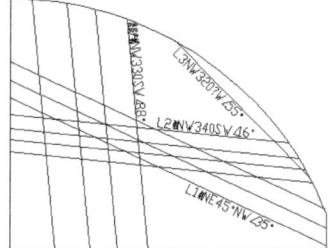

(a) Photos of the surrounding rocks at the working face (b) Record of the structural occurrence of rock layers at the working face

Fig. 1 Geological information of the working face

4.2 TGS360Pro Advanced Geological Forecast Results

The results of TGS360Pro advanced geological forecast data processed by supporting software are shown in Fig. 2, in which Fig. 2 a–h are the main eight parameters ("relative stress change," "water content probability," "dynamic Young's modulus," "Poisson's ratio," "P-wave velocity," "S-wave velocity," and "hazard level"). The data points of various parameters are mixed, and it is difficult to distinguish the potential geological risks indicated by them. These data show large overlap and redundancy in the image, and even some parameters are mutually exclusive, resulting in blurred boundaries of various geological hidden dangers, and the stratification characteristics of rock strata cannot be significantly displayed. On the other hand, the inherent similarities and differences in the data are difficult to identify, and it is difficult to accurately identify the geological structure and potential hidden danger types from various parameters.

4.3 Multiple Attribute Decision-Making of Forecast Result

The eight parameter results predicted by TGS360Pro are extracted according to the network degree of 0.5 m × 0.5 m, and a total of data (100 ÷ 0.5) × (30 ÷ 0.5) = 12,000 points are extracted for each attribute, that is, in the formula (1–6), $n = 12,000$ and $m = 8$ in this example, according to the formula (5), The weighting vectors of the eight attributes (i.e., the "relative stress change," "water content probability," "dynamic Young's modulus," "Poisson's ratio," "P-wave velocity," "S-wave velocity," and "hazard level") are (0.268, 0.241, 0.375, 0.362, 0.154, 0.217, 0.353, and 0.635), and the comprehensive attribute values calculated from formula (6) are shown in Fig. 3 after gridding.

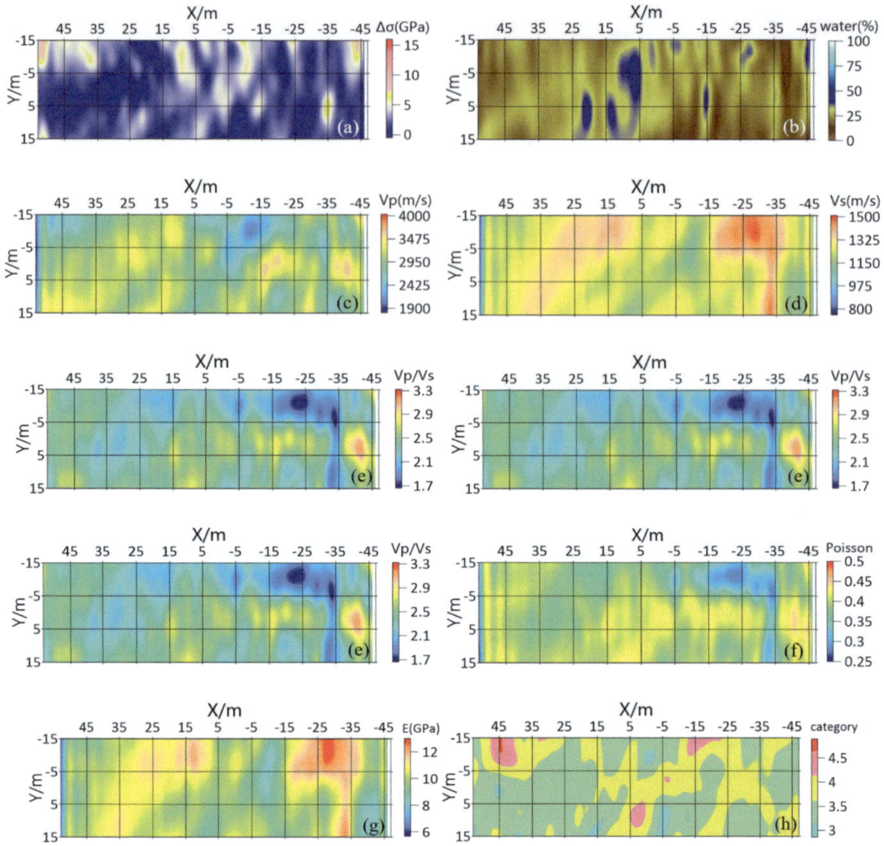

Fig. 2 TGS360Pro preliminary forecast results chart

Fig. 3 Comprehensive attribute values

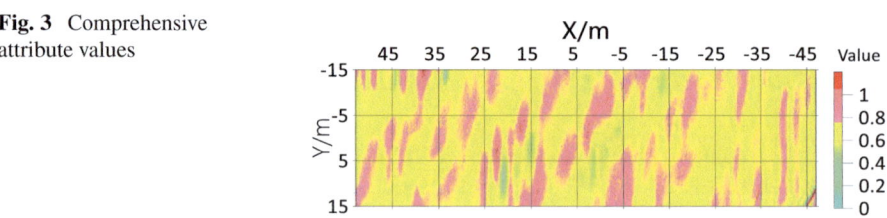

The results after updating each attribute according to formula (7) are shown in Fig. 4, where Fig. 4a–h are the relevant results of the main eight parameters (relative stress change, water cut probability, longitudinal wave velocity, shear wave velocity, longitudinal and shear wave velocity ratio, Poisson's ratio, dynamic Young's modulus, hazard level). The organization and representation of the outcome data of

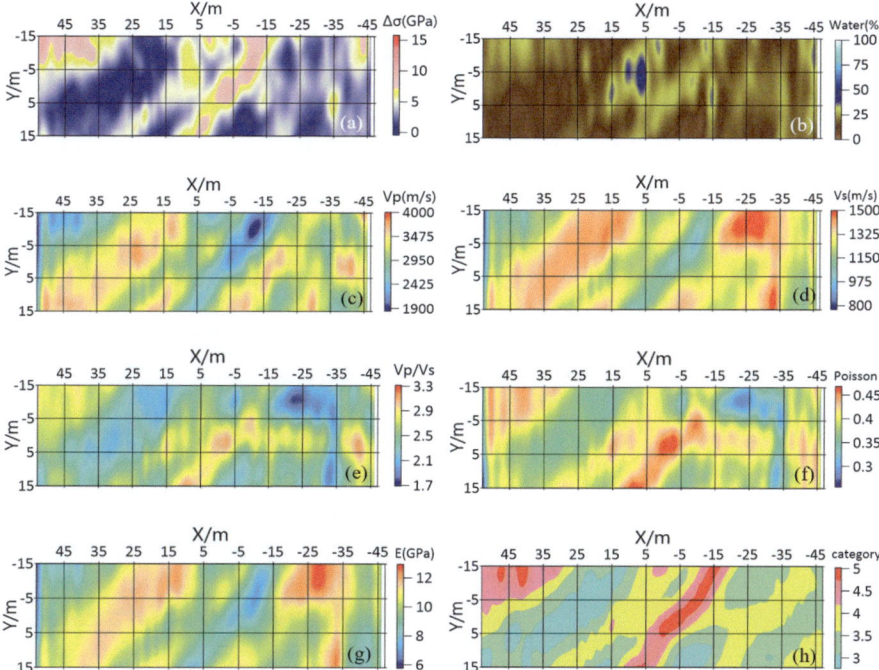

Fig. 4 TGS360Pro forecast data multi-attribute decision analysis results

each parameter have undergone significant changes. The previously disorganized data now present an orderly structure. The results of multi-attribute decision-making analysis can assign data points to their respective groups. Each group represents a unique geological phenomenon or hidden danger type. The similarities and differences between the results data have been strengthened, and the boundaries of rock strata and the scope of geological hidden dangers have been clearly revealed.

4.4 Statistic Validation of Multiple Attribute Decision Analysis Results

The updated results were statistically analyzed according to the tunnel sections, with the statistical outcomes of each parameter in different tunnel sections shown in Fig. 5. As can be seen from the figure, overall, according to the trend of change, it can be divided into four sections. The exclusivity between the eight parameters is significantly weakened, and parameters with the same attributes (cost-type and benefit-type) exhibit a strong correlation.

Fig. 5 Statistical chart of multi-attribute decision analysis results divided into tunnel sections

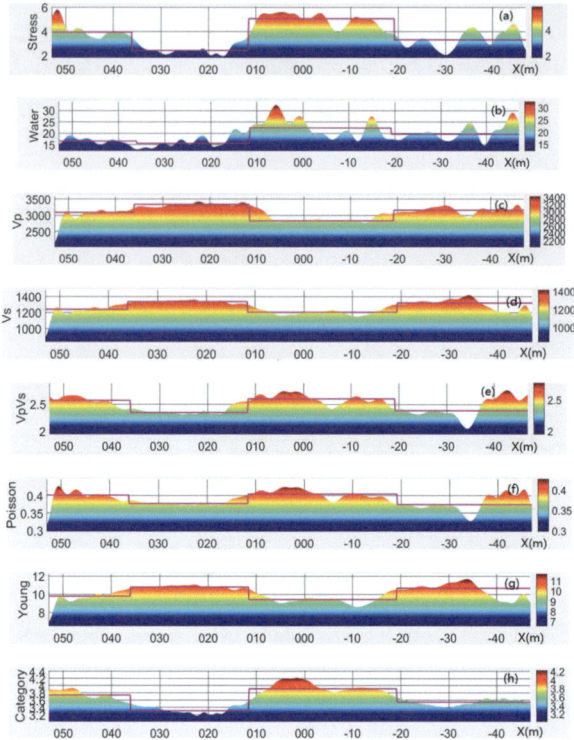

According to the statistical results of the updated multi-attribute decision-making data, the advanced geological prediction is analyzed as follows: (1) in the Sect. 0 + 053 m to 0 + 036 m from the right of the factory: The rock mass is relatively complete to poor in general, accounting for 72% of the section; Class III is the main type of surrounding rock. (2) From 0 + 036 m to 0 + 011 m to the right of the factory: The whole rock mass is relatively complete to poor integrity, and the integrity is the main part, accounting for 84% of the survey section; the structure of the rock mass is gravel-like structure to block-stone and gravel-like structure; Class IV is the main type of surrounding rock. (3) From 0 + 011 m to 0 + 019 m from the right of the factory: The rock mass is generally broken, and the rock mass structure is gravel-like structure to block-stone and gravel-like structure; Class IV is the main type of surrounding rock. (4) From 0 + 019 m to 0 + 047 m: The rock mass is relatively complete to poor in general and is mainly relatively complete, accounting for 71% of this survey section. In particular, the section from 0 + 011 m to 0 + 006 m from the right of the factory is the key risk section: The rock mass in this section is relatively broken, and the rock mass is mostly in a block-like structure; the joints are more developed and the filling material is more; Class IV is the main type of surrounding rock; the amount of water seepage at the face is serious, the fracture water in the bedrock is relatively developed, the online flow water is in the excavation range, the local flow water is strand shaped, and there is even a risk of water gushing and mud outburst, which needs to be strengthened. The quality of the rock mass is poor, there are certain risks, it is seriously affected by geological structures, and the joints are very developed.

The data of later excavation and geological cataloging show that: (1) The rock mass is generally broken in the section from 0 + 053.00 m to 0 + 035.00 m on the right of the factory, and the rock mass structure is mainly gravel-like structure; structural plane is more than three groups, mostly weathered fissures, fissures width is mainly open type, and there are many fillings; surrounding rocks are mainly IV1 type. (2) From 0 + 035.00 m to 0 + 011.00 m from the right of the plant to the left of the plant: The overall rock mass is worse than that of the broken integrity, and the integrity is the main one; Class III is the main type of surrounding rock; the quality of rock mass is average. (3) From 0 + 011.00 m to 0 + 047 m to the left of the factory: The rock mass is generally broken, the quality of the rock mass is poor, and the surrounding rock type is mainly IV2. The results of advanced geological prediction updated by multi-attribute decision-making have a high coincidence degree with the site, and the coincidence degree of local station numbers is up to 85%.

5 Conclusion

By adopting the maximal deviation method, this study has delved deeply into the complex relationships between various parameters, precisely extracted crucial information, and significantly reduced the interference effects of non-core data on decision-making. A meticulously designed weight assignment system was tailor-made for different attributes, ordered, and selected based on their pivotal role in geological forecasts. On this foundation, we constructed a decision matrix, through which the interactions and impacts between attributes were comprehensively and systematically analyzed, achieving effective data classification and aggregation. Through the integrated multi-attribute decision analysis of TGS360Pro outcome data, these data demonstrated clear structured characteristics, providing a straightforward identification pattern for previously obscure geological phenomena and hidden risk categories.

This research, through exhaustive case studies, has highlighted the marked effectiveness of multi-attribute decision analysis in handling highly complex geological forecast data, introducing an innovative data analysis tool to geological forecasting practice. By integrating the insights of multi-attribute decision analysis with the actual observations from field surveys, this method successfully predicted the critical geological conditions encountered during the construction process of hydropower station tunnels, thus providing a solid scientific decision-making foundation for tunnel design and construction, ensuring the smooth progress of the project. Moreover, this study has not only injected new vitality into the field of geological prediction but also offered insights and inspiration to related fields. Looking ahead, we believe this multi-attribute decision analysis technology will play a significant role in a broader range of application scenarios, propelling the development of geology and other related fields.

References

1. Liu B, Wang J, Ren Y et al (2022) Decoupled elastic least-squares reverse time migration and its application in tunnel geologic forward prospecting. Geophysics 87(1):EN1–EN19
2. Dhang PC (2019) Report based on tunnel seismic prediction (TSP) at a tunnel in lesser Himalaya, Jammu & Kashmir. J Geol Soc India 94(6):646–646
3. Choud K (2019) India A. 3D-tunnel seismic prediction during tunneling while underpassing an abandoned tunnel
4. Lin CJ, Li SC (2014) Tunnel seismic prediction (TSP) and its application in tunnel engineering. Appl Mech Mater 501–504:1779–1782
5. Yamamoto T, Shirasagi S, Yokota Y et al (2010) Imaging geological conditions ahead of a tunnel face using a three-dimensional seismic reflector tracing system
6. Li J, Wang Z (2015) The application of tunnel reflection tomography in tunnel geological advanced prediction based on regression analysis. Open Civ Eng J 9(1):805–810
7. Pisetski VB (2000) Method for determining the presence of fluids in a subterranean formation: USA, 6,028,820
8. Pisetski VB (2002) Method for predicting dynamic parameters of fluids in a subterranean reservoir: USA, 6498,989 B1
9. Schey C, Postma M, Krabbe P, et al (2020) The application of multi-criteria decision analysis to inform resource allocation. F1000Research 9:445
10. Ataei Y, Mahmoudi A, Feylizadeh MR et al (2020) Ordinal priority approach (OPA) in multiple attribute decision-making. Appl Soft Comput 86:105893
11. Więckowski J, Sałabun W, Kizielewicz B et al (2023) Recent advances in multi-criteria decision analysis: a comprehensive review of applications and trends. Int J Knowl-Based Intell Eng Syst 27(4):367–393
12. Dessole M, Marcuzzi F (2022) Deviation maximization for rank-revealing QR factorizations. Numer Algorithms 91(3):1047–1079
13. Wei D, Nair R, Dhurandhar A et al (2022) On the safety of interpretable machine learning: a maximum deviation approach. arXiv

Open Access This chapter is licensed under the terms of the Creative Commons Attribution 4.0 International License (http://creativecommons.org/licenses/by/4.0/), which permits use, sharing, adaptation, distribution and reproduction in any medium or format, as long as you give appropriate credit to the original author(s) and the source, provide a link to the Creative Commons license and indicate if changes were made.

The images or other third party material in this chapter are included in the chapter's Creative Commons license, unless indicated otherwise in a credit line to the material. If material is not included in the chapter's Creative Commons license and your intended use is not permitted by statutory regulation or exceeds the permitted use, you will need to obtain permission directly from the copyright holder.

Ultimate Bearing Capacity of Rigid Foundation Near Embankment Slope Subjected to Water Drawdown

Qi Wang, Boyang Xia, Gang Zheng, Zheng Wang, Boyao Zhao, and Xin Yin

Abstract This study is used to calculate the bearing capacity and failure mode of the foundation near embankment slope under different water drawdown conditions. The results exhibit when the reservoir water level rapidly drops, and the ultimate bearing capacity decreases at beginning and then remains unchanged with the decrease in water level. For drawdown of slope water level after the reservoir has been rapidly drained condition, the ultimate bearing capacity initially decreases and then increases with the water level decreasing. Finally, the bearing capacity remains unchanged when the water level exceeds a certain depth. For the slow drawdown of slope water level conditions, the ultimate bearing capacity decreases first and then rises with the decrease in water level and remains unchanged when the water level exceeds a certain depth. The most unfavorable water level exists in the mode of the drawdown of slope water level after the reservoir has been rapidly drained, and for the slow drawdown of slope water level, its location is related to failure mode and slope gradient.

Keywords Embankment slope · Ultimate bearing capacity · Failure mode · Reservoir water level

1 Introduction

According to statistics on various disaster incidents, changes in water levels lead to failure of embankment slopes and the foundations [1–5]. Emerson and Macek-Rowland [1] compiled data on landslides and slope instability events along the shores of Roosevelt Lake over a 10-year period. The statistics indicate that 49% of instability events occurred during the initial stages of water impoundment, while 30% occurred during rapid decreases in reservoir water levels. In Japan, around 60% of incidents

Q. Wang · Z. Wang
Cangzhou Qugang Expressway Construction Company Limited, Cangzhou, Hebei, China

B. Xia (✉) · G. Zheng · B. Zhao · X. Yin
School of Civil Engineering, Tianjin University, Tianjin, China
e-mail: boyang_1027@tju.edu.cn

involving landslides or instability occur during rapid water level drops, with the remaining 40% happening during water level rises. Particularly for embankments with high flood levels, when the water level on the downstream face of the slope decreases, the increased seepage forces can lead to a decrease in the slope's bearing capacity, posing a safety risk.

The drawdown of slope water level is classified into three modes: the rapid drawdown of the reservoir water level, the drawdown of slope water level after the reservoir has been rapidly drained, and the slow drawdown of slope water level [6]. Fang et al. [2] used centrifuge model tests to analyze the impact of water level changes on the overall stability of slopes. Ukritchon and Keawsawasvong [5] proposed a nonlinear regression design equation based on lower bound solutions and predicted the failure mechanisms of embankment slopes when the water levels change. Li et al. [7] introduced an uncoupled analysis system with second-order multi-directional response function to investigate the stability and reliability indicators of embankment slopes under water level drawdown. Previous studies have predominantly focused on the impact of water level drawdown on embankment slope stability, ignoring its influence on the bearing capacity of rigid foundations near the embankment slope.

This study utilized DLO computational program to investigate the variations in the ultimate bearing capacity of the foundation of adjacent slopes under three different water level drawdown scenarios. An analysis of the failure mechanism of the slope foundation is conducted. The most unfavorable water level drawdown height is determined. The findings of this research can serve as a valuable reference for the design of the bearing capacity of rigid foundations adjacent embankment slopes under water level drawdown conditions.

2 Numerical Modeling and Verification

2.1 Numerical Modeling

Discontinuity layout optimization (DLO) combined with limit analysis provides an efficient analytical approach for complex rock and soil stability issues [8–11]. The DLO calculation model, as illustrated in Fig. 1, features a rigid foundation placed at the crest of the embankment slope. The foundation has a width of B, while the slope itself has a height of H and an embankment slope angle of β. The shear strength parameters of the slope soil, c and φ, are based on effective strength indicators. Additionally, the slope soil has a unit weight of γ, and Lw represents the varying water level drop heights.

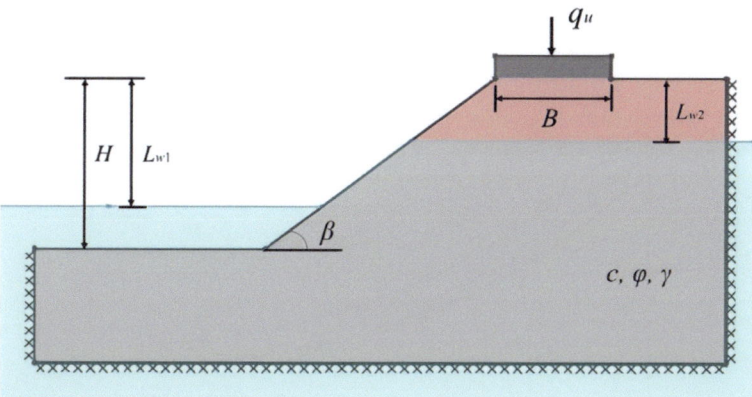

Fig. 1 DLO analysis model diagram

2.2 Verification

Safety factors obtained from the limit analysis of DLO under different working conditions are compared with the results obtained by previous literatures [3, 12], as shown in Fig. 2. By comparing the results, it is found that the trend of safety factor (Fs) variations calculated by DLO model aligns with the results of both studies.

3 Influence of Water Level Conditions on the Bearing Capacity of Foundation

3.1 The Rapid Drawdown of Reservoir Water Level Condition

Figure 3 depicts the dimensionless ultimate bearing capacity of the rigid foundation under the rapid drawdown of reservoir water level. In this scenario, the geometric dimensions of the slope are $H/B = 1$, with a slope ratio of 1:2, and soil strength parameters $c/\gamma B = 1$ (dimensionless clay cohesion), with φ of 10, 20, and 40°. For $\varphi = 10°$, a surface failure occurs, as shown in Fig. 3a. The ultimate bearing capacity of the foundation initially decreases as the reservoir water level drawdown. However, once the water level descent exceeds the sliding range at the lowest point on the slope surface, further changes in water level no longer impact the bearing capacity of the foundation. The ultimate bearing capacity of the foundation ceases to change with the water level decrease. An increase in φ of the slope soil causes the sliding failure surface to extend downwards and outwards. Finally, the failure mode transitions from slope surface failure to toe failure and base failure, resulting in a decrease in the bearing capacity of foundation as the reservoir water level decreases, as illustrated in Fig. 3b and c.

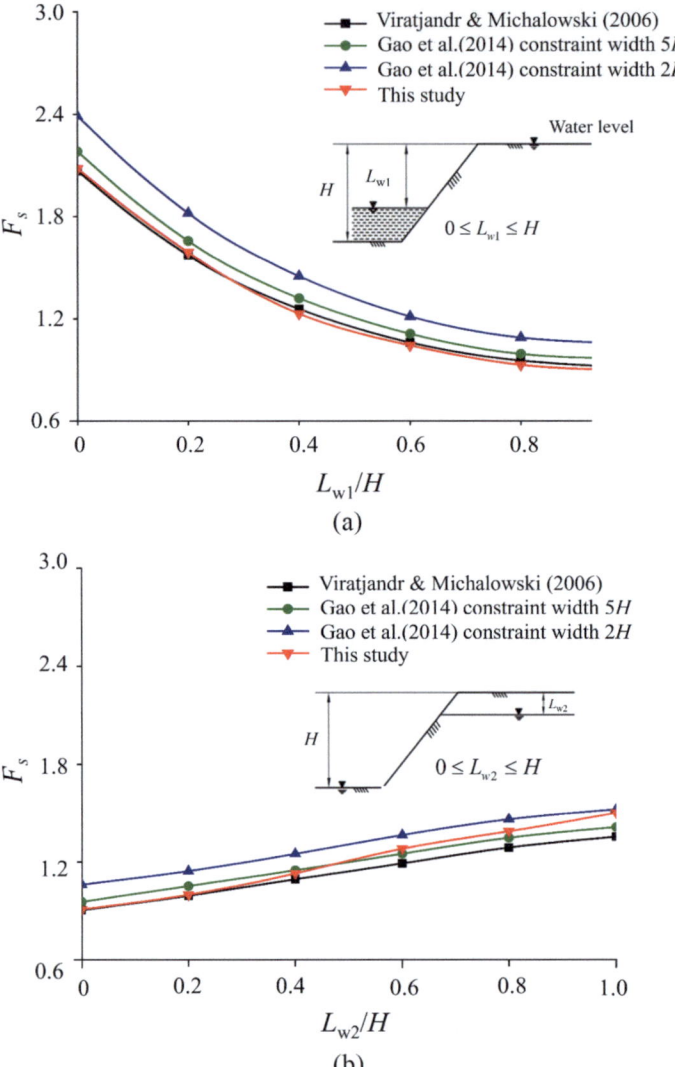

Fig. 2 Comparison verification: **a** case 1: $0 \leq L_{w1} \leq H$; **b** case 2: $0 \leq L_{w2} \leq H$

3.2 The Drawdown of Slope Water Level After the Reservoir Has Been Rapidly Drained Condition

Figure 4 shows the variation pattern of the bearing capacity under the drawdown of embankment slope water level after the reservoir has been rapidly drained. When slope surface failure occurs, the ultimate bearing capacity decreases initially with the reducing of water level and then gradually increases. Increasing internal friction

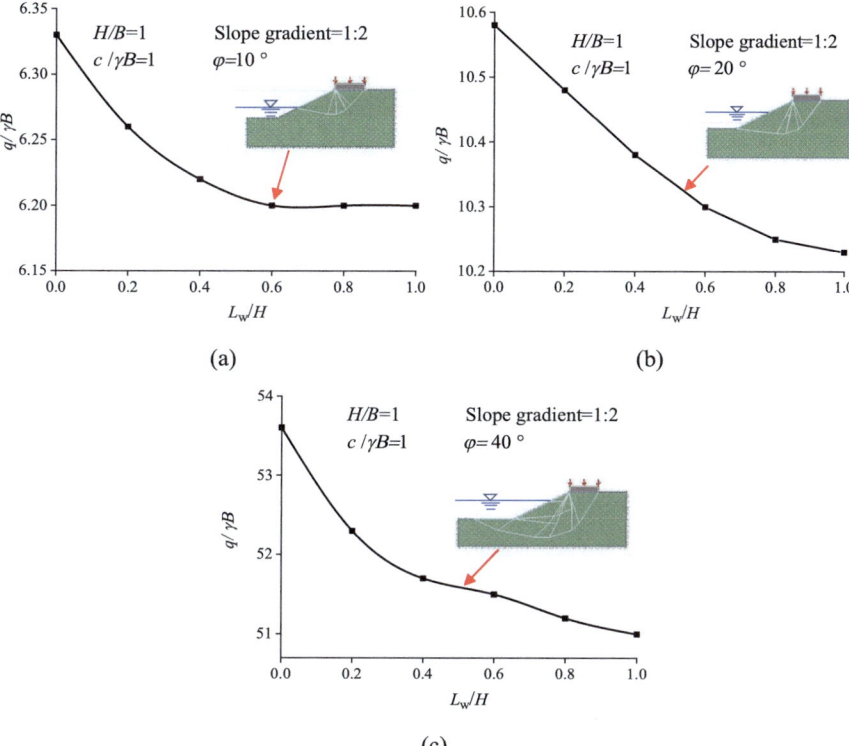

Fig. 3 The variation of bearing capacity under reservoir water rapid drawdown condition: **a** surface failure; **b** toe failure; **c** base failure

angle leads to downward extension and outward propagation of sliding failure surface of the slope. The failure mode transitions from surface failure to toe failure and base failure.

3.3 The Slow Drawdown of Slope Water Level Condition

Figure 5 shows the variation of ultimate bearing capacity of rigid foundation corresponding to different failure modes under slow drawdown of slope water level. As depicted in Fig. 5a, when the surface failure occurs, the ultimate bearing capacity initially decreases, then increases, and eventually remains constant. When the shear strength of the slope soil is relatively high, the failure mode transitions from slope surface failure to toe failure and base failure, resulting in the foundation's ultimate bearing capacity decreasing first and then increasing as the reservoir water level drops. The most unfavorable water level is located at 0.3 times of the embankment slope, as illustrated in Fig. 5b and c.

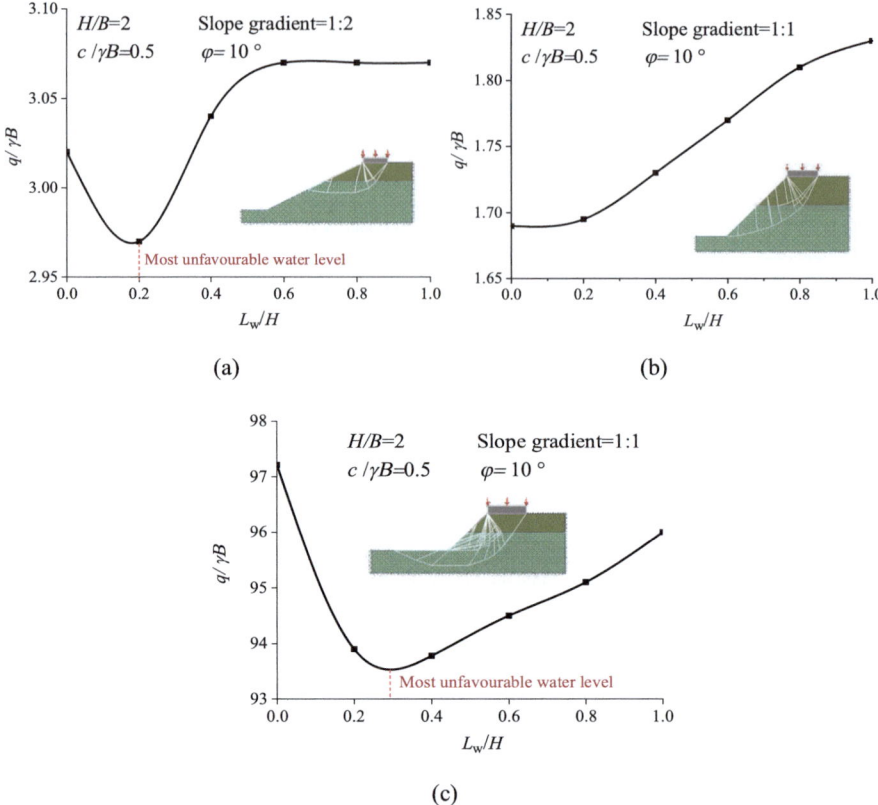

Fig. 4 The variation of foundation bearing capacity under drop of slope water level: **a** surface failure; **b** toe failure; **c** base failure

4 Influence of Failure Mode on the Critical Water Level

Based on the calculations from Sect. 4, it is exhibited that under the rapid drawdown of reservoir water level condition, the bearing capacity of the foundation decreases with decreasing water level, and there is no specific unfavorable water level. In the case of the slow drawdown of slope water level condition, the most unfavorable water level is located at 0.3 times the slope height, and the failure mode does not affect the position of the most unfavorable water level. However, under the drawdown of slope water level after the reservoir has been rapidly drained condition, the position of the most unfavorable water level is influenced by the failure mode, as shown in Fig. 6. For the surface failure mode, the most unfavorable water level drop height is 0.2 times of the slope height ($L_w/H = 0.2$). Subsequently, as the water level decreases, the ultimate bearing capacity of the foundation increases until it exceeds the depth of the sliding surface and remains constant. In the case of toe failure, the ultimate bearing capacity gradually increases with decreasing water level, growing slowly

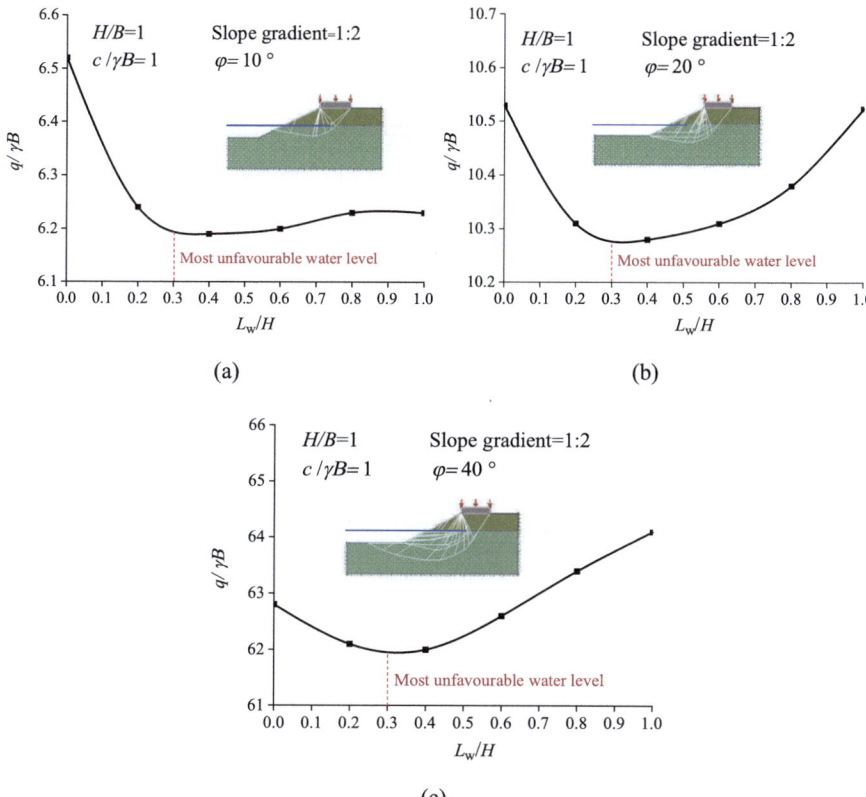

Fig. 5 The variation of foundation bearing capacity under slow drop of slope water level: **a** surface failure; **b** toe failure; **c** base failure

when $L_w/H < 0.2$ and then rapidly increasing. In the event of base failure, the most unfavorable water level drop height is 0.3 times of the slope height ($L_w/H = 0.4$).

5 Conclusions

This study employed the DLO program to calculate the variation process of the ultimate bearing capacity and failure mode of the foundation located near the embankment slope under different water level drawdown conditions. The following conclusions are drawn:

(1) In the case of the rapid drawdown of reservoir water level, the overall trend of the foundation's ultimate bearing capacity decreases with the decreasing of the water level. After the reservoir water is drained, under the condition of the water level decreasing inside the slope, the ultimate bearing capacity

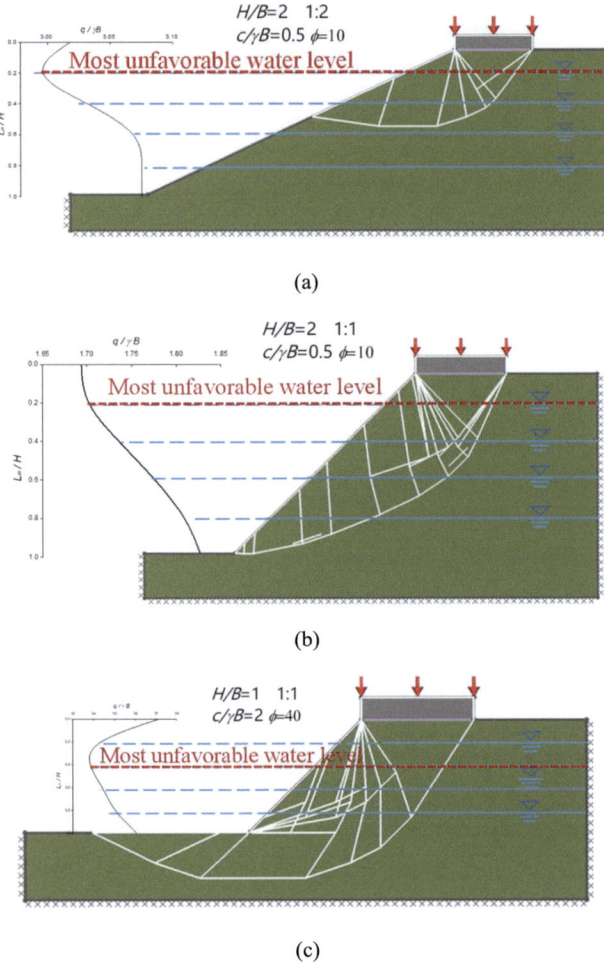

Fig. 6 The most unfavorable water level under the drawdown of slope water level after the reservoir has been rapidly drained condition: **a** surface failure; **b** toe failure; **c** base failure

of the foundation is decreasing at beginning and then proposes an increasing tendency. The most unfavorable water level decrease height is 0.2 times of the embankment height. Under the slow water level decrease mode, the change in bearing capacity is related to the slope. When the slope is steep, the ultimate bearing capacity gradually decreases with the water level, while with a lower slope, it first decreases and then increases.

(2) The failure mode of the foundation within the drawdown of slope water level after the reservoir has been rapidly drained will affect the position of the most unfavorable water level. When slope surface failure occurs, the most unfavorable depth of water level decrease is 0.2 times the slope height. When the slope surface

failure transitions to slope bottom failure, the most unfavorable depth of water level is 0.3 times the slope height.

Acknowledgements This research was supported by the Science and Technology Project of Hebei Provincial Department of Transportation (No. 201405).

References

1. Emerson DG, Macek-Rowland K (1986) Flood analysis along the little Missouri river within and adjacent to Theodore Roosevelt National Park, North Dakota (No. 86–4090). US Geological Survey. https://doi.org/10.3133/wri864090
2. Fang K, Tang H, Li C, Su X, An P, Sun S (2023) Centrifuge modelling of landslides and landslide hazard mitigation: a review. Geosci Front 14(1):101493. https://doi.org/10.1016/j.gsf.2022.101493
3. Gao Y, Zhu D, Zhang F, Lei GH, Qin H (2014) Stability analysis of three-dimensional slopes under water drawdown conditions. Can Geotech J 51(11):1355–1364. https://doi.org/10.1139/cgj-2013-0448
4. Zheng Y, Li J, Zheng X, Guo N, Yang G (2024) Pre-and post-failure behaviour of a dike after rapid drawdown of river level based on material point method. Comput Geotech 170:106269. https://doi.org/10.1016/j.compgeo.2024.106269
5. Zheng G, Guo Z, Zhou H, Tan Y, Wang Z, Li S (2024) Multibench-retained excavations with inclined–vertical framed retaining walls in soft soils: observations and numerical investigation. J Geotech Geoenvironmental Eng 150(5):05024003. https://doi.org/10.1061/JGGEFK.GTENG-11943
6. Ukritchon B, Keawsawasvong S (2018) A new design equation for drained stability of conical slopes in cohesive-frictional soils. J Rock Mech Geotech Eng 10(2):358–366. https://doi.org/10.1016/j.jrmge.2017.10.004
7. Li D, Li L, Xu L, Li C, Meng K, Gao Y, Yang Z (2022) Stability analysis of upstream and downstream dam slopes with water level drawdown using response surface function. Geotech Geol Eng 40(6):3107–3123. https://doi.org/10.1007/s10706-022-02082-0
8. Zheng G, Xia B, Zhou H, Zhao J, Yu X, Sun X, Du J (2020) Seismic bearing capacity of strip footings on ground reinforced by stone columns using upper-bound solutions. Int J Geomech 20(9):06020024. https://doi.org/10.1061/(ASCE)GM.1943-5622.0001794
9. Zheng G, Guo Z, Zhou H, He X (2024) Design method for wall deformation and soil movement of excavations with inclined retaining walls in sand. Int J Geomech 24(4):04024042. https://doi.org/10.1061/IJGNAI.GMENG-9029
10. Xia B, Zheng G, Zhou H, Yu X, Zhao J, Diao Y (2024) Stability analysis and optimization of concrete column-supported embankments in soft soil. Acta Geotechnica 19(5):2515–2531. https://link.springer.com/article/https://doi.org/10.1007/s11440-024-02302-2
11. Zhou H, Zhang J, Yu X, Hu Q, Zheng G, Xu H, Yang S (2024) Stochastic bearing capacity and failure mechanism of footings placed adjacent to slopes considering the anisotropic spatial variability of the clay undrained strength. Int J Geomech 24(4):04024022. https://doi.org/10.1061/IJGNAI.GMENG-9136
12. Viratjandr C, Michalowski RL (2006) Limit analysis of submerged slopes subjected to water drawdown. Can Geotech J 43(8):802–814. https://doi.org/10.1016/j.jrmge.2017.10.004

Open Access This chapter is licensed under the terms of the Creative Commons Attribution 4.0 International License (http://creativecommons.org/licenses/by/4.0/), which permits use, sharing, adaptation, distribution and reproduction in any medium or format, as long as you give appropriate credit to the original author(s) and the source, provide a link to the Creative Commons license and indicate if changes were made.

The images or other third party material in this chapter are included in the chapter's Creative Commons license, unless indicated otherwise in a credit line to the material. If material is not included in the chapter's Creative Commons license and your intended use is not permitted by statutory regulation or exceeds the permitted use, you will need to obtain permission directly from the copyright holder.

Rock Mass Deformation Analysis and Local Collapse Treatment of Powerhouse Slope of Wukuo Power Station

Hainian Shan, Feng Zhang, Han Zhang, and Jianhua Deng

Abstract The powerhouse slope of Wukuo Power Station presents significant engineering challenges due to its high height and fractured rock mass. During excavation, the slope has shown considerable deformation, particularly in the shallow rock layers. Furthermore, the area between the outlets of water conveyance tunnels 6 and 7 experienced local collapse due to inadequate systematic anchor support. This research work aims to design effective support measures to ensure slope stability, monitor deformation accurately, and analyze the causes of local collapse. We employed finite element analysis (FEA) and computational fluid dynamics (CFD) for safety monitoring and data analysis. These techniques were used to simulate stress distribution, deformation, fluid flow, and erosion patterns. The monitoring data exposed that deformation was predominantly in the shallow rock mass, with effective control achieved through targeted support measures. FEA and CFD analyses demonstrated the effectiveness of these measures in improving slope stability. This research highlights the importance of selecting suitable support measures for similar projects, contributing insights into reducing construction costs and time while ensuring safety. The amalgamation of FEA and CFD proved to be better when compared to traditional methods by showing significant reductions in stress, displacement, and erosion rates.

Keywords Analysis and treatment · Collapse · Computational fluid dynamics · Excavation · Finite element analysis · Power station · Powerhouse slope · Rock mass deformation · Support measures

---

H. Shan (✉) · J. Deng
NARI Group Corporation (State Grid Electric Power Research Institute), Nanjing, Jiangsu, China
e-mail: shanhainian@sgepri.sgcc.com.cn

F. Zhang · H. Zhang
China Three Gorges Corporation, Wuhan, Hubei, China

1 Introduction

The expansion project of Wuqiangxi Hydropower Station [1] mainly employs the regulation capacity of Wuqiangxi Reservoir and abundant incoming water in summer to generate electricity to improve the utilization of hydropower resources. Wuqiangxi Hydropower Station has an installed capacity of 1200 MW and a water utilization rate of 80.94%. In addition, the expansion project introduces an installed capacity of 500 MW (i.e., 2 × 250 MW), with the mean annual energy production being 558.3 million kWh, resulting in an increase in the water utilization rate to 89.38% after the expansion. Simultaneously, its water conveyance and power generation system are arranged on the right bank of the riverbed. With the water conveyance tunnel waterfront ground powerhouse serving as the overall layout, it utilizes the outlet facilities of the existing power station to transport the power generation offered by the expansion project to the outside world.

Taking into account the steep slope, high height, and sub-optimal rock mass integrity of the powerhouse slope below the road around the reservoir [2], the rock mass typically deforms to a great extent during the underlying excavation, exhibiting a continuous change in the over-alert value. During the later stage of excavation, the rock mass between the water conveyance tunnels 6 and 7 with an elevation above 33 m appears to have collapsed locally. To ensure the safety of the project, a series of supporting measures, such as anchor piles and fitting-slope concrete, are added based on the analysis of safety monitoring data [3]. During the later excavation, the deformation is effectively controlled, thus ensuring the safety of the project.

2 Literature Review

This section investigates the related works in rock mass deformation analysis and local collapse treatment of powerhouses that use computer technologies such as finite element analysis (FEA), building information modeling (BIM), and computational fluid dynamics (CFD). Zhao et al. [4] conducted an investigation in the deformation mechanisms of rock masses with WIZ at the Baihetan Hydropower Station. They utilized an array of in situ monitoring techniques that include microseismic (MS) monitoring, digital borehole cameras, and multipoint extensometers. Their findings emphasized that large deformations in rock masses were predominantly controlled by the characteristics of the WIZ, such as mineral composition, water physical properties, and weak structures. This approach provided valuable information on the spatiotemporal evolution of rock deformation and internal fracture propagation.

Ma et al. [5] focused on the stability of the surrounding rock mass at the Jinping I Hydropower Station. They considered complex geological conditions with high ground stress and significant unloading depths. They proposed an energy dissipation theory to better understand the deformation and failure mechanisms. Experimental results from borehole television, acoustic wave testing, and numerical simulations demonstrated elaborate facts about the relaxation characteristics and stability of the rock mass. Li et al. [6] emphasized the importance of MS monitoring in understanding rock mass deformation in large underground caverns. They were able to capture microcracks induced by excavation activities by establishing a MS monitoring system at the Houziyan Hydropower Station. Their dynamic analysis model with MS data proved vital in predicting rock mass deformations and ensuring the stability of the underground caverns.

Zhu and Wang [7] and Dhawan et al. [8] explored finite element analysis (FEA) for jointed rock masses and underground openings. Zhu and Wang developed an equivalent continuum model that integrates isotropic rock elements with anisotropic jointed rock elements. Their nonlinear FEA simulations effectively predicted the mechanical properties of jointed rock masses which are validated by physical model tests. Dhawan et al. compared 2D and 3D FEA for the Koyna hydroelectric project. This work concludes that 3D analysis provided more accurate deformation predictions. This is essential for evaluating the stability of underground openings in heterogeneous rock masses. Zhang et al. [9] introduced the smoothed particle hydrodynamics (SPH) method for simulating the collapse processes of rock slopes. Their mesh-free approach addressed limitations in traditional LEM and FEM methods. This offers a robust solution for analyzing damage and failure processes in rock slopes. This method makes use of the Drucker-Prager Mohr–Coulomb strength criterion for collapse process that proved effective in predicting deformation and failure in various rock-slope models.

Yang et al. [10] studied the failure characteristics of tunnels in expansive rock strata using numerical simulations based on a case study in Dugongling, China. They identified expansion and softening of the surrounding rock as primary causes of tunnel lining cracking. Their research presented a formula for calculating expansion force and proposed several reinforcement schemes to alleviate the deformation effects. Li et al. [11] proposed a BIM-based framework for the digital design and stability simulation of large underground caverns. Their automatic numerical simulation framework facilitated the efficient transformation of 3D parametric models into numerical models. This enhances the calculation efficiency and accuracy of geotechnical analyses. This method was illustrated through a case study of the Suki Kinari underground powerhouse caverns.

Qiao et al. [12] examined the collapse mechanism of a deeply buried hard rock tunnel using field investigations and theoretical calculations. They identified shear slip along structural surfaces and proposed a method that involves advance bolts, steel arch supports, and step excavation to address the collapse issues effectively. Feng et al. [13] discussed the integration of artificial intelligence (AI) in rock mechanics and engineering. This work proposed a system based on the Metaverse concept for intelligent recognition and evaluation of geological structures and stresses. The case studies presented in this work demonstrated the capability of the system to solve complex engineering problems through enhanced data-driven information and digital twin technologies.

Pu et al. [14] utilized 3D mobile laser scanning for deformation analysis of a roadway tunnel in soft swelling rock mass. Their method involved advanced point cloud data processing techniques to spot and visualize roadway deformations. This offers a reliable tool for monitoring and managing deformations in similar underground structures.

3 Methodology

The main objective of this work is to ensure the stability and safety of the powerhouse slope at Wukuo Power Station during and after excavation activities. It is critical to implement and monitor effective support measures with the given high height, steep slope, and broken rock mass characteristics. The scope of this monitoring project involves the design, implementation, and analysis of slope safety measures using advanced monitoring techniques.

Monitoring the powerhouse slope is necessary due to the following reasons:

Risk of Deformation: The excavation activities lead to a significant risk of rock mass deformation, which could compromise the structural integrity of the powerhouse slope.

Local Collapse Issues: Previous incidents of local collapses between water conveyance tunnels 6 and 7 highlight the need for continuous monitoring to prevent future occurrences.

Cost and Time Efficiency: Appropriate monitoring and timely intervention can help optimize the selection of support measures, reducing overall project costs and construction periods.

3.1 Slope Safety Monitoring Layout and Support Measures

The powerhouse slope safety monitoring project above and below the road around the reservoir primarily focuses on internal deformation and support stress. Specifically, the design of the E1-E1 typical monitoring section can be summarized as 2 sets of multipoint displacement meters for monitoring the internal deformation of slope rock mass as well as 8 anchor bolt stress meters for monitoring the stress of anchor bolt support, with 2 groundwater level holes being arranged along different elevations [15]. Figure 1 illustrates the specific slope safety monitoring layout in detail.

The support measures of the excavated powerhouse slope employ suspended-net shotcrete and system anchor bolt, encompassing C25 shotcrete with a thickness of 10 cm as well as ¢25 system anchor bolt with a length of 4.5 m and a row spacing of 2 m. On the other hand, the support measures for the slumping block with an elevation of more than 33 m include anchor piles, anchor bolts, and fitting-slope concrete. In consideration of various factors such as rock mass characteristics, cost,

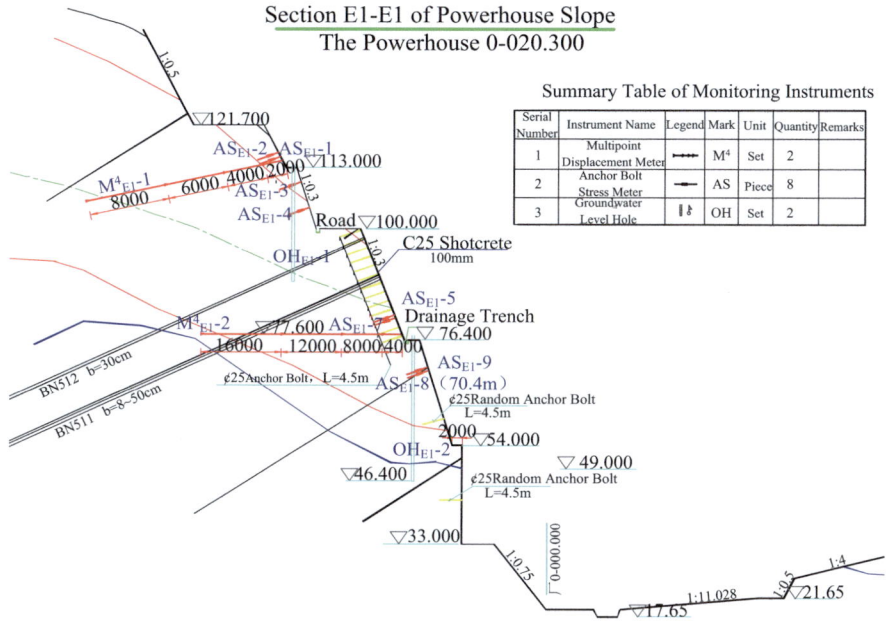

Fig. 1 Layout of powerhouse slope monitoring section

and construction period, the anchor cable support measure to be adopted in the initial design stage is finally canceled.

3.2 Safety Monitoring Data Analysis Results of Slope at Water Inlet and Outlet

Safety monitoring of slopes at water inlets and outlets is an important aspect in ensuring the stability and integrity of hydraulic structures. These slopes are subjected to various environmental and operational stresses that can compromise their stability. Implementing a robust monitoring system helps in early detection of possible slope failures that allows for timely intervention and mitigation. This process uses finite element analysis (FEA) and real-time data acquisition systems to assess and manage slope stability. The methodology for safety monitoring and data analysis is structured into several systematic steps which is illustrated in Fig. 2.

This process begins with the data collection phase, where sensors such as inclinometers, piezometers, and strain gauges are strategically installed on the slope. Initial baseline data is recorded to establish normal conditions. The process continues with continuous monitoring, where real-time data is acquired from the sensors and transmitted to a central monitoring system. This data undergoes processing that

Fig. 2 Flow of safety monitoring and data analysis

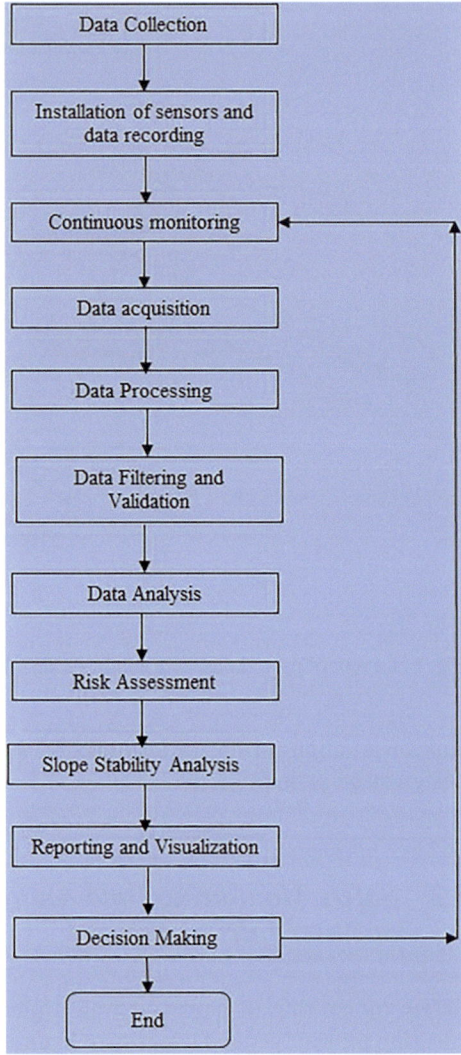

involved filtering out noise and validating for consistency and irregularities. In the data analysis stage, trends in the data are analyzed, and comparisons against predefined safety thresholds are made to identify potential risks. The risk assessment involves conducting slope stability analyses using computational models like FEA to evaluate the stability of the slopes followed by hazard identification based on the analysis results. The outcomes are compiled into detailed reports during the reporting phase, together with visualizations such as graphs, charts, and 3D models. In the decision-making step, the safety of the slope is evaluated, and necessary measures are suggested and implemented if risks are detected. Finally, a feedback loop ensures continuous improvement, where the monitoring plan is updated based on the analysis outcomes and feedback, and enhancements are made to the monitoring system and data analysis methodologies. This approach ensures monitoring and timely involvement to maintain slope stability and safety at water inlets and outlets.

3.3 Multipoint Displacement Meter

The monitoring results of the M4e1-2 multipoint displacement meter with an elevation of ▽77.6 m below the road of the E1 section demonstrate that the deformation of rock mass is closely related to the excavation in this part. Particularly, the deformation is mainly reflected in the surface layer, whereas the deformation of rock masses in the middle and deep layers is relatively limited. Along with the end of excavation and the introduction of support measures, the deformation gradually slowed down and stabilized.

Rock mass deformation was mainly concentrated from October 2020 to January 2021. More specifically, the monthly increment values of rock mass deformation from October to December in 2020 were 2.98 mm, 1.63 mm, and 1.15 mm, respectively, with the monthly increase values in the later period being all within 1 mm. Since February 2021, the rock mass deformation has shown a significant trend of stabilization. Several findings can be inferred from these observations. Firstly, the deepening of excavation depth, coupled with the implementation of support measures, led to the continuous growth of deformation. However, the deformation was significantly reduced compared with the initial excavation. Secondly, the baseplate concrete pouring into the foundation pit of the powerhouse caused a more obvious deformation to stabilize. Lastly, the cumulative deformation of the orifice is 7.78 mm after the deformation stabilizes, whereas that of the measuring point with a depth of 4 m is 0.38 mm. It implies that the deformation is mainly reflected in the surface layer, while the rock mass deformation in the middle and deep layers is relatively limited, indicating an appropriate selection of support measures [16]. Table 1 shows the statistics of rock mass deformation (Orifice) variation of powerhouse slope.

3.4 Anchor Bolt Stress

Four anchor bolt stress meters measuring points are installed in the area below the road around the reservoir, of which the ASe1-7 measuring point is located at the upper part of the M4e1-2 multipoint displacement meter (i.e., ▽78.4 m), while ASe1-8 and ASe1-9 measuring points are located at the lower part of the M4e1-2 multipoint displacement meter (i.e., ▽70.5 m). Moreover, the change in the anchor bolt stress is closely associated with the excavation process, with the stress and deformation increasing synchronously. As the excavation ends, the stress tends to be

Table 1 Statistics of rock mass deformation (Orifice) variation of powerhouse slope

Years	2020				2021					
Measuring points	September	October	November	December	January	February	March	April	May	June
Monthly increment value (mm)	0.72	2.98	1.63	1.15	0.77	0.18	0.28	0.02	0.01	0.03
Monthly deformation rate (mm/d)	/	0.10	0.05	0.04	0.03	0.01	0.01	0.00	0.00	0.00

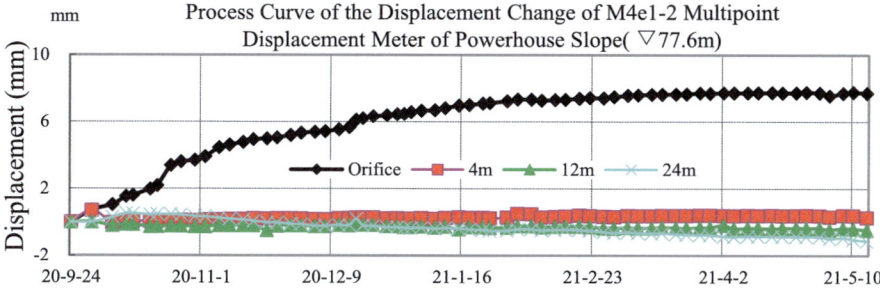

Fig. 3 Measurement process curve of M^4_{e1}-2 multipoint displacement meter of powerhouse slope

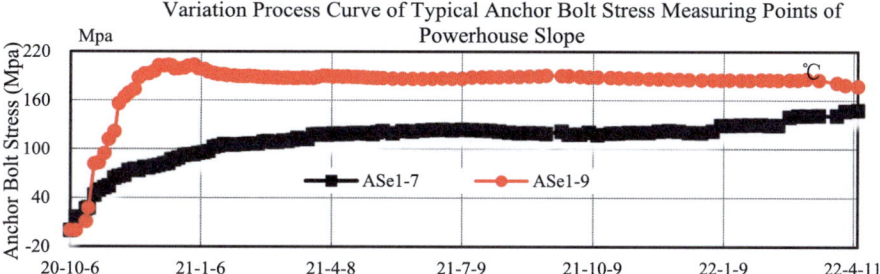

Fig. 4 Duration process curve of typical anchor bolt stress measuring points of powerhouse slope

stable significantly, which is primarily reflected at ASe1-7 and ASe1-9 measuring points. The process curve of the displacement change is illustrated in Figs. 3 and 4.

Likewise, the stress variation of the anchor bolt support was primarily concentrated from October 2020 to January 2021. Notably, the 2 measuring points exhibited significant incremental variation in October and November, with their variation being consistent with the deformation of the multipoint displacement meter measuring point. These observations reveal several findings. More precisely, for one thing, the anchor bolt support measures change synchronously with the deformation during excavation, playing a relatively effective supporting role. Secondly, the stress variation showcases a significant stabilization trend with the baseplate concrete pouring of the foundation pit of the powerhouse. Table 2 expresses the statistics of the monthly variation of stress at typical anchor bolt measuring points on the powerhouse slope.

4 Experiments

The experimentation is done to analyze the stability of water inlet and outlet. The experimental setup for the comparative analysis of the proposed work against existing models involved a series of detailed simulations. Finite element analysis (FEA) was conducted using ANSYS, where material properties such as Young's modulus,

Table 2 Statistics of monthly variation of stress at typical anchor bolt measuring points of powerhouse slope

Years	2020				2021					
Measuring points	September	October	November	December	January	February	March	April	May	June
AS_{e1}-7/MPa	/	52.35	24.36	16.30	11.78	3.49	9.96	1.31	2.84	1.64
AS_{e1}-9/MPa	27.30	141.78	33.28	− 12.99	− 4.16	1.02	− 12.44	− 0.51	− 6.72	− 8.45

Poisson's ratio, and density were specified for the soil and rock. Boundary conditions included fixed supports and gravitational loading, with external loads applied to simulate real-world conditions. Results were obtained by calculating Von Mises stress, principal stresses, total displacement, displacement vectors, and factors of safety. A fine mesh was generated to ensure precision in stress and displacement simulations, with realistic boundary conditions and loads applied, including gravitational forces. For computational fluid dynamics (CFD) analysis, ANSYS Fluent was utilized to simulate fluid flow, setting up the slope area and surrounding water bodies as the computational domain with structured mesh and appropriate inlet and outlet conditions to analyze erosion patterns and flow velocities.

The experimental results were illustrated from Fig. 5a–e. Figure 5a illustrates the comparison of maximum stress values between the proposed work and existing models. The results show that the proposed method shows a maximum stress of 50 MPa, which is significantly lower than that of the existing works. This reduction in stress indicates that the proposed support measures are more effective in distributing the stress more evenly across the slope, thereby reducing the likelihood of structural failure. Figure 5b presents the comparison of maximum displacement values. The proposed method demonstrates a maximum displacement of 5 cm, which is considerably lower than the displacement observed in existing models. Lower displacement values suggest that the proposed slope design is more stable and less prone to significant movement. Figure 5c compares the flow velocity for different slope designs. The proposed design achieves a flow velocity of 2.5 m/s, which indicates improved flow management compared to other models. Figure 5d shows the erosion rate comparison between the proposed and existing models. The proposed work has an erosion rate of 0.3 cm/year, which is significantly lower than that of the existing designs. A lower erosion rate indicates that the proposed support measures are more effective in protecting the slope from erosive forces. Figure 5e illustrates the predicted failure rates for different slope designs. The proposed method shows a much lower predicted failure rate compared to existing models. This indicates that the proposed design is more robust and reliable in preventing slope failure.

The lower maximum stress indicates the effectiveness of the support measures in distributing stress more evenly, reducing the risk of structural failure. The lower displacement values prove that the slope is more stable and less prone to significant movement, which is important for maintaining structural integrity. The reduced erosion rate highlights the effectiveness of the measures in protecting the slope from erosion, which can weaken slope stability over time. The controlled flow velocity indicates that the design is effective in managing water flow, reducing the risk of erosion and structural damage due to uncontrolled water movement.

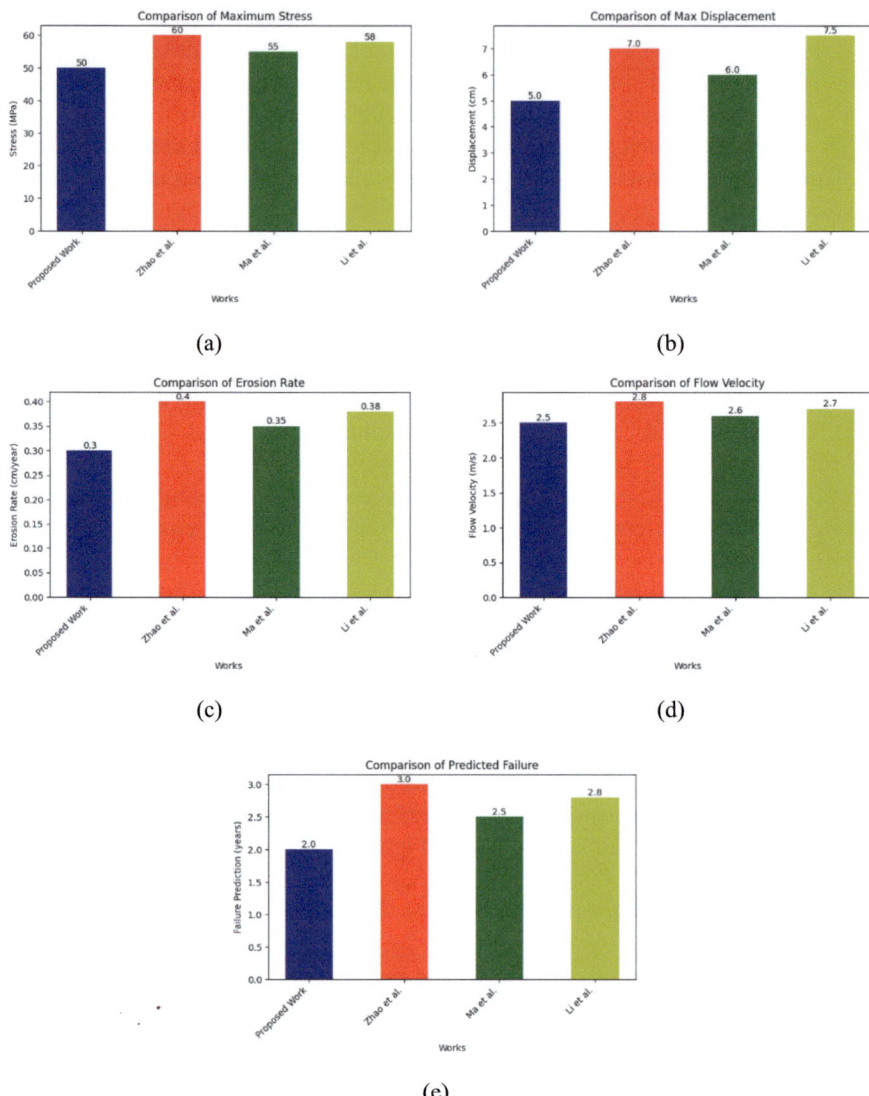

Fig. 5 Comparison of maximum stress, maximum displacement, flow velocity, erosion rate, predicted failure for stability analysis

5 Analysis of Slope Support Measures and Collapse Causes

(1) The powerhouse slope presents a broken rock mass with poor integrity. Particularly, the local rock mass exhibits a fault structure. During the excavation, the shallow layer of rock mass showcases obvious deformation to the free surface, which is greatly influenced by excavation blasting and unloading.
(2) The slope presents a high elevation, which can reach 80 m at its maximum.
(3) The suspended-net shotcrete, in conjunction with the system anchor bolt, is adopted as the main support measure, while the local shallow collapse parts are supported by fitting-slope concrete and an anchor bolt. The overall smooth excavation construction demonstrates the appropriate selection of safe and controllable support measures.
(4) The analysis of safety monitoring data reveals that the rock mass deformation is mainly manifested as the shallow rock mass within 4 m, indicating that the anchor bolt support measure with a length of 4.5 m is appropriately selected. Meanwhile, deformation is closely related to excavation. With the excavation away from the tunnel face, the deformation immediately slows down.
(5) The rock mass between the outlets of the water conveyance tunnels 6 and 7 was solely supported by shotcrete in the initial stage. Failure to design the system anchor bolt leads to a great decrease in the stability of shallow rock masses, ultimately inducing local collapse.

6 Conclusion

Taken together, the powerhouse slope of Wukuo Power Station is facing challenges posed by the high excavation height, broken rock mass, and fault structure. During the excavation construction, the shallow rock mass of the slope exhibits great deformation, whereas the middle and deep rock masses show relatively limited deformation. In this regard, the support measures of the suspended-net shotcrete and the system anchor bolt are appropriately selected to address these challenges. Moreover, the collapse of the shallow rock mass between the outlets of the water conveyance tunnels 6 and 7 can be fundamentally attributed to the failure to design systematic anchor support for the rock mass in this part. Subsequently, the rock mass in this part is supported by fitting-slope shotcrete and anchor bolts. The implementation of these support measures effectively controls the rock mass deformation, thereby ensuring the safety of foundation pit excavation. Importantly, other similar projects can learn from the construction experience of the Wukuo Power Station, further reducing the cost and saving the construction period on the premise of ensuring safety. The amalgamation of FEA and CFD in the safety monitoring and analysis of the Wukuo Power Station's powerhouse slope has proven to be highly effective. FEA simulations highlighted critical stress zones and deformation patterns, allowing for the optimization of support measures. CFD analysis unearthed fluid dynamics and erosion impacts. The empirical evaluation confirmed the superiority of the proposed system over existing methodologies with significant reductions in stress, displacement, and erosion rates.

This research work mainly focuses on short-term monitoring and immediate responses. There is a need for long-term data to understand the ongoing behavior of the slope and the effectiveness of the support measures over extended periods. The impact of environmental factors such as weather conditions and water infiltration on slope stability is not extensively analyzed. These factors can significantly influence the deformation and stability of the slope. In future, an effort will be made to explore the integration of more advanced and automated monitoring systems, such as real-time remote sensing technologies and intelligent algorithms to enhance data collection and analysis. Effort also will be made to develop a risk assessment framework that includes probabilistic methods to investigate uncertainties in geotechnical properties and environmental conditions.

Acknowledgements This research was sponsored by China Three Gorges Corporation research projects (202103469) and China Three Gorges Construction Engineering Corporation research projects (JGAJ030222003).

References

1. Li XT, Qu Y (2019) Commencement of first expansion project of Wuqiangxi hydropower station. Sichuan Hydropower 01:108
2. Wang CM (1994) Major geological problems of Wuqiangxi hydropower station. Water Power 11:19–23
3. Wang J, Li XX, Wang CM (2002) Deformation mechanism and engineering treatment of high slope in hydropower projects. J North China Univ Water Resour Electr Power 04:26–29
4. Zhao JS, Chen BR, Jiang Q et al (2024) In-situ comprehensive investigation of deformation mechanism of the rock mass with weak interlayer zone in the Baihetan hydropower station. Tunn Undergr Space Technol 148:105690
5. Ma K, Zhang J, Zhou Z et al (2020) Comprehensive analysis of the surrounding rock mass stability in the underground caverns of Jinping I hydropower station in Southwest China. Tunn Undergr Space Technol 104:103525
6. Li B, Xu N, Dai F et al (2019) Dynamic analysis of rock mass deformation in large underground caverns considering microseismic data. Int J Rock Mech Min Sci 122:104078
7. Zhu W, Wang P (1993) Finite element analysis of jointed rock masses and engineering application. Int J Rock Mech Min Sci Geomech Abstr 30(5):537–544
8. Dhawan KR, Singh DN, Gupta ID (2002) 2D and 3D finite element analysis of underground openings in an inhomogeneous rock mass. Int J Rock Mech Min Sci 39(2):217–227
9. Zhang X, Song X, Wu S (2022) Simulation of collapse failure process of rock slope based on the smoothed particle hydrodynamics method. Front Earth Sci 10
10. Yang M, Liu N, Li N et al (2021) Failure characteristics and treatment measures of tunnels in expansive rock stratum, Front Earth Sci 9
11. Li H, Chen WZ, Tan XJ et al (2022) Digital design and stability simulation for large underground powerhouse caverns with parametric model based on BIM-based framework. Tunn Undergr Space Technol 123:104375
12. Qiao S, Cai Z, Tan J et al (2020) Analysis of collapse mechanism and treatment evaluation of a deeply buried hard rock tunnel. Appl Sci 10(12):4294

13. Feng XT, Yang CX, He BG et al (2024) Artificial intelligence technology in rock mechanics and rock engineering. Deep Resour Eng 2024:100008
14. Pu J, Yu Q, Zhao Y et al (2024) Deformation analysis of a roadway tunnel in soft swelling rock mass based on 3D mobile laser scanning. Rock Mech Rock Eng
15. Wu ZR (2003) Safety monitoring theory & its application of hydraulic structures. Version 1. Higher Education Press, Beijing
16. Luo HY, Wang J (2000) Displacement monitoring and deformation characteristics of ship lock slope at left bank of Wuqiangxi hydropower station. Hydropower Autom Dam Monit 03:22–24

Open Access This chapter is licensed under the terms of the Creative Commons Attribution 4.0 International License (http://creativecommons.org/licenses/by/4.0/), which permits use, sharing, adaptation, distribution and reproduction in any medium or format, as long as you give appropriate credit to the original author(s) and the source, provide a link to the Creative Commons license and indicate if changes were made.

The images or other third party material in this chapter are included in the chapter's Creative Commons license, unless indicated otherwise in a credit line to the material. If material is not included in the chapter's Creative Commons license and your intended use is not permitted by statutory regulation or exceeds the permitted use, you will need to obtain permission directly from the copyright holder.

Research on Deformation Control Technology for Filling and Mining of High and Steep Slopes in Guizhou Mountainous Areas

Yu Wu, Xiaohu Zheng, Qing Liu, Yao Zhong, Qianyong Lv, Jie Huang, and Dandan Liu

Abstract Guizhou mountainous areas have numerous high and steep slopes with underground mineral resources, and frequent underground mining activities can easily cause high-level landslides on the slopes. Therefore, starting from the root problem of underground mining, it is proposed to use backfill mining methods to solve such disasters. This article establishes numerical models for five typical mining-induced landslides in mountainous areas of Guizhou. The mining numerical simulation software GDEM simulates the evolution of fractures and displacement changes in slopes with different filling rates and proposes a reasonable filling rate scheme. The research results show that when the filling rate is between 10 and 90%, the overall crack development scale of the slope decreases with the increase of the filling rate, and the displacement of the slope also decreases overall with the increase of the filling rate. According to the scale of crack development and monitoring point data, it can be concluded that when the filling rate reaches 50%, the deformation of the entire slope and surface can be controlled.

Keywords Slope · Underground mining · Landslide · Filling · Deformation

1 Introduction

Coal is the main energy source in China and plays an important role in ensuring energy supply and security. Among them, the underground coal resources in the mountainous areas of Guizhou are abundant and distributed on numerous steep slopes. However,

---

Y. Wu (✉) · X. Zheng · Q. Liu · J. Huang
Bijie Power Supply Bureau of Guizhou, Power Grid Co., Ltd., CSG, Bijie, China
e-mail: 512740130@qq.com

Y. Zhong · Q. Lv
Electrical Science Institute of Guizhou Power Grid Co.,Ltd., CSG, Guiyang, China

D. Liu
Powerchina Guizhou Electric Power Engineering Co., Ltd, Guiyang, China

frequent underground coal mining activities can cause geological disasters such as overburden fractures, surface cracks, and landslides on high and steep slopes. For example, the Madaling landslide that occurred in Duyun, Guizhou on May 18, 2006 [1], the Pusa landslide in Nayong County, Guizhou on August 28, 2017 [2], the Panzhihua coal mine landslide in Shuicheng County, Guizhou in August 2019 [3], the Fa'er landslide in Guizhou on September 15, 2020 [4], and the Zhijin Baiyan landslide in Guizhou on May 8, 2022 [5], all of these high and steep slope landslides have caused huge losses of personnel and property.

Many scholars have conducted research on the high and steep slope landslides caused by underground mining, and have achieved rich results. Zhao Jianjun et al. [6] used the Madaling landslide in Guizhou Province as a geological prototype and used physical simulation methods to reproduce the entire process of landslide formation from incubation to occurrence. They analyzed the mechanism of landslide formation induced by heavy rainfall and found that its deformation and failure mechanism can be summarized as the overlying rock of goaf slope deformation rainfall-induced slope overall deformation evolution sliding surface expansion landslide occurrence. Xiong Shaozhen et al. [7] used the 3DEC discrete element numerical simulation method to simulate the mechanism, instability mode, and motion trajectory of the collapse of the Pusa steep slope under underground mining and summarized the deformation and failure process of Pusa collapse. Zhang Shunbo et al. [3] used PFC2D software to simulate the deformation trend, mechanical behavior, and fracture evolution process of the slope and overlying rock mass of Panzhihua Coal Mine in Shuicheng County under underground mining. They summarized that the geomechanical model of the slope can be divided into four stages: surface modification, plastic flow tensile cracking, bending tensile cracking, and collapse. The abovementioned scholars analyzed the deformation and failure mechanisms and modes of high and steep slopes in Guizhou mountainous areas after underground mining using numerical simulation and physical simulation methods.

In response to this serious and difficult problem of underground mining deformation, many scholars have also approached it from the perspective of overburden deformation. Lian Changjun [8] used FLAC^{3D} software to analyze and study the stress response mechanism and movement deformation law of overlying rock in strip mining goaf. By studying the stress response mechanism and movement deformation law of overlying rock in strip mining goaf, accurate prediction of surface subsidence in strip mining areas can be achieved. Zhan Yapeng [9] used similar model experiments as a means to study the movement law of overlying rock under pressure mining in the 9301 mining face and 470 m horizontal coal seam of Tiexin Coal Mine. Li Xingli [10] monitored the entire mining process of the paste-filling working face in Daming Mine based on its characteristics. Through data analysis, she mastered the laws of surface movement and overlying rock damage in paste-filling comprehensive mining. After paste-filling mining, the roof of the roadway will not collapse, and surface movement and deformation will be greatly reduced.

However, few scholars have proposed new prevention and control techniques starting from the root problem of underground mining. Therefore, this article proposes filling mining prevention and control techniques for slope landslides induced by mining in mountainous areas of Guizhou and conducting relevant research. As a green method, backfill mining plays a crucial role in controlling overburden and surface deformation. It not only has significant advantages in solving the problem of "three underground" coal mining but also has good deformation control effects on civil buildings such as transmission and substation towers built on the surface of mines. Filling the goaf with filling materials can reduce the subsidence space of the roof, play a role in supporting the roof, and effectively slow down deformation and damage phenomena such as fissure development, collapse, and surface subsidence of the overlying rock. Based on the high and steep slopes in Guizhou mountainous areas, analyze the evolution of fractures and displacement changes of slopes under different filling rates, and propose a reasonable filling rate scheme.

2 Numerical Model

2.1 GDEM Continuous Non-continuous Software Calculation Principle

The continuous discontinuous element method (CDEM) of GDEM software was proposed by the Institute of Mechanics of the Chinese Academy of Sciences. It is an explicit numerical method based on the generalized Lagrangian equation. This method can deeply integrate continuous and discontinuous numerical calculations, thus achieving the unity of finite element algorithm and discrete element algorithm. The expression is shown in Eq. (1):

$$\frac{d}{dt}\left(\frac{\partial L}{\partial v_i}\right) + \left(\frac{\partial L}{\partial u_i}\right) = Q_i \tag{1}$$

Adopting an incremental display algorithm for node resultant force calculation and node motion calculation, where the expression for node resultant force calculation is shown in Eq. (2):

$$F = F^E + F^e + F^c + F^d \tag{2}$$

In the formula, F is the joint force of nodes; F^E is the external force on the node; Fe is the finite element deformation force; F^c is the nodal force generated by the contact surface; F^d is the nodal damping force. The expression for calculating node motion is shown in Eq. (3):

$$\begin{cases} a = F/m, \quad v = \sum_{t=0}^{T_{\text{now}}} a\Delta t \\ \Delta u = V\Delta t, \quad u = \sum_{t=0}^{T_{\text{now}}} \Delta u \end{cases} \quad (3)$$

In the formula, α represents the node acceleration; V is the node speed; Δu is the incremental displacement of the node; u is the node displacement; m is the node quality; Δt is the calculation time step. The alternating calculation of Eqs. (2) and (3) is the explicit solution process.

3 Establishment of a Generalized Model

This study conducted in-depth geological investigations on five typical slope collapse cases in mountainous areas of Guizhou and combined relevant literature data. These cases include the Madaling landslide [1], Pusa landslide [2], Panzhihua coal mine landslide [3], Fa'er landslide [4], and Baiyan landslide [5]. Through statistical analysis of these cases, we have drawn the following conclusions, as shown in Table 1.

The height range of these slopes is from 160 to 876 m, with an average height of approximately 400 m. The terrain features are characterized by a "steep top and gentle bottom," with a slope of 50–90 ° in the middle and upper parts, and a slope of 6–50 ° in the lower part. The dip angle of the rock layer ranges from 3 ° to 35 °, and there are usually 3–6 layers in the goaf. The thickness of the coal seam ranges from 1.23–3.5 m, while the distance between coal seams is between 6–40 m. The distance of slope collapse is 300–1500 m.

In terms of lithological composition, the upper part is mainly composed of limestone from the Yongningzhen Formation of the Lower Triassic, with a thickness ranging from 15 to 300 m. The middle part is composed of sandstone, mudstone, and argillaceous sandstone from the Feixianguan Formation of the Lower Triassic, with a thickness ranging from 100 to 400 m. The lower part is mainly composed of coal-bearing strata from the Longtan Formation of the Permian System, including

Table 1 Statistics of 5 typical mining-induced slope slide characteristics in Guizhou Province

Disasters	Slope height/m	Middle-upper part Slope/°	Underpart Slope/°	Dip angle/°	The number of gob	Thickness of coal seam /m	Seam spacing /m	Slip distance /m
MaDaling	160	80–90	20–40	3–12	2	1.70–3.50	15–25	1500
Pusa	200	80–90	20–35	5–10	3	1.23–2.12	12–40	780
Panzhihua	876	50–85	45–50	20–35	5	1.56–2.78	6–35	300
Faer	577	55–63	6–20	12–20	6	1.41–2.33	9–22	500
Baiyan	250	55–63	6–20	6–15	3	1.45–2.30	9–21	1000

mudstone, argillaceous sandstone, and coal seams, forming a geological structure with a "steep top and gentle bottom."

Based on the above analysis, a generalized model is established as shown in Fig. 1, with a height of 400 m, a length of 800 m, and a rock dip angle of 15 °. The lithology of the upper part of the slope is limestone, the middle part is argillaceous sandstone, and the lower part is mudstone. The thickness of limestone, mudstone and siltstone is 100 m, 150 m and 150 m respectively. The upper, middle, and lower slopes are 80 °, 50 °, and 10 °, respectively, highlighting the characteristic of steep slopes at the top and gentle slopes at the bottom in mountainous areas of Guizhou. The spacing between coal seams is set at 25 m, with 3 layers and a coal seam thickness of 3 m. The physical parameters of each lithology are shown in Table 2.

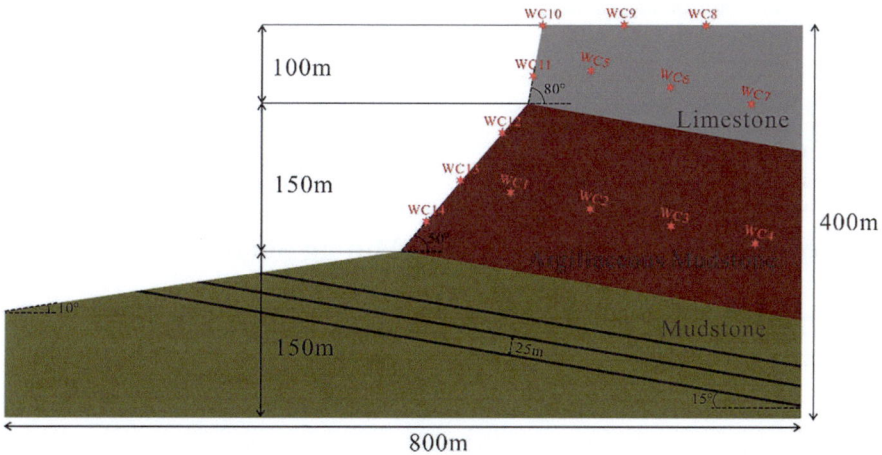

Fig. 1 Generalized slope model

Table 2 Physical and mechanical parameters of each rock layer

Name	Density (g/cm^3)	Compressive strength (MPa)	Tensile strength (MPa)	Poisson's ratio	Elastic modulus (GPa)	Cohesion (MPa)	Internal friction angle (°)
Limestone	2.85	84.20	9.36	0.29	16.1	13.2	36
Argillaceous siltstone	2.32	63.58	6.25	0.35	11.9	16.6	32
Mudstone	1.93	21.15	1.93	0.42	1.7	3.1	24
Coal	1.72	19.70	1.81	0.22	1.5	2.9	23

Table 3 Physical parameter table of filling material

Name	Density (g/cm^3)	Compressive strength (MPa)	Tensile strength (MPa)	Poisson's ratio	Elastic modulus (GPa)	Cohesion (MPa)	Internal friction angle (°)	Dilation angle (°)
Backfill	1.21	9.53	0.15	0.38	3.2	0.6	22	15

4 Filling Plan

After establishing the corresponding numerical model, simulation studies were conducted on a coal seam mining height of 3 m and a filling height of 2 m. Analyze the fracture evolution and displacement characteristics of high and steep slopes under different filling rates (10–90%, with each 10% increment), and determine a reasonable filling rate. The filling method is to fill from the middle of the goaf to both sides. The physical parameters of the filling material refer to relevant literature, as shown in Table 3.

5 Deformation Results of Backfill Mining Slope

5.1 Development Law of Slope Fractures Under Different Filling Rates

The crack development diagram of the slope under different filling rates is shown in Fig. 2. As shown in the figure, with the increase of filling rate between 10 and 40%, there is no significant change in the development degree of limestone fractures in the upper part of the slope within this range; however, within this filling rate range, the deep fractures of limestone will extend from the bottom to the front edge, which may lead to the cutting of the front rock mass and the formation of potential landslide bodies. Especially at filling rates of 30 and 40%, this phenomenon is more pronounced. Between 50 and 90% of the filling rate, as the filling rate increases, the degree of fractures in limestone increases slightly, but the fractures do not extend to the front edge of the slope. The number of developed fractures in the sandstone and mudstone in the middle and lower parts of the slope generally decreases with the increase of filling rate between 10 and 90%. At a filling rate of 60%, the number of fractures in the middle and lower parts of the slope decreases more significantly.

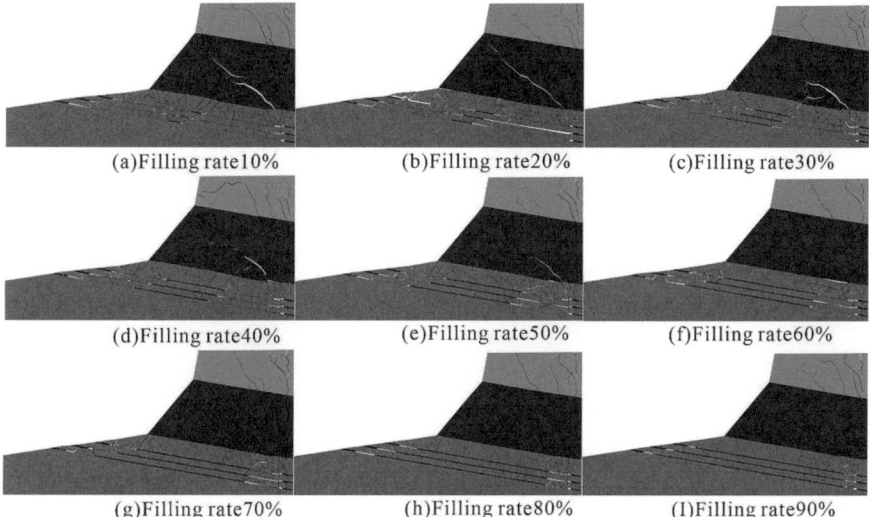

Fig. 2 Development law of slope fractures under different filling rates

5.2 Characteristics of Slope Displacement Change Under Different Filling Rates

In Fig. 3, as the filling rate increases, the maximum displacement value of the slope shows a decreasing trend. When the filling rate is 10%, the displacement value reaches its maximum of 9.312 m. When the filling rate increases to 90%, the displacement value significantly decreases to 4.227 m. This indicates that an increase in the filling rate helps to reduce the displacement that occurs on slopes. Although the height of the filling body is lower than the mining height of the coal seam, the displacement range has not significantly expanded due to the increase in filling rate. On the contrary, the displacement values within the displacement-affected area generally decrease with the increase in the filling rate. The results indicate that by adjusting the filling rate, the displacement of the slope can be controlled to a certain extent, thereby improving the safety of mining operations.

This article deeply analyzes the displacement characteristics of slopes under different filling rates by setting monitoring points on the overlying rock, slope top, and slope surface of goaf, as shown in Fig. 1. As shown in the monitoring curve in Fig. 4a, when the filling rate increases from 10 to 20%, the combined displacement values of WC1 and WC2 monitoring points significantly decrease, from 5.6 m and 4.8 m to 3.4 m and 3.3 m, respectively, while the displacement values of other monitoring points remain relatively stable. As the filling rate further increases to 30%, the displacement values of most monitoring points in the overlying rock generally decrease. However, as the filling rate increased from 30 to 40%, the displacement values of WC4 and WC7 continued to decrease, while the displacement values of

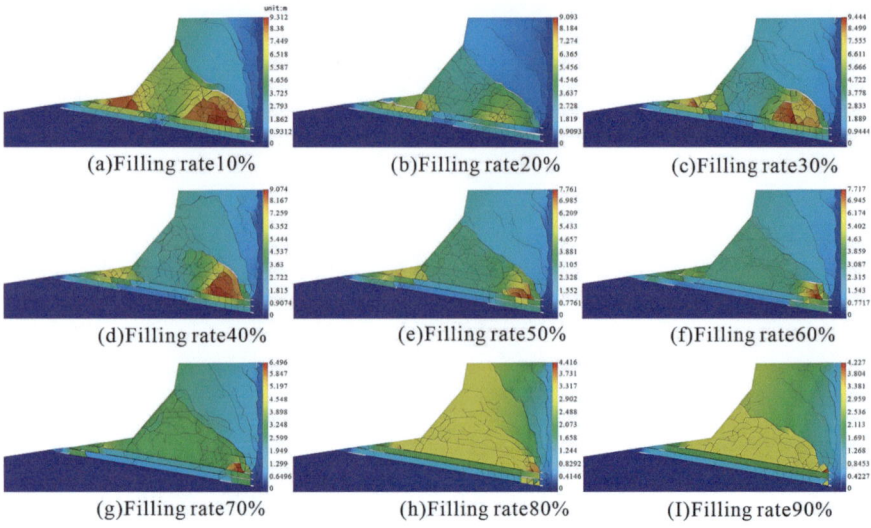

Fig. 3 Cloud map of slope displacement under different filling rates

WC1, WC3, WC5, and WC6 increased. This may be related to the stress concentration caused by the supporting effect of the filling body, resulting in uneven local stress distribution and increased displacement at these positions. When the filling rate increases to 50%, the displacement values of most monitoring points reach their lowest point. When it continued to increase to 60%, there were slight fluctuations in the displacement values of WC3 and WC6, while other monitoring points showed little change. During the process of increasing the filling rate from 60 to 80%, the displacement values of most monitoring points increased. Finally, when the filling rate reached 90%, the displacement values of WC2, WC3, and WC5 showed a significant decrease, while the displacement values of the remaining monitoring points remained relatively stable. These results indicate that the increase in filling rate has multiple effects on slope displacement, and it is necessary to comprehensively consider filling rate, stress distribution, and geological conditions to achieve slope stability and safe mining.

Figure 4b shows the displacement monitoring curves of the top and surface of the slope, revealing the influence of changes in filling rate on displacement. In the initial stage, when the filling rate increased from 10 to 20%, the displacement values of slope monitoring points WC12, WC13, and WC14 significantly decreased, from 5.8 m, 5.8 m, and 6.7 m to 3.4 m, 3.5 m, and 3.8 m, respectively. Meanwhile, other monitoring points also recorded a slight decrease in displacement. As the filling rate continues to increase to 50%, the displacement values of most monitoring points at the top and surface of the slope decrease overall, although some monitoring points have experienced relative increases in displacement due to uneven stress distribution. Especially when the filling rate is 50%, the displacement values of most measuring points reach their lowest point. Further, increase the filling rate to 70%, and the displacement

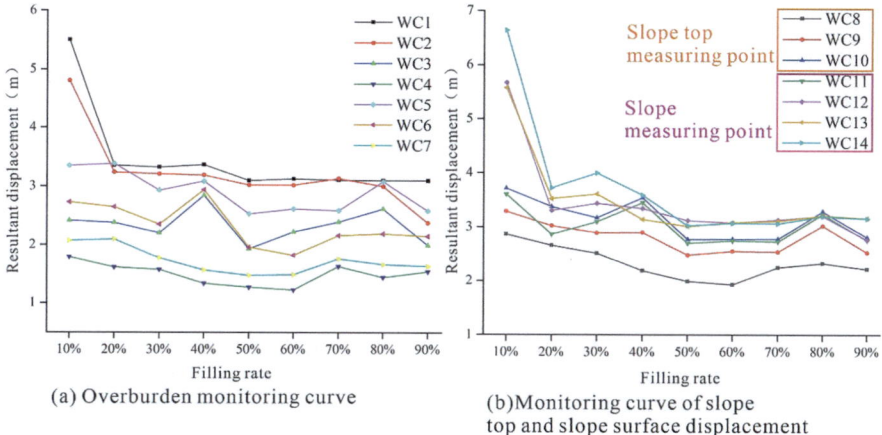

Fig. 4 Monitoring curves of displacement of overlying rock, slope top, and slope surface under different filling rates

values of most monitoring points remain stable, with only the WC8 monitoring point recording a small increase. When the filling rate increases from 70 to 80%, there is a slight increase in displacement values at monitoring points WC8, WC9, WC10, and WC11, while there is little change at other monitoring points. At a filling rate of 90%, the displacement values of monitoring points WC8, WC9, WC10, WC11, and WC12, which had previously increased at 80%, decreased, while the displacement values of monitoring points WC13 and WC14 remained unchanged. Comprehensive analysis shows that as the filling rate increases, the displacement trends of the overlying rock, slope top, and slope surface monitoring points are roughly the same. During the process of increasing the filling rate from 10 to 50%, the supporting effect of the filling material may lead to uneven local stress distribution, resulting in an increase in displacement values at certain measuring points. However, the displacement values of most measuring points reach their lowest point when the filling rate is 50%. When the filling rate increases from 50 to 60%, the displacement value of the measuring point does not change significantly. Subsequently, the displacement values of most measuring points increased from 60 to 80%. Finally, when the filling rate reached 90%, the displacement values of the previously added measuring points decreased, while the other measuring points remained stable.

5.3 Determination of Reasonable Filling Rate

Based on the displacement values of all monitoring points under different filling rates mentioned above, sum them up and take the average to make a fitting curve, as shown in Fig. 5. The fitting value R^2 is 0.9382, indicating good reliability of the fitting curve; the fitting formula is:

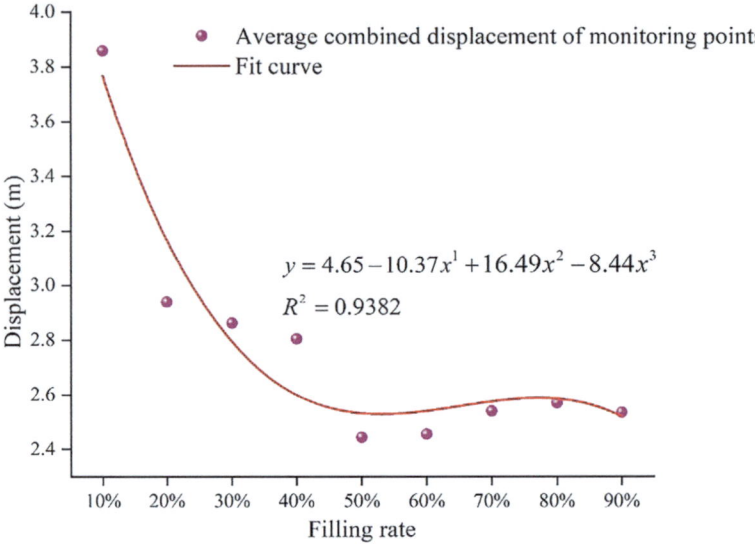

Fig. 5 Fitting curve of average displacement of monitoring points under different filling rates

$$y = 4.65 - 10.37x^1 + 16.49x^2 - 8.44x^3 \qquad (4)$$

As shown in the figure, the displacement value of the fitted curve shows a small increase at filling rates of 70 and 80%, but its numerical variation is within the range of 0.1 m. The overall fitting curve decreases with the increase of filling rate. The fitting curve shows a negative correlation between the filling rate of 10 and 50%, and the curve is steep, indicating that the filling effect is good within this filling rate range. The average settlement displacement significantly decreases from the filling rate of 10 to 20%, and the curve is steepest. The average settlement displacement reaches its lowest value at the filling rate of 50 to 60%. From the above crack development pattern, it can be seen that when the filling rate is 50%, the scale of crack development at the top of the slope is small, and the cracks at the top of the slope have not developed to the leading edge position. Therefore, the possibility of slope collapse is relatively small. Summarizing the above analysis, it can be concluded that in the mountainous areas of Guizhou, with a slope terrain of "steep on the top and gentle on the bottom" and a slope structure of "hard on the top and soft on the bottom," when using filling mining technology for underground mining, a filling rate of 50% can achieve a good filling effect. After reaching 50%, if the filling rate is further increased, the improvement in filling effect is small and the cost is high.

6 Conclusion

This chapter takes the high and steep slopes in the mountainous areas of Guizhou as the background, and summarizes the deformation results of the slope model analysis under different filling rates. The main conclusions are as follows:

(1) At a filling rate of 10–40%, there is no significant change in the degree of crack development in the upper limestone of the slope, but the cracks will extend to the rock mass at the front edge of the slope. At a filling rate of 50 ~ 90%, the number of cracks developed in the upper part of the slope slightly increases with the increase of filling rate, but the cracks do not extend to the front edge of the slope. The number of fractures in mudstone and mudstone in the lower part of the slope generally decreases with the increase of filling rate.
(2) The displacement value of the slope decreases with the increase of filling rate, and the range of displacement influence does not show significant changes. According to the monitoring data, it can be seen that the displacement values of most measuring points reach their lowest values when the filling rate is 50%.
(3) According to the displacement mean fitting curve of the monitoring points, there is a negative correlation between the filling rate of 10–50%, and the slope of the curve is steep. The filling effect is better within this range.

Acknowledgements This work is supported in part by the Research on Monitoring and Evaluation Technology for Power Transmission and Distribution Lines in Mining Collapse Areas of China Southern Power Grid Co., Ltd. (No. GZKJXM20222498).

References

1. Zhao JJ, Lin B, Ma YT et al (2016) Physical simulation study on deformation characteristics of overlying rock mass in goaf of gently inclined coal seam. J Coal Min 41(06):1369–1374
2. Tao TW, Shi WB, Xiong SZ et al (2023) Experimental study on deformation and failure of a mining slope under the action of rainfall. Can Geotech J 00:1–18
3. Zhang SB, Shi WB, Wang Y et al (2022) Study on the deformation and failure process of ultra-high, steep, medium, and gentle reverse slopes under underground mining action. J Eng Geol 1–15
4. Li HJ, Dong JH, Zhu QQ et al (2019) Characteristics and genesis mechanism of Jianshanying landslide in Fa'er coal mine, Guizhou. Sci Technol Eng 19(26):345–351
5. Li B, Zhao CY, Li J et al (2023) Mechanism of mining-induced landslides in the karst mountains of Southwestern China: a case study of the Baiyan landslide in Guizhou. Landslides 20:1481–1495
6. Zhao JJ, Li JS, Ma YT et al (2020) Experimental study on failure process of mining landslide induced by rain fall. J China Coal Soc 45(2):760–769
7. Xiong SZ, Shi WB, Peng XW et al (2022) Study on the Pusa collapse process based on discrete element method. J Nat Disasters 31(05):202–211
8. Lian CJ (2020) Study on partial extraction overburden stress response mechanism and moving deformation pattern. Coal Geol China 32(05):32–37

9. Li YP (2020) Research on overburden movement and failure law based on mining under safe water pressure of aquifer. Shandong Coal Sci Technol 02:138–141
10. Li XL (2014) Law of surface movement and overburden failure for paste filling fully mechanized mining. Saf Coal Mines 45(08):60–63

Open Access This chapter is licensed under the terms of the Creative Commons Attribution 4.0 International License (http://creativecommons.org/licenses/by/4.0/), which permits use, sharing, adaptation, distribution and reproduction in any medium or format, as long as you give appropriate credit to the original author(s) and the source, provide a link to the Creative Commons license and indicate if changes were made.

The images or other third party material in this chapter are included in the chapter's Creative Commons license, unless indicated otherwise in a credit line to the material. If material is not included in the chapter's Creative Commons license and your intended use is not permitted by statutory regulation or exceeds the permitted use, you will need to obtain permission directly from the copyright holder.

Study on the Settlement Change Law of Adjacent Buildings Caused by Deep Excavation Construction in Saturated Soft Loess Geology

Hanjuan Yao, Wenyao He, Man Wang, and Liudi Yang

Abstract The geological conditions in the loess region are complex and varied, and the applicability of existing research is limited. In order to study the influence of deep foundation pit excavation on the deformation of adjacent buildings under saturated soft loess geological conditions, this paper takes a deep foundation pit in Xi'an as an example to analyze the deformation monitoring data of buildings during construction. The results indicate that the pile anchor support system and precipitation measures can effectively ensure the safety of buildings; the settlement deformation of buildings can be divided into three stages: uniform deformation, differential deformation, and stable change, and the deformation perpendicular to the long side of the foundation pit is significant; the sealing measures between interlocking piles are effective, reducing geological losses and suppressing significant settlement of outer layers and buildings outside the foundation pit.

Keywords Saturated soft loess · Deep foundation pit · Pile anchor support · Building subsidence

1 Introduction

In densely populated urban areas, deep excavation projects adjacent to buildings face many safety risks [1, 2]. Although existing studies have explored the deformation effects of deep excavation on adjacent buildings [3–6], most of them are limited to specific working conditions, and the applicability of the conclusions is limited. The engineering geological conditions in loess areas are highly complex and variable, especially those containing collapsible and saturated soft loess, which pose high construction risks and are prone to safety accidents. Therefore, engineering monitoring has become an essential safety control measure in construction. Based on this, this article takes deep foundation pit engineering in saturated soft loess strata

---

H. Yao (✉) · W. He · M. Wang · L. Yang
Shaanxi Construction Engineering Railway Construction Engineering Co., Ltd., Xi'an, China
e-mail: 18710570639@163.com

in loess areas as the research background. Through on-site monitoring methods, the impact of deep foundation pit construction on existing buildings is analyzed, aiming to provide practical reference for building pre-reinforcement and safety control for similar projects.

2 Project Overview

A certain project is located at a convenient intersection with many old residential buildings in the surrounding area, which requires high requirements for disturbance control during construction. The main buildings of the project are the arc suppression coil and capacitor room, the comprehensive distribution building, and the main transformer room. The comprehensive distribution building adopts a semi-underground layout, and the foundation pit is a rectangular pit with a length of 76.9 m, a width of 36.4 m, and a depth of 19.1 m. The support structure adopts a bite pile + anchor cable support system, and the structural design service life is 24 months. The geomorphic unit of the construction area belongs to Huangtuliangwa, and the terrain of the site is relatively flat, with a maximum height difference of 0.8 m and an elevation of 413.0 ~ 413.8 m. According to geological survey data, the groundwater depth is 6.9–7.0 m, and the soil in the construction area is composed of artificial fill, loess, ancient soil, silty clay, silt, medium sand, and silty clay from top to bottom.

3 Monitoring Plan

3.1 Monitoring Purpose

To ensure construction safety and reduce disturbance to the surrounding environment, a comprehensive and rigorous monitoring system should be established during construction. At the same time, the monitoring results should be used to guide construction and monitor and warn the safety of the foundation pit enclosure structure and surrounding buildings to avoid safety accidents.

3.2 Monitoring Content

Given that the excavation site of the foundation pit is surrounded by numerous old buildings, and these buildings generally adopt shallow foundation designs, strict monitoring of the settlement of surrounding buildings is crucial to ensure the safe excavation of the foundation pit. This article aims to illustrate the impact of foundation pit construction on surrounding buildings in saturated soft loess areas. Therefore,

Fig. 1 Layout of monitoring points for foundation pit

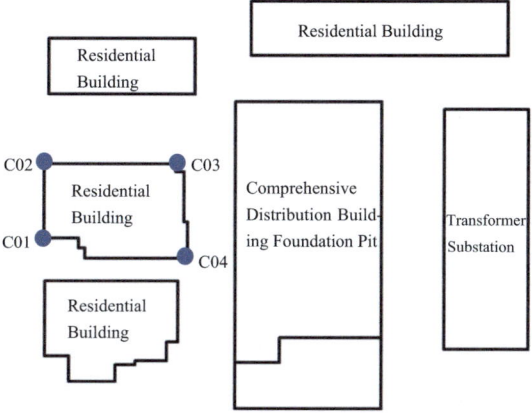

a brick concrete residential building located in the middle of the west side of the foundation pit is selected as an example to monitor the four corner points of the building and analyze the impact of foundation pit construction on the building. The relative position of the foundation pit and the building, as well as the layout of monitoring points, are shown in Fig. 1.

3.3 Monitoring Methods and Frequency

The settlement deformation of the building is monitored using the SZYL-A1000C intelligent settlement instrument, and the technical parameters of the SZYL-A1000C intelligent settlement instrument are shown in Table 1.

The monitoring frequency is once per (2–3) days when excavating from 0 to 6.4 m; when excavating 6.4–12.8 m, once per (1–2) days; when excavating from 12.8 to the bottom of the foundation pit, (1–2) times per day; within 7 days after pouring the bottom plate, once a day; bottom plate pouring takes 7–14 days, once every 3 days;

Table 1 Technical parameter table

Category	Parameter
Product range	1000 mm
Measurement accuracy	± 0.1%FS
Sensitivity	0.01 mm
Working voltage	DC12V
Environmental temperature for use	− 10°C ~ 50°C
Temperature measurement accuracy	± 0.5°C
Signal output	RS485
Long-term stability	0.05%/Year

bottom plate pouring takes 14–28 days, once every 5 days; and 28 days after pouring the bottom plate, once every 7 days.

4 Monitoring Results

4.1 Analysis of Building Settlement and Deformation

A detailed analysis was conducted on the settlement data of the four corner points of the residential building on the west side of the foundation pit. Based on the relationship between the settlement amount and time, as shown in Fig. 2, the settlement process of the residential building can be clearly divided into three stages:

(1) Uniform deformation stage. At this stage, from initial excavation to 20 days of excavation construction, the settlement of residential buildings did not change significantly. Among all the measuring points, the settlement of point C04 was the most significant, with a cumulative settlement value of up to 2.14 mm. At this time, the excavation depth of the foundation pit is relatively shallow and no dewatering operation has been carried out inside the pit. The settlement of the building is mainly attributed to the excavation and unloading of the soil in the foundation pit.

(2) Differentiated deformation stage. At this stage of excavation construction for 20–90 days, the settlement of each measuring point began to show unevenness, and the phenomenon of differential settlement became particularly evident. At this point, the unloading amount of soil in the foundation pit increases, resulting in increased stress on the supporting piles and deformation. At the same time, the dewatering operation carried out during this stage resulted in the loss of

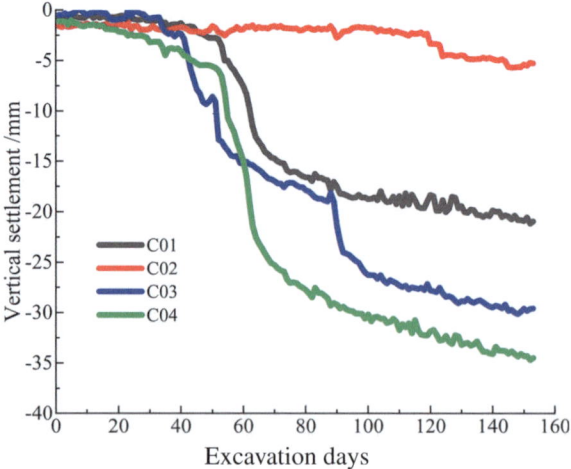

Fig. 2 Deformation trend of buildings with excavation time

groundwater mixed with the soil outside the foundation pit, and further problems such as water leakage between interlocking piles and anchor holes led to the loss of outer layers of the foundation pit, exacerbating the settlement phenomenon of buildings.

(3) Stable stage. At this stage, after 90 days of excavation construction, the anchor cable construction of the foundation pit has been fully completed and gradually exerted its reinforcement effect. Therefore, the settlement rate of the building has significantly slowed down, indicating that the settlement deformation of the building is gradually stabilizing, and the soil pressure on the supporting structure and its soil facing side is also maintained at a relatively stable level.

4.2 Analysis of Differential Settlement of Buildings

To more intuitively reflect the specific impact of foundation pit excavation on buildings, analyzing the differential settlement of buildings can more accurately evaluate the degree of impact of foundation pit excavation on building structural safety, timely identify potential problems, and take corresponding measures for prevention or correction. The differences in settlement measured from different directions of the building are shown in Fig. 3.

It can be seen that the rapid growth period of differential settlement in different directions of the building occurs during the excavation construction period of 40–90 days. After the overall excavation of the foundation pit is completed, the differential settlement values of measuring points C02-C03 show a steep slope like growth, while the differential settlement values of measuring points C03-C04 show a cliff like decrease, and then gradually tend to stabilize. The main reason for this phenomenon is that the pile body at the bottom of the biting pile in the middle of the left side of

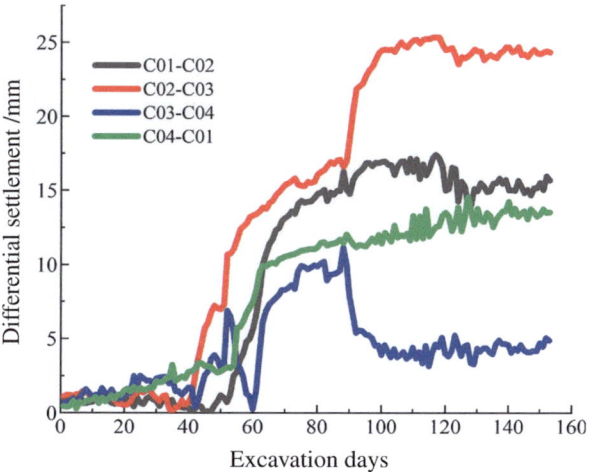

Fig. 3 Differences in settlement of buildings in different directions

the foundation pit splits, and a large number of soil particles are carried out by the groundwater at the bifurcation, causing a sudden increase in soil settlement near the side of the foundation pit. This also indicates that when the building is affected by the construction of the foundation pit, the force effect perpendicular to the long side of the foundation pit is more significant, exceeding the influence parallel to the long side of the foundation pit.

To address the water leakage caused by the bifurcation between interlocking piles, methods such as heavy loading, filling with recycled cotton, post pile grouting, reinforcement grouting, and hanging mesh spraying are used to seal the leakage. For areas with multiple bifurcations between interlocking piles, vertical deep hole grouting is also required to increase the seepage path of groundwater outside the foundation pit. With the plugging work and the construction of the last layer of anchor cables, the deformation of the soil outside the foundation pit has been effectively controlled, and the settlement of the building has gradually stabilized. The differential settlement of each measuring point has also gradually reached a relatively stable state.

5 Summarize

(1) The maximum settlement of buildings caused by excavation of foundation pits in saturated soft loess areas is 34.48 mm, indicating that the impact of excavation on surrounding buildings is within a controllable and safe range. The use of interlocking piles and anchor cable support system combined with internal and external dewatering measures can effectively ensure the safety of buildings.

(2) Under complex geological conditions, the construction of deep foundation pits should strengthen the control of the quality of supporting structures to avoid the phenomenon of increased settlement of buildings caused by geological losses due to water leakage from supporting structures. The inevitable leakage of water should be promptly addressed through leak sealing construction, such as heavy loading, filling with recycled cotton, post pile grouting, reinforcement grouting, hanging net spraying and mixing, and deep hole grouting.

(3) By analyzing the differential settlement of buildings, the force effect perpendicular to the long side of the foundation pit is more significant and tends to tilt towards the interior of the pit. It is recommended that in future similar construction projects, the entire construction process of buildings near the side of the foundation pit should be monitored, and the quality of the supporting structure and groundwater level should be strictly controlled. If necessary, grouting reinforcement can be carried out on the strata and the load-bearing components of the building can be reinforced to avoid safety accidents.

References

1. Guan YL, Li YH, Li JY et al (2024) Impact analysis of deep foundation pit construction on deformation of adjacent buildings. Geotech Eng Tech 1–5. http://kns.cnki.net
2. Liu GX, Liu Y, Feng XX et al (2023) Analysis on deformation of adjacent buildings induced by deep foundation pit construction in water-rich strata: taking the deep foundation pit of Jinan Metro line 2 as an example. Sci Technol Eng 23(36):15665–15672. http://kns.cnki.net
3. Boone SJ, Westland J, Nusink R (1999) Comparative evaluation of building responses to an adjacent braced excavation. Can Geotech J 36(2):210–223. https://doi.org/10.1139/cgj-36-2-210
4. Rechea C, Levasseur S, Finno R (2008) Inverse analysis techniques for parameter identification in simulation of excavation support systems. Comput Geotech 35(3):331–345. https://doi.org/10.1016/j.compgeo.2007.08.008
5. Dai Z (2023) Full process analysis of the impact of deep excavation construction on surrounding buildings in subway stations. J Liaoning Prov Coll Commun, 25(05):21–25. http://kns.cnki.net
6. Liu MN (2022) Analysis of settlement of surrounding buildings caused by construction of subway station foundation pit. J Shijiazhuang Inst Railw Technol 21(03):29–34. http://kns.cnki.net

Open Access This chapter is licensed under the terms of the Creative Commons Attribution 4.0 International License (http://creativecommons.org/licenses/by/4.0/), which permits use, sharing, adaptation, distribution and reproduction in any medium or format, as long as you give appropriate credit to the original author(s) and the source, provide a link to the Creative Commons license and indicate if changes were made.

The images or other third party material in this chapter are included in the chapter's Creative Commons license, unless indicated otherwise in a credit line to the material. If material is not included in the chapter's Creative Commons license and your intended use is not permitted by statutory regulation or exceeds the permitted use, you will need to obtain permission directly from the copyright holder.

Influence of Subgrade Excavation on Vertical Deformation of Collinear Metro Tunnels in Soft Soil Area

Jianzhao Li and Qingyuan Zeng

Abstract With the development of urban construction, excavation of new projects adjacent to metro tunnels will affect soil and existing metro tunnels. Based on a highway construction project collinear with metro tunnels in Shanghai, the impact of different subgrade excavation conditions on tunnels was investigated using the FDM software FLAC 3D. The results show that the vertical deformation of tunnels is less sensitive to the increase in the width of subgrade excavation area, compared to the area length in the longitudinal direction. Doubling the longitudinal length of excavation area will increase vertical displacement of tunnels by 90%, compared with 17% when doubling its width. In the case of excavating simultaneously in two adjacent areas, the additional vertical displacement of tunnels decreases rapidly with the increase of distance between two excavations. When the spacing of two pits is 30m, the extra displacement is less than 13% of a single excavation.

Keywords Metro tunnel · Subgrade replacement · Numerical simulation · Modified Cam-Clay model

1 Introduction

With the rapid development of urbanization and urban construction, the metro can greatly alleviate the pressure of urban ground transportation. Meanwhile, excavation of new construction projects adjacent to metro tunnels will cause the release of in-situ stress and thus soil displacement, which inevitably affects existing metro tunnels [1]. In recent years, research on the impact of construction on adjacent metro tunnels has been conducted by many scholars. For example, theoretical analysis was used in references [2, 3] to study the rule of deformation of tunnels under different cases;

J. Li (✉)
Tongji University, Shanghai, China
e-mail: sicklelee@tongji.edu.cn

Q. Zeng
China MCC5 Group Shanghai Corp. Ltd., Shanghai, China

numerical simulation methods were used in references [4–6] to investigate the impact of various construction on the adjacent metro tunnels; centrifuge model tests were used in references [7, 8] to analyze the problem.

Generally, in the case of the highway construction project collinear with metro, the excavation area of subgrade replacement is characterized by large scale and being located directly above tunnels, which exerts excessive deformation and hazards the safety of metro operation. However, there is relatively little research on the impact of large-scale excavation and unloading as roadbed replacement on collinear metro tunnels in operation. In this study, based on the construction project of Luxiang Road in Shanghai, the finite difference method (FDM) is used to analyze the vertical displacement of tunnels under various excavation conditions, and guidance and suggestions of subgrade excavation and replacement zoning plan for similar projects are provided.

2 Project Overview

The Luxiang Road construction project is located in Baoshan District, Shanghai, with a total length of 2452 m and a width of 40 m. The length of project collinear with Shanghai Metro line 7 is 1670 m, including 4 sections of roads with a total length of 1146 m and 4 bridges with a total length of 524 m. The subgrade is constructed using the EPS replacement method, with an excavation depth ranging from 1.47 m to 2.10 m.

As is shown in Fig. 1, the outer diameter of two parallel tunnels of Shanghai Metro line 7 is 6.2 m, and its inner diameter is 5.5 m. The horizontal distance between the centerlines of twin tunnels is about 14.0 m on average, and the burial depth of tunnel crown varies from 6.0 m to 10.0 m.

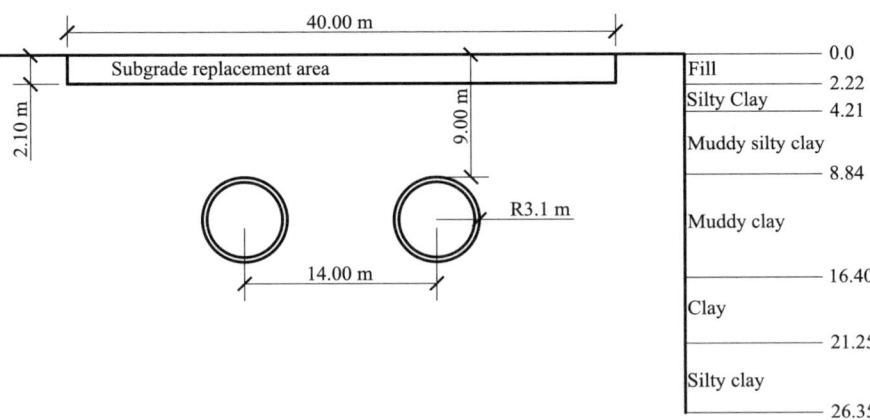

Fig. 1 Profile of subsurface soil layers and relative position between the subgrade replacement area and twin tunnels [9]

The metro tunnels and roads were constructed in typical soft soil strata. Thick sedimentary soft soils characterized by high water content, high compressibility, and low strength are widely distributed in the construction site and are shown in Fig. 1 [9].

3 Numerical Simulation Method

In this study, the FDM software FLAC 3D (Fast Lagrangian Analysis of Continua in 3 Dimensions) was used for simulation. A detailed introduction to model parameters and simulation cases is shown as follows.

To eliminate the influence of the model scope on the results, the domain of model is determined to be 120 m × 140 m × 26.35 m, which is extended to 4 times of burial depth of tunnels. The burial depth of twin tunnels is determined to be 9 m based on a typical subgrade replacement area of the construction site. The outer diameter of tunnels is 6.2 m, and the lining thickness is 0.35 m. The distance of the centerline of tunnels is 14.0 m, as shown in Fig. 2. The tunnel is simulated by linear elastic model, with an elastic modulus of 3.45MPa, Poisson's ratio of 0.20, and density of 2400 kg/m^3 [10].

The choice of soil model has a significant impact on the results. Compared with the Mohr-Coulomb model, the modified Cam-Clay model can reflect the deformation characteristics of soft soil more accurately, due to the modulus of soil being related to its stress. Thus, the modified Cam-Clay model is used in this study. The parameters of soil are shown in Table 1.

To take the safety and economy of various excavation schemes into account, 2 variables are taken into consideration in this study: length of excavation in the longitudinal direction of tunnels and its width. In addition, to study the impact of simultaneous excavation in multiple areas on tunnels, the distance of two adjacent

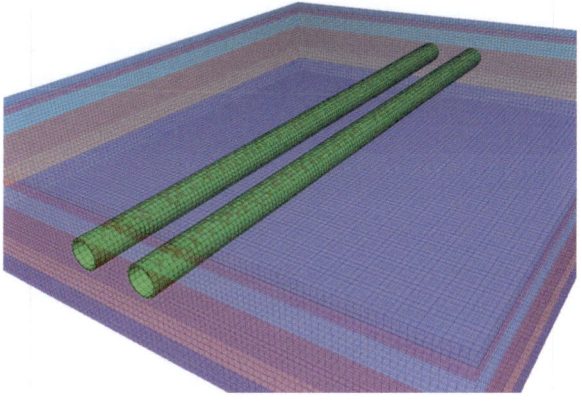

Fig. 2 Mesh of the FD model

Table 1 Soil parameters for modified Cam-Clay model

Soil layer	Thickness (m)	μ	M	κ	λ	ρ(kg/m^3)	G(MPa)
①Fill	2.22	0.3	1.278	0.0182	0.0728	1750	6.923
②Silty clay	1.99	0.3	1.264	0.0143	0.0573	1860	6.923
③Muddy silty clay	4.63	0.35	1.259	0.0253	0.1011	1750	5.696
④Muddy clay	7.56	0.35	1.101	0.0228	0.0912	1680	8.026
⑤Clay	4.85	0.3	1.106	0.0242	0.0970	1760	16.53
⑥Silty clay	5.10	0.3	1.264	0.0099	0.0397	1950	21.91

Note: M = friction constant, based on the effective internal friction angle φ' and empirical formulas [11]. λ = slope of normal consolidation line, κ = slope of elastic expansion line, both are based on compressive index C_c [9]. G = unload shear modulus, based on stress path controlled triaxial test of on-situ soil or empirical formulas [12]

Table 2 Simulation cases

Cases	Width	Length (m)	Spacing of excavations	Cases	Width	Length (m)	Spacing of excavations (m)
SH1	Half	10	Only a single excavation	DH1	Half	10	10
SH2	Half	20		DH2	Half	10	20
SH3	Half	30		DH3	Half	10	30
SH4	Half	40		DF1	Full	10	10
SF1	Full	10		DF2	Full	10	20
SF2	Full	20		DF3	Full	10	30
SF3	Full	30					
SF4	Full	40					

Note Half = Half width of the road (20 m); Full = Full width of the road (40 m)

pits is also taken as a variable. The difference of all simulation cases is shown in Table 2.

4 Results and Discussions

The vertical deformation curve of the tunnel with greater deformation is shown in Fig. 3. The maximum vertical displacement of tunnels is shown in Fig. 4.

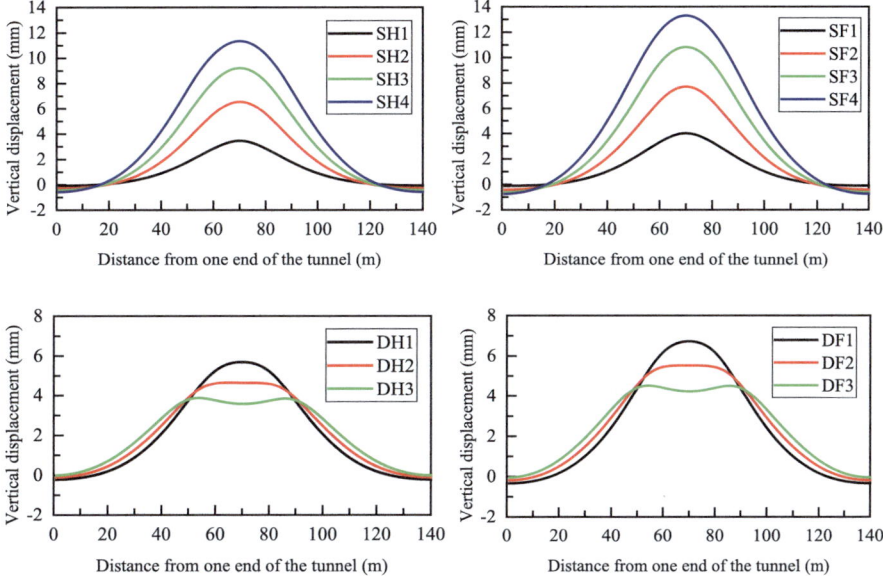

Fig. 3 Vertical deformation curve of the tunnel

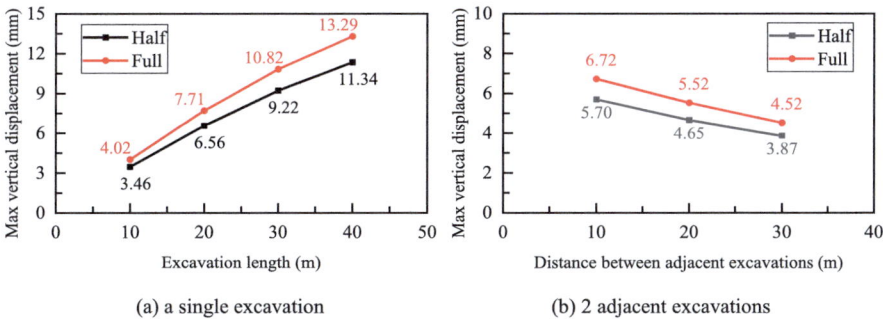

(a) a single excavation　　　(b) 2 adjacent excavations

Fig. 4 Maximum vertical displacement of tunnels

4.1 The Impact of Excavation Length and Width

As is shown in Fig. 4a, as the length of the excavation area increases, the rate of increase in vertical deformation of tunnels slows down. When the length of the half-width excavation is 20 m, 30 m, or 40 m, the maximum vertical displacement of the tunnel will be 190%, 268%, and 330% respectively, compared with the 10m-length excavation. In the case of full-width excavation, it will be 192%, 269%, and 331% respectively.

If the length of excavations is the same, the maximum displacement of the tunnel under the full-width excavation will be 15.8% to 17.0% greater than that under the half-width excavation. It can be seen that doubling the width of excavation area has a significantly minor impact on the tunnel than doubling its length. Thus, as the horizontal distance between the centerline of the excavation area and the tunnels, the impact of soil unloading on the tunnel will decrease sharply.

4.2 The Impact of the Distance Between Adjacent Excavations

As is shown in Fig. 4b, as the distance between adjacent excavations increases, their mutual effect decreases. When the spacing between half-width excavations is 10 m, 20 m, and 30 m, the maximum displacement of tunnels will be 166%, 136%, and 112% of simulation case SH1, respectively. In the case of full-width excavation, it will be 167%, 137%, and 113% respectively. Thus, when the spacing of adjacent excavation areas is more than 30 m, the additional vertical displacement of tunnels caused by mutual effects is only about 12–13%. In addition, the impact of excavation width follows the same rules as in Sect. 3. The vertical displacement of tunnels under a full-width excavation is about 16.4% greater than that under a half-width excavation.

5 Conclusion

A series of 3D-FDAs were conducted to investigate the deformation of tunnels due to a collinear subgrade excavation. The conclusion of this study is as follows:

1. Compared with the length of excavation area in the longitudinal direction, tunnel deformation is less sensitive to increasing the width of the excavation area. Doubling the width of the area only increases the vertical displacement of tunnels by less than 17%, as a comparison, 90% when doubling its length. Thus, to increase the efficiency of construction, the width of a single excavation area can be widened while implementing reinforcement and monitoring measures. But increasing its length should be cautious.
2. To improve efficiency and shorten the construction period, multiple subgrade replacement areas can be excavated simultaneously in the longitudinal direction of metro tunnels. The additional vertical displacement of tunnels caused by mutual effects will be less than 13% when the spacing of adjacent excavations is over 30 m.

From the above findings, engineers and designers can improve the subgrade excavation zoning in similar construction conditions as investigated in this study. Specifically, widening the excavation area to the full width of the road rather than extending its length and excavating simultaneously at intervals of more than 30m would be an effective way to improve construction efficiency while reducing the tunnel deformation. However, in this study, the computer performance and modeling accuracy were limited, and thus, the FD model of tunnels cannot accurately reflect the deformation characteristics of real metro tunnels, which needs further research.

References

1. Cheng K (2022) Overview on the influence and control measures of vibration induced by shield construction on adjacent buildings. J Water Resour Arch Eng 20(3):95–101
2. Wu H, Shen S, Liao S et al (2015) Longitudinal structural modeling of shield tunnels considering shearing dislocation between segmental rings. TunnLing Undergr Space Technol 50:317–323
3. Shuan C, Huaina W, Renpeng C et al (2021) Deformation of a collinear tunnel induced by overlying long-distance excavation. J Shanghai Jiao Tong Univ 55(6):698–706
4. Chen R, Liu M, Meng F et al (2023) Circumferential forces and deformations of shield tunnels due to lateral excavation. Chin J Geotech Eng 45(1):24–32
5. Prateep L, Pornkasem J, Pattaramon J et al (2017) Numerical investigation of tunnel deformation due to adjacent loaded pile and pile-soil-tunnel interaction. Tunn Undergr Space Technol 70:166–181
6. Bing W, Chen K, Weng T et al (2024) Refined numerical model analysis of the influence of adjacent foundation pit excavation on shield tunnel structure. J Ningbo Univ 37(0), 1–10
7. Ng CWW, Shi J, Hong Y (2013) Three-dimensional centrifuge modelling of basement excavation effects on an existing tunnel in dry sand. Can Geotech J 50(8):874–888
8. Liang F, Chu F, Song Z et al (2012) Centrifugal model test research on deformation behaviors of deep foundation pit adjacent to metro stations. Rock Soil Mech 33(3):657–664
9. CCCC First Highway Consultants (2019) Detailed geotechnical engineering investigation report of luxiang road (Poyang Lake Road—Yangnan Road), Shanghai
10. Ministry of Housing and Urban Rural Development of the PRC (2015) Code for design of concrete structures. China Architecture & Building Press, Beijing
11. Shihua W, Wang X (1997) The relationship between angle of internal friction and plastic index of clay soil in Shanghai. Dam Obs Geotech Tests 21(2):41–43
12. Liu G, Hou X (1996) Unloading modulus of the shanghai soft clay. Chin J Geotech Eng 18(6):18–23

Open Access This chapter is licensed under the terms of the Creative Commons Attribution 4.0 International License (http://creativecommons.org/licenses/by/4.0/), which permits use, sharing, adaptation, distribution and reproduction in any medium or format, as long as you give appropriate credit to the original author(s) and the source, provide a link to the Creative Commons license and indicate if changes were made.

The images or other third party material in this chapter are included in the chapter's Creative Commons license, unless indicated otherwise in a credit line to the material. If material is not included in the chapter's Creative Commons license and your intended use is not permitted by statutory regulation or exceeds the permitted use, you will need to obtain permission directly from the copyright holder.

MIX
Papier aus verantwortungsvollen Quellen
Paper from responsible sources
FSC® C105338

If you have any concerns about our products,
you can contact us on
ProductSafety@springernature.com

In case Publisher is established outside the EU,
the EU authorized representative is:
Springer Nature Customer Service Center GmbH
Europaplatz 3, 69115 Heidelberg, Germany

Printed by Libri Plureos GmbH
in Hamburg, Germany